I0875077

Brain and Race

Nuncius Series

Studies and Sources in the Material and Visual History of Science

Series Editors

Marco Beretta (*University of Bologna*)
Sven Dupré (*Utrecht University / University of Amsterdam*)

VOLUME 4

The titles published in this series are listed at *brill.com/nuns*

Brain and Race

A History of Cerebral Anthropology

By

Claudio Pogliano

BRILL

LEIDEN | BOSTON

Cover illustration: *Negerköpfe*, Four Studies of a Male Head, Workshop of Peter Paul Rubens (about 1617-1620), Oil on panel, The J. Paul Getty Museum, Los Angeles. Digital image courtesy of the Getty's Open Content Program.

The Library of Congress Cataloging-in-Publication Data is available online at http://catalog.loc.gov
LC record available at http://lccn.loc.gov/2020016564

Typeface for the Latin, Greek, and Cyrillic scripts: "Brill". See and download: brill.com/brill-typeface.

ISSN 2405-5077
ISBN 978-90-04-42933-8 (hardback)
ISBN 978-90-04-43188-1 (e-book)

Copyright 2020 by Koninklijke Brill NV, Leiden, The Netherlands.
Koninklijke Brill NV incorporates the imprints Brill, Brill Hes & De Graaf, Brill Nijhoff, Brill Rodopi, Brill Sense, Hotei Publishing, mentis Verlag, Verlag Ferdinand Schöningh and Wilhelm Fink Verlag.
All rights reserved. No part of this publication may be reproduced, translated, stored in a retrieval system, or transmitted in any form or by any means, electronic, mechanical, photocopying, recording or otherwise, without prior written permission from the publisher.
Authorization to photocopy items for internal or personal use is granted by Koninklijke Brill NV provided that the appropriate fees are paid directly to The Copyright Clearance Center, 222 Rosewood Drive, Suite 910, Danvers, MA 01923, USA. Fees are subject to change.

This book is printed on acid-free paper and produced in a sustainable manner.

Contents

Illustrations

Introduction

> The brain of our species is, as we know, made up largely of potassium, phosphorus, propaganda, and politics [...][1]

∴

The exergue is from the last book published by James Thurber, an American journalist, humorist, and cartoonist who assiduously contributed to *The New Yorker*. I came across it by reading a work by the South African paleoanthropologist Phillip v. Tobias, and it seemed to me that it conveys approximately the first lesson of the research I am introducing. To those four terms that have in common the initial letter *p* a fifth could be added, *prejudice*. Brain is indeed plainly an organ of the body, with its complex structure and electrochemical functioning, but it is also an object of fierce disputes, besides outlining a field of continuous scientific, cultural, and social practices. "Ein symbolisch kontaminiertes Organ", as aptly stated by Michael Hagner in 2004.[2] I am aware I have not said anything new so far, given the existence of a historical-critical literature that has thus portrayed the brain during the last couple of decades.[3]

---

1 James Thurber, *Lantern and Lances* (London: Hamish Hamilton, 1961), p. 112. The quote is in Phillip v. Tobias, *The brain in hominid evolution* (New York-London: Columbia University Press, 1971), p. 15.

2 Michael Hagner, *Geniale Gehirne. Zur Geschichte der Elitegehirnforschung* (Göttingen: Wallstein, 2004), p. 8.

3 I list here only a few of the many possible titles: Michael Hagner, *Homo cerebralis. Der Wandel vom Seelenorgan zum Gehirn* (Berlin: Berlin Verlag, 1997); Michael Hagner (Hrsg,), *Ecce Cortex: Beiträge zur Geschichte des modernen Gehirns* (Göttingen: Wallstein, 1999); Daniel Lord Smail, *On Deep History and the Brain* (Berkeley: University of California Press, 2008); Suparna Choudhury, Jan Slaby (eds.), *Critical Neuroscience: A Handbook of the Social and Cultural Contexts of Neuroscience* (Oxford: Wiley-Blackwell, 2011); Nikolas Rose, Joelle M. Abi-Rached, *Neuro: The New Brain Sciences and the Management of the Mind* (Princeton: Princeton University Press, 2013); "Focus: Neurohistory and the History of Science," *Isis*, 2014, *105*:100-154; Andrew P. Wickens, *A History of the Brain. From Stone Age Surgery to Modern Neuroscience* (London-New York: Psychology Press, 2015); Stephen T. Camper, Delia Gavrus (eds.), *The History of the Brain and Mind Sciences* (Rochester, University of Rochester Press, 2017); Fernando Vidal, Francisco Ortega, *Being Brains. Making the Cerebral Subject* (New York: Fordham University Press, 2017; Mattia Della Rocca and Claudio Pogliano (eds.), "Different Histories from 20th Century Neuroscience," *Nuncius. Journal of the Material and Visual History of Science*, 2017, *32*/2:253-500.

© KONINKLIJKE BRILL NV, LEIDEN, 2020 | DOI:10.1163/9789004431881_002

Since the second half of the eighteenth century, generations of scientists and philosophers have persevered to look for an answer to the question about the relationships between quantitative and qualitative characters of the brain, on the one hand, and the degree of 'intelligence' – the most nebulous of concepts – and morality on the other. During an initial period, which lasted about fifty years, a few agents triggered some of those discursive and operational devices that would have gained more and more consensus during the nineteenth century. At first human diversity was the actual object of application, seen through the postulated existence of races on earth, the last and secluded part of which had recently been approached. It didn't take long, however, before another object, the female of the human species, attracted a search for further traits – in addition to the sexual organs – which could explain her specific being. Although the cerebral status of the woman represents a very secondary thread of my narrative, nevertheless it needs to be reported. In the early decades of the nineteenth century, new arguments reinforced a somatic and stigmatizing *cliché*, and her 'racialization' – as it were – easily took hold. Black male is to white male what in general the female is to the male: this was stated in 1847 during a *séance* of the Société d'Ethnologie de Paris. The time when the Cartesian François Poullain de La Barre declared the entire anatomical similarity of the brain "entre les hommes et les femmes" was in the distant past.[4]

More or less simultaneously, a further kind of figure began to solicit scientific attention. The existence of individuals, whose lives and achievements revealed an outstanding level, had always aroused much curiosity and attempts at explanation, but from a certain moment onwards it seemed that the cerebral secret of their excellence was within reach of a scalpel, or at least of a craniological scrutiny.[5] At the opposite end, there would be the delinquent, except that the historiography relating to criminal anthropology – as a cultural phenomenon both before and after Cesare Lombroso – is already gigantic enough to advise me to add almost nothing in this regard. Since the beginning

4 François Poullain de La Barre, *De l'egalité de deux sexes, discours physique et moral, où l'on voit l'importance de se défaire de préjugez* (Paris: Jean Du Puis, 1673), p. 112. In a book that has become a reference, Londa Schiebinger pointed out a deep intellectual shift in the public consideration of women – from recognition to exclusion – occurred at the beginning of the nineteenth century: see her *The Mind Has No Sex? Women in the Origins of Modern Science* (Cambridge, Mass.: Harvard University Press, 1989). See also Elizabeth Fee, "Nineteenth-Century Craniology: the Study of the Female Skull", *Bulletin of the History of Medicine*, 1979, *53*:415-433; Nicole C. Karafyllis, Gotlind Ulshöfer (eds.), *Sexualized Brains. Scientific Modeling of Emotional Intelligence from a Cultural Perspective* (Cambridge, Mass.: The MIT Press, 2008). More bibliographic information and a particular interpretation are provided by Gina Rippon, *The Gendered Brain. The New Neuroscience that Shatters the Myth of the Female Brain* (London: The Bodely Head, 2019).

5 See Hagner, *Geniale Gehirne* (cit. note 2).

of his career, Franz Joseph Gall had in mind to write a treatise on the natural history of the criminal, which was never published, even though this was always a big bee in his bonnet.[6] And it is worth just recalling here, *en passant*, that five years before the first edition of his *L'uomo delinquente*, the polygenist Lombroso had gathered some popular lectures for ladies about *L'uomo bianco e l'uomo di colore*, with no restraint in expressing all the stereotypes of scientific racism. He could not fail to commiserate the poor brains of Blacks, so uncomfortable (*a disagio*) in their long, narrow, and heavy skulls: a cerebral development suspended in simian stupidity, not very different from what could be found in the head of some delinquent, given the tenets of the atavistic theory.[7]

For a while, the grandiose program announced by Gall in 1798, aimed at founding a new science called *organology*, was carried out in Europe, and then disseminated like a cultural epidemic, in the form of phrenology, throughout the western world, which began either to embrace it as a faith or to fight it. Sort of a scientific ideology, it acted as the precondition and facilitator of a process that soon went well beyond its boundaries, and invested a series of interests at the intersection of various fields.[8]

Human races do constitute the backbone of this research, which investigates a long period when the species seemed to come naturally partitioned into a number of discrete types, each furnished with its own distinctive, innate, and hereditary characters. Once the globe had been progressively explored, these types were easily identified with the continents they lived in. The stream of classification that so much energy had been spent on since the days of Linnaeus – with variations increasingly ramifying since the nineteenth century – does not directly affect my story, which takes it for granted. It should also be said that scientific interest did not regard all the classified races to the same extent. From the beginning, it has been the brain of Africans (or of their

6 See Sigrid Oehler-Klein, "Mordsinn und Diebsinn. Franz Joseph Galls Erklärung des Verbrechens und die Anfänge der Neuropsychologie zu Beginn des 19. Jahrhunderts," in Fayet, Roger (Hrsg.), *Die Anatomie des Bösen: Ein Schnitt durch Körper, Moral und Geschichte* (Baden: Hier & Jetzt, 2008), pp. 107-122.

7 Cesare Lombroso, *L'uomo bianco e l'uomo di colore. Letture sull'origine e le varietà delle razze umane* (Padova: Tip.Editrice F. Sacchetto, 1871), pp. 28-29.

8 In the last decades the history of phrenology has raised a great scholarly interest, producing a considerable mass of books and essays. Suffice it to indicate here just a few titles: Marc Renneville, *Le langage des crânes. Une histoire de la phrenologie* (Paris: Sanofi-Synthélabo, 2000); John van Wyhe, *Phrenology and the Origins of Victorian Scientific Naturalism* (Aldershot-Burlington: Ashgate, 2004), pp. 27-29; Stanley Finger, Paul Eling, *Franz Joseph Gall. Naturalist of the Mind. Visionary of the Brain* (Oxford: Oxford University Press, 2019); James Poskett, *Materials of the Mind. Phrenology, Race, and the Global History of Science, 1815-1920* (Chicago: University of Chicago Press, 2019).

descendants in America) that captured the eye and the study, mostly in comparison with Europeans. Attention was only sporadically paid to individuals of other origins, due to the scarce availability of their brains, but also because Blacks were – as Julien-Joseph Virey decreed in 1817 – "en quelque sorte l'inverse de l'Européen".[9]

Having often dealt in the past, separately, with both elements that appear in the title of the book – *Brain & Race* – I thought it might be suggestive to make them react with each another, so that the result, whether successful or not, will describe the effects of this chemical mixture.[10] Needless to say, it is not the first time that this operation is performed[11], but I have hopefully coped with it in a more extensive way, without judging the past as if it were the present, nor separating the good from the bad. By the way, I remember how quickly I devoured, long ago, the first edition of Steve Gould's *Mismeasure of Man*, which might serve here as a kind of ideal *introibo*, despite its being quite dated by now, and criticized for its open partisanship. I must also anticipate that the four chapters of this book have frequent recourse to citations from the original sources, to convey the flavor of their language, which I have also used in my text – a language that today can sound obsolete, sometimes.

Almost like in a chronicle, the thread of time drives the narrative from the mid-eighteenth century to the mid-twentieth century. The chronological order has not only helped the treatment of a conspicuous amount of sources, but also served to highlight the continuity of the racial study of skulls and brains. Given the questionable advancement of knowledge that has gradually occurred along the way, this continuity sometimes acquires the features of resilience: i.e., to still keep playing a game, the dubious outcomes of which did not fully satisfy the players. A game that is intrinsically political, even though it has to do with bones and nervous matter. Never leading to certain and definitive conquests, their search often sowed doubts and disillusions, expressed at least by the most serious and prudent among them. In the midst of his numerous brain studies, which inspired a host of followers, Paul Broca surprisingly and after much reasoning claimed that "il ne peut donc venir à la pensée d'un

9 Julien-Joseph Virey, "Homme," in *Nouveau Dictionnaire d'Histoire naturelle, appliquée aux Arts, à l'Agriculture, à l'Économie rurale et domestique, à la Médecine, etc. par une Société de Naturalistes et d'Agriculteurs* (Paris: Deterville, 1817), vol. 15, p. 168.

10 As a sample see Claudio Pogliano, "Between Form and Function: A New Science of Man," in Pietro Corsi (ed.), *The Enchanted Loom. Chapters in the History of Neuroscience* (New York-Oxford, Oxford University Press, 1991), pp. 144-203; Id., *L'ossessione della razza. Antropologia e genetica nel XX secolo* (Pisa: Edizioni della Normale, 2005); Id., *Storie di cervelli, Dall'antichità al Novecento* (Milano: Editrice Bibliografica, 2017).

11 For instance, see Gavin Evans, *Black Brain White Brain. Is Intelligence Skin Deep?* (London: Thistle, 2014), a book with a distinctly journalistic approach.

homme éclairé de mesurer l'intelligence en mesurant l'encéphale." A little further in the same text he nevertheless added that "il serait difficile de ne pas admettre l'existence d'une relation bien déterminée entre la masse du cerveau et la puissance de l'intelligence."[12] Does it seem a contradiction? In part it is, but the collective enterprise I am about to describe is full of obvious contradictions. Significantly, as early as 1888, all the jumble (*fatras*) of data on brain weight, which only served to clutter up science, annoyed Léonce Manouvrier, one of Broca's last students. He himself, however, did not cease attending the field.[13]

So to speak, one side of the coin shows an obstinacy that appears boringly repetitive: from this point of view the whole, daring polycentric initiative might be judged, all in all, as a sort of a failure. For this outcome, the confused array of the measurement systems and units – which had to be translated into one another, thus inevitably impairing their accuracy – was also responsible. On the other side, however, such dedication, in the attempt to snatch from the brain the secret of inter-individual and racial variation, stemmed from the positivistic belief that better methods and more data would always correct any shortcomings and errors. Consequently, the transnational industry, which had been tirelessly working for decades to discover something new on the same subject, was occasionally forced to experiment with updated guidelines and tools. So, what at first sight may look like a static picture actually contains successive small changes of direction.

In concluding the sixth book of his *Ideen zur Philosophie der Geschichte der Menschheit*, in 1785 Herder confessed the desire to have at disposal a magic wand (*Zauberrute*), with the help of which he would transform all previously given indefinite verbal descriptions of men into images, and thus produce a gallery of drawn shapes of fellow brothers on earth. But how far – he complained – one was from fulfilling this anthropological wish! For centuries, its surface had been scored by sword and cross, coral and liquor barrels: the peaceful pencil instead neglected. Just the "marvelous" had been mainly sought. Afterwards, even when better drawings had been provided here and there, the intent was still to beautify, whereas no true zoologist would ever alter strange animal figures in depicting them. Visual representations of non-European peoples were needed, for both scientific and philanthropic purposes: that is why

12 Paul Broca, *Sur le volume et la forme du cerveau suivant les individus et suivant les races* (Paris: Typographie Hennuyer, 1861), pp. 10, 16.

13 Léonce Manouvrier, "Recherches d'anatomie comparative et d'anatomie philosophique sur les caractères du crâne et du cerveau. Sur l'interprétation de la quantité dans l'encéphale et dans le cerveau en particulier," *Mémoires de la Société d'Anthropologie de Paris*, II série, 1888, *3*:137-326, p. 289.

Herder called for an anthropological map, modeled on Zimmerman's zoological geography. Of the same mind, in January 1787, from Vilna Georg Forster informed his friend Soemmerring that he planned to write more about human *Verschiedenheiten*, but that he lacked the sources, first of all good drawings, like those he had seen during one of his stays in England.[14]

The nineteenth century would try to fulfill these desiderata through a mass of anthropological visualizations that grew over time. Also the whole business of Brain & Race had its own iconography, without which its expanding constellation would not have been possible. The demand for images, to be stored as in an archive, and shown in books or articles is perfectly understandable in craniology or cephalometry, where scholars had the necessary expertise to decipher, and mutually verify the peculiar elements of each skull or head, sometimes quarreling about divergent interpretations. In particular, the brain began to require drawings or photographs when, from the simple weighing of it – a methodologically controversial operation – a shift occurred towards the observation of the cortical surface and to the strict analysis of convolutional patterns.

The retrieval and collection of specimens, or the irresistible impulse to pile data on data, fueled all this ingenuity, put at the service of science, and deposited a considerable amount of material in various centers of Europe and the United States, which became pilgrimage destinations. A thousand times complaints resonated because of their scarcity, provenance, anatomical treatment or conservation techniques. Many of those who spent their time and energy to compare racial brains had almost always something to reproach to their predecessors, possibly intending to remedy their alleged flaws. In a sense, the entire exercise might even seem self-correcting, if only actual and positive breakthroughs had been reached. There was something really important at stake: on the one hand, the racial question that became increasingly explosive; on the other hand, the coeval transformation of the human brain into an epistemic object, with multiple facets and political implications. These two threads intertwined and uninterruptedly fed each other in a process that never really reached a conclusion. Doubts and disillusion came along with the intensification of research, which nonetheless did not die out, and tried each time to modify some of its procedures.

The nineteenth century proved to be a "long" one, also as for the topic of this book. Researches in 'excellent' (or otherwise criminal) brains still flourished in

14 Johann Gottfried Herder, *Ideen zur Philosophie der Geschichte der Menschheit* (Riga: Hartknoch, 1785), zweiter Teil, sechstes Buch, pp. 244-245; von Georg Forster, Vilna, den Januar 1787 in Samuel Thomas Soemmerring, *Briefwechsel 1784-1792. Teil II Januar 1787 – Oktober 1792* (hrsg. von Franz Dumont) (Stuttgart: Gustav Fischer, 1998), pp. 507-508.

the first decades of the new century, at least until Lenin's comrades commissioned the German neurologist Oskar Vogt to study his preserved precious organ, and a Brain Institute was founded in Moscow to investigate the biological basis for greatness of mind.[15] Statistics began to work more systematically on a selection of already collected data, while research also began to focus more frequently on the brains of some other human group, for instance North American Indians, Chinese, and Australians. In the meantime the practice rarely included the microscopic observation of the cortex: since the turn of the century, the advent of neurocytology has only indirectly influenced the slow decline of cerebral anthropology. Moreover, it cannot be forgotten that the first decades of the century corresponded to the rise of Mendelian genetics that so consistently changed the game, and to the golden age of eugenics, the spread of which managed to infiltrate legislation in many countries. It's no coincidence that measuring (and improving) intelligence had been one of the main enigmas to which Francis Galton had devoted himself, and psychologists who chose to tackle the challenge soon employed reaction times and tests.

This is obviously another story, which fully developed in the twentieth century and does not come up in my book since it belongs to the psychological domain, adjacent to the territory, in itself quite vast, which I have explored here. However, the scheme of tests was originally very similar to that invented by cerebral anthropology, which had paved the way. By comparing reaction times in three small groups composed of Whites, Indians, and Africans, respectively, in 1895 R. Meade Bache had already argued that the Negro was "more of an automaton than the white man is", because of the quickness of his sensory responses. A slowing of these – from a Spencerian viewpoint – accompanied the intellectual evolution of the species. Two years later, George R. Stetson ran memory tests on a large sampling: one thousand children in the Washington (DC) area, equally divided between black and white. And here too, a marginally better performance of Blacks was cunningly explained as due in part to their deficiency in reasoning power.[16] Alfred Binet's comment on Stetson's experiment is noteworthy:

> Je me demande seulement quelle signification on peut bien attribuer à ces chiffres, puisque l'auteur nous apprend d'autre part que les noirs parlent mal l'anglais, ne le comprennent pas exactement et s'obstinent dans leur patois; dès lors, n'est-il pas probable que, pour expliquer les

15 See Hagner, *Geniale Gehirne* (cit. note 2), pp. 249-264.

16 R. Meade Bache, "Reactions Time with Reference to Race," *The Psychological Review*, 1895, 2:475-486; George R. Stetson, "Some Memory Tests of Whites and Blacks," *The Psychological Review*, 1897, 4:285-289.

> différences précèdentes dans le rang des élèves dans la classe, il faut tenir grand compte de la langue qu'ils parlent et de la cause d'infériorité qu'il existe pour les noirs? Cela suffit pour mettre en suspens toute conclusion.[17]

After dealing for a while with cephalometry and craniometry, Binet became disillusioned with their results and moved towards new horizons. In collaboration with his psychiatrist colleague Théodore Simon, and until his death in 1911, he developed three versions of "échelles" by administering tests to assess school-children activities. The following events are well known, after in 1912 William Stern had defined the *Intelligenzquotient*, while many others, less cautious than Binet, ended up reifying intelligence in the IQ. Conclusions were not at all put "on hold", as had been recommended by the French psychologist, but rather dictated a variety of practical applications there and then.

In 1997, Graham Richards called all this "an undead controversy", which resurfaced from time to time: the notorious 1969 article by Arthur Jensen aroused a durable, daunting mass of reactions, supported by Hans Eysenck who stressed genetic factors in 1971.[18] At the end of the century, the massive tome *Bell Curve* compiled by another two psychologists, Richard J. Herrnstein and Charles Murray, again stirred up polemics. Here it would be really too much to dwell on the racist, unflagging campaign that the Canadian J. Philippe Rushton had pursued since the 1980s, also as a director of the white supremacist Pioneer Fund until his death in 2012.[19] After all, biological determinism is always lurking about – a recurrent temptation both in popular science and in academic research, promising an easy shortcut to tackle economic, social, and political problems. Even more so in an epoch that is so prone to the cult of the gene.

---

17 Alfred Binet, review of Stetson, *L'Année psychologique*, 1897, *4*:575-576.

18 Arthur Jensen, "How Much Can We Boost IQ and Achievement?," *Harvard Educational Review*, 1969, *39/1*:1-123; Hans Eysenck, *The IQ Argument: Race, Intelligence, and Education* (New York: Library Press, 1971).

19 The critical literature on the history of IQ exploitation is quite rich: see for instance William H. Tucker, *The Science and Politics of Racial Research* (Urbana, Ill.: University of Illinois Press, 1995); Stephen Jay Gould, *The Mismeasure of Man* (New York-London: Norton & Company, 1996, revised and expanded edition); Graham Richards, *'Race,' Racism and Psychology* (London-New York: Routledge, 1997); William H. Tucker, *The Funding of Scientific Racism. Wickliffe Draper and the Pioneer Fund* (Urbana, Ill.: University of Illinois Press, 2002); Jonathan M. Kaplan, "Race, IQ, and the Search for Statistical Signals Associated with So-Called 'x'- Factors: Environments, Racism, and the 'Hereditarian Hypothesis'," *Biology and Philosophy*, 2014, *30*:1-17.

CHAPTER 1

Eighteenth-century Onset

1 Darker Skin and Brain

In 1665 Marcello Malpighi investigated the structure of the skin by first removing the outside layer of a boiled hoof of a pig, then also the reticular layer to reach the papillary pyramidal bodies, which – observed at the microscope – he held responsible for the sense of touch. Incidentally (*transeunter*) he happened to suggest that it was not incongruous to place the cause of *nigredo Aethiopum* in the mucous body that connects and protects the *papillae*. When in 1684 the médecin-philosophe François Bernier, a libertine and a great traveler, divided the Earth according to the different species or races of its inhabitants, he explained the dark color of the skin of the Egyptians or Indians with an accidental cause, namely the exposure to the powerful rays of the sun. On the contrary, with the exception of the northern coast of the continent, most Africans owed their essential *noirceur* to the inner structure of their body, to the seed or to the blood that would be dark like their exterior. And indeed – Bernier ventured to write – if one of their couples were transported to a cold country, their descendants would be born of the same color until a crossing with a white individual occurred.[1]

In Paris, the anatomist and surgeon Alexis Littré exploited the opportunity of a dissection carried out on the corpse of a black man to test and amend Malpighi's assumption in 1702. The Académie des Sciences gave an account of his experiments, from which he got the certainty that the dark skin was to be attributed partly to the tissue of the reticular membrane and partly to the action of an air "très-échauffé".[2] In 1724 Gian Domenico Santorini, chief physician of the Republic of Venice, collected his *Observationes anatomicae*, with a dedication to Peter the Great: the first pages of the first chapter discussed *De Aethiopum cutis nigrore*, about which anatomists had not yet established a true consensus. In fact, some of them placed the seat of dark color in the epidermis,

1 Marcello Malpighi, *De externo tactus organo anatomica observatio* (Neapoli: apud Aegidium Longum, 1665), pp. 20-21; [François Bernier], "Nouvelle division de la Terre par les différentes Especes ou Races qui l'habitent, envoyèe par un fameux Voyageur à N. l'abbé de la ***** à peu près en ce termes," *Journal des sçavans*, 1684, pp. 133-140.

2 The report of Littré's text is in *Histoire de l'Académie Royale des Sciences. Année MDCCII. Avec les Mémoires de Mathématiques et de Physique* (Paris: chez Gabr. Martin, Jean Bapt. Coignard & les Frères Guerin, 1743), pp. 30-32.

© KONINKLIJKE BRILL NV, LEIDEN, 2020 DOI:10.1163/9789004431881_003

others in the reticular body, and others in both. After macerating the skin of a black man for many days, Santorini agreed with Malpighi but also noticed greater darkness in the outer part of the epidermis and surmised that the cause was to be identified in the bile.[3]

Since then, an observational and discursive field took shape and developed about the color of "Negro" skin, its origins and natural history, as Europeans' geographical expansion proceeded. Until the early nineteenth century, at least thirty-eight dissections of Africans were performed just to study their cutaneous pigmentation, while there is no evidence of Amerindian or Asiatic bodies subjected to the same kind of investigation.[4] Let's just mention, here, that a peak of interest in the matter came with the prize offered in 1739 by the Académie royale de sciences de Bordeaux "à tous les Sçavans de l'Europe" for the best response to the question "Quelle est la cause physique de la couleur des nègres, de la qualité de leur cheveux, et de la dégénération de l'un et de l'autre?".[5] Among the several competitors, the Perpignan-born naturalist and physician Pierre Barrère – who had spent a few years in Guyana – anonymously published a dissertation in 1741, in consonance with Malpighi's hypothesis, which maintained that the darkness of the skin – surely not "une couleur d'emprunt" – was produced by the blood and the black bile.[6] Barrère brought new reasons in favor of the bile argument that had been already put foward by ancient authors and repeated by Renaissance sources then reappeared everywhere in the first half of the 18th century.

To get an idea of how prominent the issue was in that period, just browse the seventh volume of the *Teatro crítico universal* (1736) by the enlightened

3 *Observationes anatomicae Jo. Dominici Santorini* (Venetiis: Apud Jo. Baptistam Recurti, 1724), pp. 1-4.

4 Renato G. Mazzolini, "Skin Color and the Origin of Physical Anthropology," in *Reproduction, Race, and Gender in Philosophy and Early Life Sciences*, edited by Susanne Lettow (Albany, NY: State University of New York Press, 2014), p. 138. See the chart 6.1 (p. 137) showing "the chronological distribution, between 1640 and 1849, of the first editions of four hundred works that either sought to explain why the skin color of sub-Saharan Africans was black or used skin color as a criterion for the classification of humankind". The high number of texts recorded by Mazzolini does not include entries in dictionaries and encyclopedias, and would demonstrate "that the problem of the pigmentation of Africans was a major and constant subject of inquiry."

5 The prize was announced in various locations, for example in the *Journal des Sçavans pour l'année M. DCC. XXXIX, Juin,* p. 1131. Sixteen essays, now available for consultation at the Bibliothèque municipale de Bordeaux, were submitted but none of them rewarded: see Andrew S. Curran, *The Anatomy of Blackness. Science & Slavery in an Age of Enlightenment* (Baltimore: The Johns Hopkins University Press, 2011), pp. 82-87.

6 M.***[Pierre Barrère], *Dissertation sur la cause physique de la couleur des nègres, de la couleur de leurs cheveux, et de la degeneration de l'un et de l'autre* (Paris: Pierre-Guillaume Simon, 1741).

Benedictine monk Benito Jerónimo Feijóo y Montenegro, and find its third part dedicated to *Color etiópico*, with a review of past and present opinions: a question that, as stated by him, would be of equal interest to natural philosophy and religion. It discussed, in detail and from a rationalistic point of view, whether the biblical curse of Noah was still a reliable explanation for the dark pigmentation, or if it was the burning heat of the African sun, or even the power of imagination in the Ethiopian's female ancestor, or water and food consumed by the inhabitants. According to Feijóo, the true, unique cause of the color is the influence of the climate, i.e. the inhabited country. The general influential cause may also consist not in a single thing, but in a combination, or complex of multiple things. Vapors, exhalations, corpuscles of the atmosphere – effluvia of the earth – or the bodies from which they exhale, must be reckoned as the particular good or bad qualities of a country. Different countries, by their different features, induce a distinctive variety in color, and even in the configuration of their dwellers.[7] From the second half of the 18th century onward, the topic became obligatory for anyone concerned with the natural history of man.

Probably dating back to the period when Montesquieu was working on the draft of the *Esprit des Lois* – between 1736 and 1743 – his *Essai sur les causes qui peuvent affecter les esprits et les caractères* contains interpretative cues that were subsequently recast in his major work.[8] The text is divided into two parts, dedicated to physical and moral causes. The general thesis presented in the first part is well known and shared by many in the eighteenth century: the climate and the kind of food determine the physical constitution of individuals and nations. In order to communicate impressions to the soul through the nervous juice, it is necessary for the nerve fibers to be flexible and susceptible to move and be moved. Using a musical metaphor, the smaller the string of an instrument, the more acute the sound from more vibrations; and, vice versa, the bigger the string, the deeper the sound. In the same way, therefore, when the fibers of the soul are coarse, the vibrations become slower and less frequent. Their stiffness or coarseness can produce slowness of spirit; but from their great flexibility, together with relaxation, a weakness might come. It was almost unknown what particular arrangement of the brain would be required to have a lively spirit, but some conjecture seemed possible to Montesquieu.

7 Benito Jerónimo Feijóo y Montenegro, *Teatro critico universal, ò Discursos varios en todo genero de materias, para desengaño de errores communes* (Madrid: Imprenta de Francisco del Hierro, 1736), vol. VII, pp. 69-94.

8 The *Essai* remained unpublished for a long time, until it was included in *Mélanges inédits de Montesquieu publiés par le Baron de Montesquieu* (Bordeaux-Paris: G. Gounouilhou-J. Rouam, 1892), pp. 109-148.

Surely liveliness is lacking in peoples living in cold countries because their brains suffer from superfluous moisture. On a similar basis, by the way, when presenting his "art de former les hommes" in 1762, Jean-Jacques Rousseau decided to set Émile in a temperate climate, because the organization of the brain is less perfect in extreme environments. A Frenchman adapts rather easily to life in Guinea or Lapland, but an African cannot survive in northern Finland, nor a Samoyed in Benin.[9]

A surprising inter-individual variety was revealed by anatomical observation: perhaps there had never existed two persons whose organic parts were arranged in the same way under every aspect. The differences were not negligible in the very small vessels of the brain. Here Montesquieu proposed another metaphor, of Eraclitean inspiration: in the human body, the soul is like a spider in its web. It cannot move without moving the thread that stretch out, and no one can touch the threads without the spider moving. The more these threads are extended, the better the spider is warned. If some thread is loose, the communication to the spider or to another thread will be lesser, and the spider's ability to interfere will be suspended in its own web. In other words, the state of the soul depends not only on the arrangement of the brain, but also on the whole bodily "machine", even on those parts that one would never think of. However, Montesquieu admonished to be "extrêmement ménagers" of the cerebral fibers with an example that should not be followed. The Orientals usually reach euphoria with decoctions of hemp, which give them such intense pleasures that, for a few hours, they are beside themselves. This condition is followed by a total abatement and a state close to lethargy. For its part, the abuse of wine slowly degrades the drunkards, whose fibers are excited for a while, then collapse and need increasing alcoholic doses to produce the same effect. In view of a wise administration of the brain, other forms of abuse should be avoided, whether it is sleep or waking, fast or loneliness.[10]

The second part of Montesquieu's *Essai* deals with moral causes. When an individual begins to make use of his own reason, this can happen either in a barbaric or in a civilized people. In the first case, the individual has only ideas of self-preservation and ignores everything else. It is not by chance that the barbarous languages are sterile: few words are enough to express just as few things. The cerebral fibers of those who speak them, little accustomed to being flexed, have become rigid. Montesquieu believes that the savage Americans are

9 Jean-Jacques Rousseau, *Émile, ou de l'éducation* (La Haye: Jean Néaulme, 1762), I, pp. 56-57.

10 *Mélanges inédits de Montesquieu* (cit. note 8), pp. 109 ff.

indocile, incorrigible, and unable to reflect and learn. To teach them something – that is, to flex the fibers of their brain – would be like asking paralytics to walk. Their roughness can reach such a point that they differ little from animals. So unused brains end up losing their functions: they almost do not benefit from their soul, nor from its union with the body, made perfect only by the education of which civilized nations are capable.[11] In a conference held at the Sorbonne in 1750, the young Turgot, just graduated, likewise recalled a "heureux arrangement" of the cerebral fibers, more or less strength or delicacy in the sensory and memory organs, a certain speed in blood circulation as the only natural differences in the human species: their causes would always remain unknown, but the "dispositions primitives" act equally in savage and civil peoples, in every place and time. All the rest would come from the power of education: if Corneille had been raised in a village, tied to a plow all his life, if Racine had been born among the Hurons of Canada, they would have never unfolded their genius.[12]

Brain is not even mentioned in the second part of Buffon's *Histoire naturelle de l'homme* (1749), where the geographical varieties of the human species known at that time are described, starting from the extreme north. It should be added, however, that he had entrusted his collaborator Louis Jean-Marie Daubenton with the description, included in the same volume, of over four hundred items from the rich and constantly growing collection of the Cabinet du Roy. Deformed skeletons, mummies, monsters, anatomical preparations: eight years of work had been spent distributing the confused *bric-à-brac* in an order including classes, genera and species, the three kingdoms of nature separately organized. The environmental conditions in which the objects were preserved had a particular importance, depending on their character. The continuous view of them – according to the curator – struck with much more force and truth than the most exact words or perfect images.

Daubenton began his illustration with the part of the Cabinet concerning man. Bones and cartilages, first of all, appropriately prepared. The object number one is the skeleton of a fetus two and a half inches tall, followed by a long series of 280 other bone elements, normal, abnormal, and pathological. Afterwards, the anatomical pieces injected, dried, etc. are examined. In particular, Daubenton highlighted the difficulties of the recent art of injecting various substances into the blood vessels, and appreciated Frederik Ruysch's

11 Ibid., pp. 129-131.

12 "Plan du second discours sur l'histoire universelle, dont l'objet sera les progrès de l'esprit humain," in *Oeuvres de Turgot. Nouvelle édition classée par ordre de matières, avec les notes de Dupont de Nemours* (Paris: Guillaumin, 1844), II, pp. 645-646.

wonderful preparations, bought in 1717 by Tsar Peter I. Nor were the excellent procedures developed in Edinburgh by Alexander Monro neglected. Thus four injected heads of children are presented, one of them "nègre de deux ou trois ans"; in addition, the "tégumens injectez de la tête d'un nègre", about which Daubenton only says that the peculiar features of his physiognomy can be better distinguished than in the previous one. Nothing about the brain, probably extracted before proceeding to the preparation.[13]

Against the background of the crucial climate, physiognomy and morphology, character and behavior were the parameters chosen by Buffon to identify each individual variety. In his own words: *couleur, forme et grandeur, le naturel.*[14] Value judgments – aesthetic and ethical – do abound in over a hundred and fifty pages that account for the various areas of the world. Africa deserved the largest number of pages in Buffon's anthropological geography, due to the range of human varieties that inhabit it. All the shades of pigmentation from brown to black are present across the continent, similarly to what happens in Europe, from white to brown. Not only the skin color varies a lot, but also the type of hair, smell and customs change too, as described by travel reports.[15] From some of these, Buffon obtains information, *inter alia*, on the limited *esprit* of the Negroes of Guinea, who can only count to three, do not think and have no memory, but on the other hand they are human, docile, gullible and superstitious, ready to become good soldiers. Intense feeling (*sentiment*) is what characterizes them: cheerful or melancholic, hardworking or idler, friends or enemies depending on how they are treated. Buffon is moved by

13 *Histoire naturelle, générale et particulière, avec la description du Cabinet du roy* (Paris: De l'Imprimerie royale, 1749), III, pp. 149-150. According to Curran, "La description de ces spécimens, sevrés de leur corps et identifiés par leur couleur et leur morphologie plutôt que leur provenance géographique, illustre clairement la matérialité et la mesurabilité croissantes appliquées vers 1750 aux catégories humaines, en particulier celle du *Nègre*" (Andrew S. Curran, "Buffon et l'histoire naturelle des Africains," *Dix-huitième siècle*, 2012/1, *n. 44*:183-199, p. 183.

14 *Histoire naturelle, générale et particulière* (cit. note 13), p. 371. A little later, dealing with the topic *humaine espèce,* did nothing but synthesize Buffon's text: see *Encyclopédie ou Dictionnaire raisonné des sciences, des arts et des métiers, par une société des gens de lettres* (Neufchastel: Samuel Faulche & Compagnie, 1765), 8, pp. 346-348.

15 It is known that Buffon left France only as a young man, to travel with the Duke of Kingston in Italy, Switzerland and England – even though biographers and historians have provided different versions of his experience. Therefore all the news on foreign countries included in his *Histoire naturelle de l'homme* come from a great quantity of works composed more or less recently by travelers, missionaries, explorers. As for his portrait of Africa, see Sylviane Albertan-Coppola, "Des récits des voyageurs à l'*Histoire générale des voyages*," *Dix-huitième siècle*, 2012/1, *n. 44*: 165-181 and Curran, *The Anatomy of Blackness* (cit. note 5), pp. 67-69.

their long history of slavery and declares that humanity turns against the brutality of their masters and the hideous condition into which greed of profit has forced them.[16] He also refers to the almost centuries-old research for possible causes of the dark skin and quotes the opinion of those who claim to have seen the blood of Africans blacker than that of Whites, nevertheless without having been able to personally confirm the truth of their statement. It is enough for him to have noticed that, among Whites, a darker complexion carries a darker blood.[17]

Needless to say, the most beautiful and best-made humanity lives between 40 and 50 degrees of latitude. Only the temperate zone nurtures the 'true' color of the human species, model or unit with which to compare all the other shades of pigmentation and beauty. Buffon never doubted that the human varieties belonged to the same species – originally white – which by multiplying and spreading over the entire surface of the Earth, would undergo significant changes due to climate, nutrition, ways of life, epidemics and crossings. Adapting to the sands of the desert or the ice of the north, diversification had reached such an extent as to give an excuse for allocating "le Nègre, le Lappon & le Blanc" to different species, if it were not for their possibility to mate and propagate the great and unique human family. The alterations of their nature are nothing but superficial and all come from the same dynamic stock, representing a single type. Buffon was one of the main advocates of a naturalistic dogma that was bound to last, despite the challenges and refutations that would later become more and more aggressive.[18]

Another, imaginative perspective came from Germany around mid-century. In 1744, Johann Friedrich Meckel was just twenty years old, when he began a correspondence with Albrecht von Haller, whose lessons at the University of Göttingen he had followed for a while. In fact, at the request of his father, he then moved to Berlin to take anatomical courses under Augustin Buddeus and to learn medical practice. In the capital of the Kingdom of Prussia, he stayed for a couple of years, exchanging letters with his favorite mentor, and managed to gain the qualification of a prosector. His main interest was then in the nervous system, so that his 1748 dissertation in Göttingen dealt with the fifth pair of cranial nerves, which serve the muscles of the face. Although dedicated to Maupertuis, recently appointed president of the new Königliche Akademie der Wissenschaften, and to Johann Theodor Eller, *Leibarzt* of the King Frederick II,

16 *Histoire naturelle, générale et particulière* (cit. note 13), pp. 469-470.

17 Ibid., pp. 524-526.

18 *Histoire naturelle, générale et particulière, avec la description du Cabinet du Roi,* (Paris: De l'Imprimerie royale, 1766), XIV, pp. 311-313.

Meckel's *Tractatus* fully acknowledged his great debt to Haller, prepared the trigeminal nerve with great precision, and achieved great recognition in Germany and beyond.[19]

Back in Berlin, Meckel settled down as a practicing physician, but was already admitted into the Akademie der Wissenschaften in 1749, and soon appointed assistant professor of anatomy. After Buddeus' death in 1753, he succeeded him on the main chair at the Collegium medico-chirurgicum (Charité) – where body materials to be dissected were always available – and turned into an excellent *Präparator*.[20] That same year, Meckel presented to his academic fellows the results of an anatomical research into the corpse of a twelve-year-old *Nègre* at the service of the Count of Neale, who had died on July 26th. The first question was posed by the color of the skin: given the different opinions on the subject, direct observation could have added something to what other anatomists had already said. So, despite the unsuitable season, the dissection had been promptly and carefully completed, because similar occasions were quite rare. In previous cases – Meckel noted – only the black mask or the set of bones of "cette espèce d'hommes" had been uselessly preserved, rather than studying their body in order to discover some new truth.[21]

---

19 From November 1744 to December 1748 Meckel informed Haller in eleven letters about his activities in Berlin: *Epistolarum ab eruditis viris ad Alb. Hallerum scriptarum Pars I. Latinae. Vol. II – Epistolae 195 ad 404* (Bernae: Sumptibus Societatis Typographicae, 1773); for his dissertation see Johann Friedrich Meckel, *Tractatus anatomico-physiologicus de quinto pare nervorum cerebri, duabus figurarum tabulis illustratus* (Gottingae: Apud Abram Vandenhoeck, 1748). After listing in the first section the contribution of his predecessors (from Galen on), in the second he exposed *De vera origine et divisione nervi quinti paris*, while the following ones deal with the three branches of the nerve. The sixth section contains a *Physiologica explicatio*. Six figures in two tables visualize the anatomical details.

20 Senta Berner, *Die fünf Anatomen Meckel* (Freiburg im Brisgau: Universität, Medizinische Fakultät, Diss., 1963). His son Philipp Friedrich Theodor and especially his grandson Johann Friedrich Meckel (called the Younger), further increased the fame and content of the collection he had started: see Sabine Schwarz, *Die anatomische Privatsammlung der Anatomenfamilie Meckel unter besonderer Berücksichtigung ihres präparationstechnischen Profils. Dissertation zur Erlangung des akademischen Grades Dr.med. vorgelegt dem Rat der Medizinischen Fakultät der Martin-Luther-Universität Halle-Wittenberg*, pp. 3-23. URL: <http://webdoc.sub.gwdg.de/ebook/m/2002/pub/medizin/00H110/prom.pdf>. More information in Rashid M. Janjua (et al.), "The Legacy of Johann Friedrich Meckel the Elder (1724-1774): A 4-Generation Dinasty of Anatomists," *Neurosurgery*, 2010, *66*: 758-771.

21 Johann Friedrich Meckel, "Recherches anatomiques. I. Sur la nature de l'epiderme, et du réseau, qu'on appelle malpighien; II. Sur la diversité de couleur dans la substance médullaire du cerveau des nègres; III. Description d'une maladie particulière du peritoine," *Histoire de l'Académie Royale des Sciences et Belles Lettres. Année MDCCLIII* (Berlin: chez Haud et Spener, 1755), pp. 79-113.

The skin of the dead boy turned out to be chromatically different depending on the part of the body, sprinkled here and there with scars left by smallpox and useful to explain the origin of the epidermis and of its color. Using the microscope and following a common technique – arm and foot macerated in water for three weeks – Meckel concluded that putrefaction dissolves the Malpighian *rete mucosum* while leaving the firm tissue of their structure to the skin and epidermis. Hence, the just completed dissection allowed him to draw some clues on the differences between the various layers of the body surface and, in particular, to define the epidermis as "la croute extérieure de la membrane muqueuse desséchée et endurcie", to which the compression and action of the air gradually give greater or lesser thickness and hardness in the various areas of the body.[22]

The second part of Meckel's presentation at the Akademie der Wissenschaften is not only more original than the first one, but it contains alleged observational data that had a wide resonance and later were often repeated by a number of authors. To begin with, the dissection of the *Nègre*'s head manifested a singular diversity in the occipital bone: its upper part, up to the middle, seemed formed by a rhomboidal bone with some morphological features that were far from the "état naturel". His death dated from the day before, so that – after the opening of the skull – the brain appeared intact and solid, with a fresh smell. Cutting and peeling away slices from top to bottom, Meckel realized that the color of the medullary substance, thicker than the cortical one, was not as white as usual, but more and more bluish as the dissection proceeded downward. However, after exposing the slices to the air, their color quickly whitened. Instead of the common ashen color, the pineal gland was blackish blue too: something surely not caused by decomposition, the structure being firm and without a bad smell.[23] If externally the striated body and the layers of the optic nerve had the color of the cortical substance, the dissection showed they contained a darker medullary substance.

As for the rest, Meckel could not find any real difference in the figure and structure of the boy's brain. Assuming that the particular color of the medullary substance was generally found in black people, he hypothesized that it could represent a constitutive character, although a single case was not enough and only further dissections could support his observation. It seemed

22 Meckel, "Recherches anatomiques" (cit. note 21), pp. 79-97.

23 Later Meckel's attention turned to the pineal gland which, due to "la fiction de Descartes", had been the object of research without satisfactory results: see Johann Friedrich Meckel, "Observations anatomiques sur la glande pinéale, sur la cloison transparente, et sur l'origin du nerf de la septième paire", *Histoire de l'Académie Royale des Sciences et Belles-Lettres. Année MDCCLXV* (Berlin: chez Haude et Spener, 1767), pp. 91-101.

to him that until then, being in front of a black corpse, the anatomists had focused on the color of the skin, without really going any further. Santorini, for instance, had exclusively sought the cause of blackness in the liver, by simple analogy with jaundice. Scientific prudence did not prevent Meckel from suggesting that the evaporating liquid, which dyes the medullary substance, might also help, through the brain stem and the skin nerves, to determine the color of the subcuticular mucous membrane.[24]

Two years later, another and darker *Nègre*, a drum player, had died of a tragic accident in Berlin. Meckel could anatomize his body: again, the skin was investigated, with the maceration of its Malpighian, black mucosity and the microscopic observation of fragments of dermal tissue. And again the organ under strict examination was the brain, during a more favorable season (February) than in the previous case; the doubt that heat had produced the blackish color of the medullary substance was thus excluded. Then Meckel compared it with the brain of a European of the same age who had died the same day, being thus able to notice, in the former's brain a cortical substance more ashy and a yellowish medullary substance verging to gray, in contrast to the perfect whiteness of the latter's brain. To make the dark color fade, a longer exposure to the air was needed in winter, while the summer temperature had facilitated the evaporation of the volatile parts.

Not trusting his own eyes, Meckel had also decided to subject the two brains, side by side, to people unaware of both their internal structure and the purpose of his research. Without hesitation they had distinguished the chromatic difference, which increased as the dissection reached the *corpora striata*, of a brown similar to that of a tree bark. The color of the respective pineal glands differed greatly, as did the inside of the spinal cord. Consequently a fluid could supposedly be transmitted by means of the nerves to the skin surface and determine the color of the subcuticular mucosa and then that of the epidermis. In conclusion, Meckel also emphasized a further, dramatic difference: the *Nègre*'s blood was so dark as to blacken the sheets instead of reddening them. From all this he deduced that Blacks, in terms of internal composition, formed "almost" a different species of men.[25]

---

24 Meckel, "Recherches anatomiques" (cit. note 21), pp. 97-102. The third part of the text accounts for the illness – hardening of the peritoneum – the black boy had died of (pp. 102-113).

25 Johan Friedrich Meckel, "Nouvelles observations sur l'epiderme et le cerveau des Nègres," *Mémoires de l'Académie Royale des Sciences et Belles-Lettres. Classe de philosophie expérimentale*, 1757, *13*: 61-71. During the last twenty years of his life, Meckel's medical practice committed him more and more, not infrequently at the service of the king himself.

In the following years, a few periodicals reviewed Meckel's *Mémoire* and amplified its echo. After the news was released among *Gens des Lettres*, not surprisingly there were those who tried to ascertain its verisimilitude. At the Hôtel-Dieu in Rouen, François-Nicolas Melaure, a seventeen-year-old black boy, died on March 4, 1757, due to the rare dislocation of the odontoid apophysis. The Master surgeon Claude-Nicolas Le Cat, a local authority who also enjoyed a reputation in the European scientific community[26], dissected his corpse and presented the results at a session of the *Académie des Sciences, Belles-lettres et Arts* of the Norman town, which he had been instrumental in founding around 1744.

At first glance Melaure's brain seemed to him quite similar to that of a White, but, at a closer look, he noticed a bluish tinge in both the cortical and medullary matter, and an even bluer, almost black, pineal gland. As a *rendez-vous* of nerves and arterioles, this was necessarily provided with a large amount of *oethiops*, namely the combined liquid – a mixture of mercury and sulfur, extracted from those two kinds of vessels – that Le Cat had postulated to exist in all organisms in variable quantities: an extract of the nervous fluid, a yeast (*levain*) more copious in the African body. In general, the union of the nervous fluid with that of the arterial extremities would be necessary for all organs to function. By imitating the test already carried out by Meckel – whose experiences he had learnt about by reading the Parisian *Annonces, affiches et avis divers* in November 1756 – Le Cat put his designer in front of two pieces of brain, respectively belonging to Melaure and to a white man, to receive confirmation of their different color and get them painted. A second dissection of a *Nègre*'s corpse, made the following year, showed him the same phenomena. Curiously, evidence apparently came from the animal world: three times, between 1761 and 1764, Le Cat had examined the brains of three rabbits – black, white and gray specimens – and found, in the first one, an *oethiops* similar to that of black men. Unfortunately, a fourth experience did not give the same clear result, nor did another one related to the brains of two sheep, white and black.[27]

26 Gérard Hurpin, "Claude-Nicolas Le Cat ou de la notoriété médicale au XVIIIème siècle," *Histoire des sciences médicales*, 2001, 25:151-162; Jean-Pierre Lemercier, "Claude-Nicolas Le Cat et l'Académie des Sciences, Belles-Lettres et Arts de Rouen," ibid:163-168. Le Cat had made himself known above all for his combative ardor, which led him to argue also with prominent figures such as Haller and Rousseau. In 1739 the Royal Society had elected him a foreign member.

27 A report on Melaure's dissection and on the experiences on animals is included in Claude-Nicolas Le Cat, *Traité de la couleur de la peau humaine en général, de celle des nègres en particulier, et de la métamorphose d'une de ces couleurs en l'autre, soit de naissance, soit accidentellement* (Amsterdam, 1765), pp. 51-62.

Although the brain was reputed to be the first laboratory or source of the alleged *oethiops*, nonetheless the whole nervous system seemed to participate in secreting it. Le Cat had previously postulated the existence of this black substance in the second edition of his *Traité des Sens*: at that time called *aethiops*.[28] In the 1765 treatise, he even referred to the fifteenth book of Strabo's *Geography*, to corroborate his own idea of a color contained in the parents's *semence*: nobody would doubt that the brain is a "spermatic part", comparable to a sort of fruitful almond that produced all the rest of the animal body. The engraving next to the title page of the book is extremely eloquent, with the image of the seated white lady, surrounded by three characters that represent the other three fundamental races, and the words *non vultus, non color unus*, taken from Virgil's Aeneid but with a different meaning.[29]

In his historical work on the *Anatomy of Blackness*, Andrew S. Curran rightly comments about the tendency to search for deeper, organ-based racial differences that took place in those decades.[30] Despite the fact that the young Dutch clergyman – a philosopher and a geographer – Cornelis de Pauw had never visited the American continent, he happened to be considered one of the highest authorities on the populations of the New World, a reputation built second hand, on his readings of travel stories.[31] It was 1768-1769 when the first edition of his *Recherches philosophiques sur les Américains* in two volumes was published in Berlin, where for several months he had been a guest at the court of Sanssouci. His systematic devaluation of overseas land was in consonance with the anti-emigration policy of Frederick the Great, who might have appreciated de Pauw's warning against the disadvantages caused by colonial conquests. Buffon's denigration of all the American nature – brought out in 1766 by the 14th volume of his *Histoire naturelle* – was pushed to its extreme: de Pauw described the Amerindian as a degenerate living in a weak and corrupted nature, a *minus habens* who is incapable of any improvement, the opposite of the good savage cherished by some missionary or *philosophe*.[32]

28 Claude-Nicolas Le Cat, *Traité des sens. Nouvelle edition* (Amsterdam: J. Wetstein, 1744), pp. 159-160)

29 Le Cat, *Traité de la couleur* (cit. note 27), p. 58.

30 Curran, *The Anatomy of Blackness* (cit. note 5), p. 125.

31 "His books passed through edition after edition, and were translated from the original French into English, Dutch, and German." (Henry Ward Church, "Corneille de Pauw, and the Controversy over his *Recherches philosophiques sur les Américains*," PMLA, 1936, 51:178).

32 See the chapter *De la dégénération des animaux*, in *Histoire naturelle, générale et particulière, avec la description du Cabinet du Roi. Tome Quatorzième* (Paris: De l'Imprimerie Royale, 1766), pp. 311- 374. About the controversial thesis of a 'minor' New World, the classic historical narration is that by Antonello Gerbi, *The Dispute of the New World. The*

FIGURE 1.1 Engraving in Claude-Nicolas Le Cat, *Traité de la couleur de la peau humaine en général, de celle des nègres en particulier, et de la métamorphose d'une de ces couleurs en l'autre, soit de naissance, soit accidentellement* (Amsterdam 1765).

In the first volume of his work, even Africans come on stage for a while. The effects of heat on the human constitution under the equinoctial line – de Pauw reminded the reader – had been discovered thanks to the anatomy of Blacks and the analysis of their bodily fluids. Thus, he took for granted that the medullary substance of their brain was blackish, their pineal gland almost entirely black, their optic chiasm brownish, with a sperm colored of the same principle that is widespread in their mucous membrane. He was surprised that modern authors had ignored the relationship between blackness and seminal matter. The coloring material would be so tenacious as to require at least four generations of racial crossing to disappear altogether. Of course, the sources cited by de Pauw are Meckel's *Recherches anatomiques* and Le Cat's book of 1765, whose *Aethiops animal* is directly recalled in the second volume. He also hypothesizes that the most delicate organs of black people's brain must have been destroyed or obliterated by the torrid climate, and their faculties impaired. They would differ from Europeans both for the limits of their memory and for the impotence of their spirit, no less than for their color and physiognomy. No doubt, white men transplanted in places burnt by sunrays would see their skin darken and their qualities change for the worse, in the long run.[33]

Among the adepts of the Meckelian belief, in 1775 the French vitalist physician Pierre Roussel had already presented a *Système physique et moral de la femme*, while a dozen years later he dealt with the *Physique de l'homme* for a fortnightly *Bibliothèque universelle des dames*, conceived as a collection of works to give a general, easily available education, to women of a certain class. On that occasion, he referred to the disputed question of skin pigmentation, only to emphasize that a Black is such in every part of his body, except for the teeth. All his organs – he thus reiterated an almost common place – bring more or less the imprint of the dark color: the medullary substance of the brain is blackish, and this tint prevails in the different parts of the organ, in sperm and blood, and reaches the greatest darkness in the bile.[34]

History of a Polemic, 1750-1900. Revised and enlarged edition translated by Jeremy Moyle (Pittsburgh: University of Pittsburgh Press, 1973).

33 Mr. de P.*** [Cornelis de Pauw], *Recherches philosophiques sur les Américains, ou Mémoires intéressants pour servir à l'Histoire de l'Espèce humaine* (Berlin: George Jacques Decker, 1768-1769), tome I, pp. 179-184; tome II, p. 41.

34 Pierre Roussel, *Système physique et moral de la femme, ou Tableaux philosophique de la Constitution...* (Paris: Vincent, 1775); [Pierre Roussel], "Bibliothèque universelle des dames. Physique de l'homme. Tome premier," (Paris: Rue et Hötel Serpente, 1787), pp. 209-212. Later the two texts were brought together and posthumously expanded: see Id., *Système physique et moral de la femme, suivi du Système physique et moral de l'homme, et d'un fragment sur la sensibilité* (Paris: Crapart, Caille et Ravier, 1805).

In August 1785, two ships sailed from Brest – the Astrolabe and the Boussole – under the command of Jean-François de la Pérouse, an expert navigator and geographer commissioned by Louis XVI to explore the Pacific Ocean. The "expédition de découverte" was quite unfortunate, as the two crews disappeared in 1788, probably due to a tropical storm. Only forty years later were the shipwrecks found at Vanikoro, in the Solomon Archipelago. However, information on the long voyage was published, also containing some *Questions proposées par la Société de Médecine, aux savans qui accompagnent M. de la Pérouse*, which recommended them to use the utmost observational and descriptive accuracy. Fourteen points listed in detail the human objects on which they had to direct the scientific gaze, and the twelfth one asked some questions the *savans* should have answered:

> Le rapport de la couleur de la peau avec celle des humeurs. La liqueur spermatique des hommes plus ou moins basanés, la pulpe cérébrale et le sang, répondent-ils à la teinte de leur peau? Cette couleur varie-t-elle parmi les noirs, dans quelques individus, tels que les nègres-blancs, les blafards, &c. Cette variation est-elle le produit d'une maladie, ou d'une constitution altérée par l'influence du climat, comme on le pense des nègres transportés dans les pays froids?[35]

2 Qualitative and Quantitative Differences

The question concerning the duplicity of the human species, viz. the relationship between the material and the spiritual side, became more and more crucial in the second half of the 18th century. Buffon had faced it in a traditional, rather Cartesian way, while other *philosophes* risked more courageous hypotheses. In Germany, Ernst Platner – a young professor of medicine in Leipzig – defined *Anthropologie* as a medical-philosophical science of the whole individual; in other words, the study of body and soul in their mutual relations and limits, a harmony made possible by the mediating nerve fluid. Written in a sort of aphoristic style, his 1772 textbook, *für Aerzte und Weltweise*, became widely mentioned during the last quarter of the century and used as a manual by many university courses in Germany.

35 *Voyage de La Pérouse autour du monde pendant les années 1785, 1786, 1787 et 1787, publié conformément au décret du 22 avril 1791, et rédigé par M. L. A. Milet-Mureau* (Paris: de l'Imprimerie de la République, 1797), I, p. 184.

Platner borrowed from Albrecht von Haller the model of a nervous system consisting of a system of canals and, although placing the soul in the brain, he continued to conceive it as immaterial, in a rather aporetic way.[36] The last chapter of Platner's *Anthropologie* examines the contribution given by the body to "genius" and tries, first of all, to determine the requirements of this loose concept. Four of them are listed: an ability to be attentive to the surrounding reality; a particular vivacity and distinctiveness of the cerebral impressions, of those both arising from the sensory objects, and depending on memory; a broad connection of ideas, and an easy transition from one to the other; a distinctive speed in all these operations. If the specificities of genius are brought back to their physical causes, then the contribution of the body is found in the following qualities: a soft and irritable brain marrow, a lively movement of spirits, a multiple connection and special fluidity of the brain canals. Furthermore, Platner associates some individual or collective figures with the presence of that type of brain: much more in the young than in the old, but also nations, whose body is not hardened either by the sky, or by a raw way of life.[37]

A few years later, the young Friedrich Schiller graduated in medicine at the Karlsschule (Militärakademie) of Stuttgart with two dissertations, one of them discussing the connection between the animal and the spiritual nature of man. He set out to highlight more clearly the conspicuous contribution of the body to the activities of the soul (*Seele*), the great and real influence of the animal sensory system. The material association is the ground on which thought rests, the guide of the creative mind. Through it alone, ideas can be assembled, compared, and can then direct the will. This claim may seem dangerous to freedom, but Schiller offered a way out: the soul has an active influence on the *Denkorgan*, and all human morality is grounded in this activity. So, the *Denkorgan* is the true tribunal of the soul, and vice versa, by virtue of a constitutive reciprocity, entirely dependent on attention (*Aufmerksamkeit*). It follows that the confusion of the spirits in case of disease, when it is propagated into the *Denkorgan*, turns the wisest man into the most ridiculous fool, the thinker into the simpleton, the meekest into a fury.[38]

---

36 Ernst Platner, *Anthropologie für Aerzte und Weltweise* (Leipzig: Dyck, 1772); see John H. Zammito, *Kant, Herder, and the Birth of Anthropology* (Chicago and London: The University of Chicago Press, 2002), pp. 250-253; in general, see also the essays by several authors to "Themenschwerpunkt: Ernst Platner (1744-1818). Konstellationen der Aufklärung zwischen Philosophie, Medizin und Anthropologie," *Aufklärung*, 2007, *19*: 7-378.

37 Platner, *Anthropologie* (cit. note 36), pp. 282-291.

38 Johann Christoph Friedrich Schiller, *Versuch über den Zusammenhang der thierischen Natur des Menschen mit seiner geistigen. Eine Abhandlung welche in höchster Gegenwart*

That was a time when many thought that the brain might forever torment the ingenuity of naturalists, physicians and philosophers. Of all the bodily systems, it presented the most unknown composition, functions and pathologies. For some decades, the urgency of giving rigor and relative certainty to its study was often stressed. Among those who complained about this condition and worked to overcome it, Félix Vicq d'Azyr – a young anatomist and a perpetual secretary of the Société Royale de Médecine – made the brain the main object of his investigations. Knowing it better was essential in order to formulate an idea about the nature of each animal species, as he wrote in a *Discours sur l'anatomie en général*, which opened his excellent treatise of 1786, dedicated to Louis XVI.[39] In fact, the main cerebral provisions appear to be constantly linked with the general sensibility, the energy or weakness of instinct, the vehemence of appetites, the strength of affections, the extension of intellectual faculties. Based on this assumption – Daubenton being his "maître vénéré" – Vicq d'Azyr conceived the great project of aligning man and all animals in natural sequence, through a series of anatomical tables accompanied by appropriate captions. Above all, he wanted to show how that sequence was punctuated by a decreasing degree of cerebral complexity, along the zoological scale from top to bottom.

Although the brain of the quadrupeds contained almost all the parts present in that of man, the differences between them were undeniable. The main one consisted, of course, in the smallness of the cerebral hemispheres, together with the lacking fissure of Sylvius; furthermore, in most animals, the convolutions of one hemisphere resemble those of the other, whereas in man they have their own individuality. A long series of other peculiarities followed, as explained in a *Mémoire* read by Vicq d'Azyr in 1781 and illustrated by four

Sr. Herzoglichen Durchlaucht, während den öffentlichen akademischen Prüfungen vertheidigen wird Johann Christoph Friedrich Schiller Kandidat der Medizin in der Herzoglichen Militair-Akademie (Stuttgart: Christoph Friedrich Cotta, 1780). In view of his medical degree, Schiller had already tried to submit a text (*Philosophie der Physiologie*), refused by the commission which found too easy his way of dealing with the medical doctrines of the time. Only the initial chapter (*Das geistige Leben*) of the five planned has been saved, where the task of mediating between body and mind is assigned to a *Mittelkraft* located in a thin and mobile entity that flows through the nerves: see *Schiller's Werke. Nach den vorzüglichsten Quellen revidirte Ausgabe. Vierzehnter Theil. Kleinere prosaische Schriften* (Berlin: Gustav Hempel, 1869/70).

39 Félix Vicq d'Azyr, *Traité d'anatomie et de physiologie avec des planches coloriées. Représentant au naturel les divers organes de l'homme et des animaux* (Paris: François Didot l'Aîné, 1786), Tome premier, p. 4.

plates.[40] The organs of the nervous system – his *Discours* claimed – have relationships, which are still unknown, with the soul, but some of them could be determined in the living bodies of different orders. Comparing the pattern of physical differences with those of the intellect or instinct, sentiment or passions, movements or needs of each class of animals, there was some hope of getting a grip on the hidden agent that rules the matter. To achieve such goal, the most delicate observation and the force of the most exact logic (*en géomètre*) were necessary, avoiding the vain and dangerous speculations on the seat of the soul that had been put forward for centuries. Vicq d'Azyr's circumspection was shared by others, at that time, and yet it was somewhat suspicious: the scruple in setting the limits of the anatomical enterprise, as if it were extraneous to the mind-body problem, hardly concealed that this was at stake.

Primary physician of Marie Antoinette, Vicq d'Azyr, suffering of tuberculosis, died of a collapse in 1794, barely forty-six, immediately after taking part in the annual festival of the Supreme Being: an indirect victim of the Revolution, given his fear of ending guillotined like his friends Bailly and Lavoisier. Therefore he could not achieve the objectives of his project, but left anatomical documents on the brain that remain among the best of the time.[41] In his capacity as *secrétaire perpetuel*, Vicq d'Azyr read the *éloges* of over fifty deceased members of the Société Royale de Médicine. Among them, the Dutch Petrus Camper had his own when he died in 1789. Born in Leiden in 1722, the son of a Calvinist pastor who had served in Batavia, since childhood he had learned to draw and paint under the guidance of Carel de Moor and his son Carel Isaak. Most naturalists –Vicq d'Azyr commented – are forced to hire artists or artisans with the task to visualize the results of their work. Camper, instead, had been able to show himself what he had observed, so both his writings and his *planches* were the work of the same hands. Studies in philosophy, mathematics, and medicine led him to discuss a double doctoral thesis in 1746, and then to take an extended *Bildungsreise* through England, France, Switzerland and Germany, where he had the opportunity to meet the leading scholars of the

40 Félix Vicq d'Azyr, *Suite des recherches sur la structure du cerveau. Quatrième Mémoire sur la structure du cerveau des Animaux comparé avec celui de l'Homme*, in *Histoire de l'Académie Royale des Sciences. Année MDCCXLXXXIII. Avec les Mémoires de Mathématiques et de Physique* (Paris: De l'Imprimerie Royale, 1786), pp. 468-504.

41 A recent secondary source is in Paul Mazliak, *Félix Vicq d'Azyr, créateur révolutionnaire de l'anatomie comparée* (Paris: Hermann, 2017), however connoted by a certain presentism. See also Rafael Mandressi, *Félix Vicq d'Azyr: l'anatomie, l'État, la médecine*, in *Bibliothèque numérique Médic@*. 2006 and Yves Pouliquen, *Félix Vicq d'Azyr, les Lumières et la Révolution* (Paris: Odile Jacob, 2009), who speculates about the possible reasons for his early death (pp. 217-219).

time and to start weaving a solid network of correspondents. Vicq d'Azyr's eulogy emphasized the aptitude for travel and contacts, the vast, non-bookish culture that distinguished Camper, his permanent tendency to verify facts by observation.[42]

Among the countless themes of interest for the Dutch naturalist, Vicq d'Azyr did not forget to mention his study of the heads of Blacks – received from different parts of Asia and Africa – whose facial line he measured and found more inclined than that of Whites. A dissertation conceived in 1768, later revised and only published after his death, announced the application of this new geometric device, the facial angle. A professor of anatomy and surgery at the Athenaeum Illustre of Amsterdam, and later at the Academy of Groningen, Camper was intrigued very early by the physical racial differences and by the visual methods for representing them. In the past and until then, artists had tended to paint black people by simply darkening the skin of familiar white figures. To him, the oval of the face did not seem adequate to determine its distinguishing features with certainty. Rather, by splitting the human head horizontally straight down the middle, Camper noticed that the cavity containing the brain remained relatively constant, while the position of the upper and lower jaws was the real cause of the surprising variety of physiognomies. It was helpful therefore to trace an angle on the cranium, formed by the two lines that join the base of the nose and the auricular foramen, the apex of the incisors and the *os frontalis*, respectively. By giving different degrees of obliquity to the line, he could draw a human head, the head of a quadruped or even of a bird. All possible animal and human forms were thus distributed along a scale of perfection from 0 to 100 degrees of the facial angle, culminating – quite unsurprisingly – in Greek statuary.

From the *Cercopithecus* (45 degrees), the chain would rise through the orangutan, the Black, the Kalmuck, up to the European, increasing in their pleasant proportions, symmetry, and beauty. The reputation enjoyed by Camper all over Europe facilitated the reception of his facial angle by the scientific community. In 1770, it was presented at the Drawing Academy (*Stadstekenacademie*) in Amsterdam and seven years later at the Académie de sciences, when he visited Paris for four months. During his lifetime, other public or private lectures, together with his vast correspondence, had spread the theory of the graduated angle. Only in 1786 was his treatise completed and posthumously

42 Félix Vicq d'Azyr, *Éloge de Pierre Camper*, in *Oeuvres de Pierre Camper, qui ont pour objet l'histoire naturelle, la physiologie et l'anatomie comparée* (Paris: H. J. Jansen, 1803), vol. 1, pp. lxi-xci. In the same volume see also *Notice de la vie et écrits de Pierre Camper*, written by his son Adrien-Gilles, and a second *Éloge*, by Condorcet, pp. xcii-ciii.

published in 1791 by his son, very soon translated into French, German, and English.[43]

"Why did European scientists suddenly collect and measure human skulls?" is the question posed by Miriam Claude Meijer, one of the foremost experts on Camper, whom she qualifies as "the initiator of craniology" who would later become "infamous for two of his engravings that seemingly suggested that protruding jaws (*prognathism*) indicated an 'objective' racial hierarchy in nature".[44] If his cranial sequence was bound to act "like the central visual icon of all subsequent racism", a few authors – including Meijer – have defended Camper from the accusation of having fueled hierarchical thoughts. Suffice it to say that he inferred nothing about intelligence from his aesthetic comparisons: in fact, the facial angle was morphologically oriented and could not calculate either the cranial volume or the brain mass. Its real racist exploitation began only after Camper's death.[45]

As early as 1764 he had given a speech *De l'origine et de la couleur des nègres* at the anatomical theater in Groningen, exhibiting a few specimens and giving prominence to a necessary division of roles: ingenious and subtle philosophers are responsible for reasoning on the faculties of the soul; it is up to the anatomists to treat solely the material parts, the *composition organique* of the human being. From this point of view, it was quite clear that all men have a more or less black skin, and that the difference in color could not serve, according to Camper, to deny a common descent. He recalled that the question had been discussed since antiquity. Herodotus, blamed by Aristotle, had even argued that the sperm of the Ethiopians was as black as their body. Was it not astonishing that a great scholar like Meckel had dared to write that the *Nègres* were a quite different type of men because – besides their skin – their brain and blood would also be black? The lack of familiarity with them had probably

43 *Verhandeling van Petrus Camper, over het natuurlijk verschil der wezenstrekken in menschen van onderscheiden landaart en ouderdom; over het schoon in antyke beelden en gesneedene steenen. Gevolgd door een voorstel van eene nieuwe manier om hoofden van allerleye menschen met zekerheid te tekenen* (Utrecht: B. Wild en J. Altheer, 1791).

44 Miriam Claude Meijer, "Cranial Varieties in the Human and Orangutan Species", in Nicolas Bancel et al. (ed.), *The Invention of Race. Scientific and Popular Representations* (New York: Routledge, 2014), p. 33. See also her remarkable *Race and Aesthetics in the Anthropology of Petrus Camper (1722-1749)* (Amsterdam-Atlanta: Rodopi, 1999). Several contribution on the manifold activity of that "restless man" are in Klaas van Berkel, Bart Ramakers (eds.), *Petrus Camper in Context. Science, the Arts, and Society in the Eighteenth-Century Dutch Republic* (Hilversum: Verloren, 2015).

45 Robert Visser, "Die Rezeption der Anthropologie Petrus Campers (1770-1850)," in *Die Natur des Menschen. Probleme der Physischen Anthropologie und Rassenkunde (1750-1850)*, herausgegeben von Gunter Mann, Franz Dumont (Stuttgart-New York: Gustav Fischer, 1990, pp. 325-335.

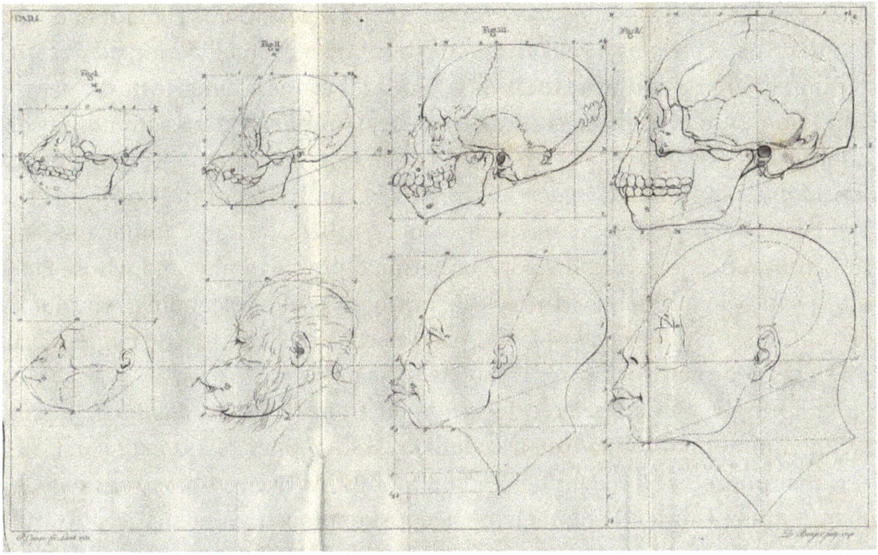

FIGURE 1.2 Tab. 1, In Peter Camper, *Über den natürlichen Unterschied der Gesichtszüge in Menschen verschiedener Gegenden und verschiedenen Alters... Übersetzt von S. Th. Soemmerring* (Berlin 1792).

inspired a sort of repugnance in him about their color. Meckel would have certainly expressed less revolting and more reasonable ideas if he could have had daily contact with black people, as was easily the case in Holland.

Camper claimed to consider them "des êtres de notre espéce" – worthy of trust and love – to whom a fraternal hand should be extended. Moreover, what idea did the unfortunate inhabitants of America, so barbarously treated, get of the white skin of Europeans? Could they not imagine that the creator had dyed the skin of those cruel invaders with a sinister color, as the eternal sign of his righteous anger? As for the rest, Camper faithfully followed Buffon's lesson: without being able to determine whether Adam was created brown, black or white, his descendants – once dispersed on the surface of the earth – had to see their own traits and colors altered according to climate, food, disease, and crossings. On April 16, 1766, he had publicly dissected the corpse of a Black in Groningen and underscored how the medullary substance of his brain was white, and the cortical one even paler than that of the Europeans. Camper had repeated the same demonstration two years later, finding a darker blood color in a young black man, a phenomenon already encountered in some corpses of Whites.[46]

46 *Oeuvres de Pierre Camper* (cit. note 42), pp. 451-476, p. 459n. It must also be noted that, since 1770, Camper was the first European to dissect orangutan specimens from the Dutch

In his article on *humain espèce* for the *Encyclopédie*, Denis Diderot had written that the *nègre* was a sensitive being but not very intelligent (*peu d'esprit*), having suffered greatly due to slavery: "Nous les avons réduits, je ne dis pas à la condition d'esclaves, mais à celles de bêtes de somme; & nous sommes raisonnables! & nous sommes chrétiens!".[47] Twice, in 1773 and 1774, Diderot spent a few months in Holland, a guest of the Russian ambassador in The Hague, and visited the small nation far and wide. Among others, he met Camper, to whom he attributed a perfect knowledge of national physiognomy, and whose facial angle captivated him by virtue of that aesthetic scale descending, without interruptions, from the faces of gods "jusqu'à la tête de l'homme, du nègre, à celle du singe; et de celle-ci jusqu'à la tête de l'oiseau."[48]

In Spring 1778, Camper received a letter from the Swiss *Gehlerter* Johann Georg Zimmermann – a former student and biographer of Albrecht von Haller – which praised and recommended an ambitious twenty-three year old German. Samuel Thomas Soemmerring had just graduated in medicine at Göttingen's Academia Georgia Augusta, determined to do research and teach anatomy in some university instead of practicing a profession that did not attract him at all. Camper invited him to his retreat of Klein Lankum, in the Frisian countryside. Three weeks spent there strengthened Soemmerring's aspiration to a scientific career: he gave proof of his expertise in making anatomical preparations and with Camper he studied skulls, both of them also interested in understanding the position of brain nerves.[49] Not surprisingly, to get his medical degree he had discussed about the base of the encephalon and the origin of the cranial nerves, not without drawing himself three accurate *ad naturam* plates with different views of the brain, and his dissertation earned

colonies and to give an initial report of his studies in 1779, much expanded later. His main objective was to differentiate with anatomical precision anthropoid apes from humans, starting from their quadruped gait, and he firmly opposed the belief that the Black was originally generated by the mating between the white man and the orangutan: see "De l'orang-outang, et de quelques autres espèces de singes," ibid., pp. 1-196.

47 *Encyclopédie ou Dictionnaire raisonné* (cit. note 14), p. 347.

48 *Oeuvres inédites de Diderot. Le neveu de Rameau. Voyage de Hollande* (Paris: Brière, 1821), pp. 283-284.

49 On their mutual respect, despite belonging to different generations, see Antonie M. Luyendijk-Elshout, "«Les beaux esprit se rencontrent». Petrus Camper und Samuel Soemmerring," in Gunter Mann, Franz Dumont (Hrsg.), *Samuel Thomas Soemmerring und die Gelehrten der Goethezeit* (Stuttgart-New York: Gustav Fischer, 1985), pp. 57-73. Soemmerring was also the translator of the posthumous *Peter Camper über den natürlichen Unterschied des Gesichtszüge in Menschen verschiedener Gegenden und verschiedenen Alters...* (Berlin: in der Vossischen Buchhandlung, 1792).

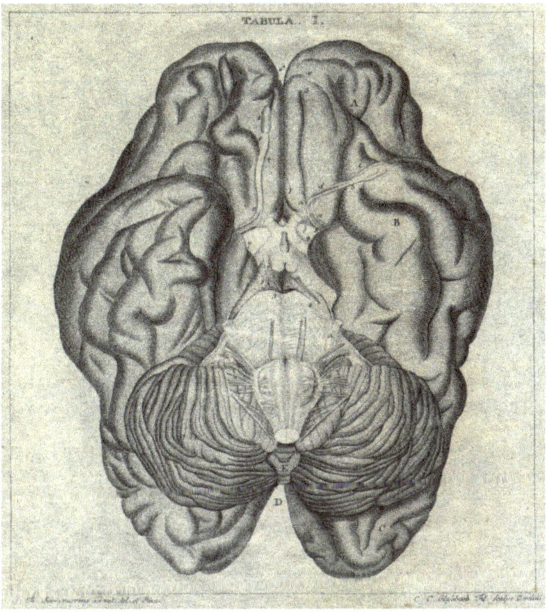

FIGURE 1.3
Tabula I, *"exhibet basin cerebri"* (exhibiting the brain base). In Samuel Thomas Soemmerring, *De basi encephali et originibus nervorum cranio egredientium* (Goettingae 1778).

him an early fame.[50] After the Dutch experience, Soemmerring went to England and Scotland, with many letters of introduction signed by Camper, and had the opportunity to get to know some prominent scientific figures. In London he also met Georg Forster, his own age, with whom he joined the Freemason's Hall, recently opened and already invested with good social prestige, professing tolerant and enlightened values, and especially appealing to young people who wanted to make a career.

Back to Germany, Soemmerring received an appointment to teach anatomy at the Collegium Carolinum in Kassel, and quickly took up his task, which included taking care of the collections for a new natural history museum. His ideal model of anatomist and physiologist was Bernhard Siegfrid Albinus – one of Camper's teachers – who had worked at the University of Leyden since 1721, and, in *Tabulae sceleti et musculorum corporis humani* (1747), had drawn the figure of a *homo perfectus* subject to physical and mathematical laws, approachable in the way an architect knows the structure of a building from the inside. Of that abstract and normative figure all human beings were merely variants. On the other hand, the physiological work of Albrecht von Haller continued to represent an example for Soemmerring, among others, as evidenced

50 Samuel Thomas Soemmerring, *De basi encephali et originibus nervorum cranio egredientium* (Goettingae: Apud Abr. Vandenhoeck Viduam, 1778).

by his 1788 German translation and commentary of the fourth edition of *Primae lineae physiologiae*.[51]

In his new Hessian residence Soemmerring also acted as a freemason – again with Forster – at the "Zum gekrönten Löwen" lodge and as a Rosicrucian: they went so far as to engage together in unsuccessful alchemical experiments. Significantly, he was welcomed as a member of the local Société des Antiquités, to which he presented an introductory lecture in December 1779 about the beauty of childish heads in ancient art, recalling Camper's work on the same subject and a speech he had given shortly before in Göttingen. Four child profiles visualized Soemmerring's arguments on the subject that intertwined anatomy and art history. In 1844, his biographer, Rudolph Wagner, described him as one of the few men of that time who, in an interaction of science and art, recognized the true source of general education (*allgemeiner Bildung*) and opened it up from both sides.[52]

During his five years in Kassel Soemmerring came into contact with a few living *Neger* in the *Mohrenkolonie* established in Wilhelmshöhe by Friedrich II, the Landgrave of Hesse-Kassel, who collected Blacks as servants and ornaments thanks to his connections in the Flanders, Portugal, the Near East, and with slave traders. Soemmerring could also dissect some of their corpses. Just before moving to Mainz, where a renewed university had called him to teach anatomy[53], he was given the opportunity to perform a further autopsy.

It was 1784 when he published a short text, *Über die körperliche Verschiedenheit des Mohren vom Europäer*, just thirty-two pages that tripled the following year in a new edition with a slightly different title – where the more 'scientific' *Neger* replaced the exotic *Mohren* – and some drawings made by the artist

51 See Reinhard Hildebrand, "Attic perfection in anatomy: Bernhard Siegfrid Albinus (21697-1770) and Samuel Thomas Soemmerring (1755-1830)," *Annals of Anatomy*, 2005, *187*: 555-573; Alberts von Haller, *Grundriss der Physiologie für Vorlesungen…durch Herrn Hofrath Soemmerring in Mainz* (Berlin: Hauder und Spener, 1788).

52 See the in-depth contributions by Ulrike Enke, "Soemmerings erste Professur am Collegium Carolinum zu Kassel"; Irmtraut Sahmland, "Soemmerring als Freimaurer und Rosenkreuzer in Kassel"; Sigrid Oehler-Klein, "Anatomie und Kunstgeschichte. Soemmerrings *Rede über die Schönheit der antiken Kinderköpfe* vor der Société des Antiquités in Kassel (1779)," in Manfred Wenzel (Hrsg.), *Samuel Thomas Soemmerring in Kassel, 1779-1784: Beiträge zur Wissenschaftsgeschichte der Goethezeit* (Stuttgart-New York: Gustav Fischer, 1994), pp. 75-141, 353-422, 189-239. Rudolph Wagner's judgment in his *Samuel Thomas Soemmerrings Leben und Verkehr mit seinen Zeitgenossen. Zweite Abtheilung* (Leipzig: Voss, 1844), p. 58.

53 About the new phase of his activity in Rhineland-Palatinate see Franz Dumont, "«Teutschlands Hippokrates». Der Anatom Samuel Thomas Soemmerring in seinen Mainzer Jahren (1784-1792/97)," in Id. (*et alii*), *Moguntia medica. Das medizinische Mainz. Vom Mittelater bis ins 20. Jahrhundert* (Wiesbaden: Wyličil, 2002), pp. 55-64.

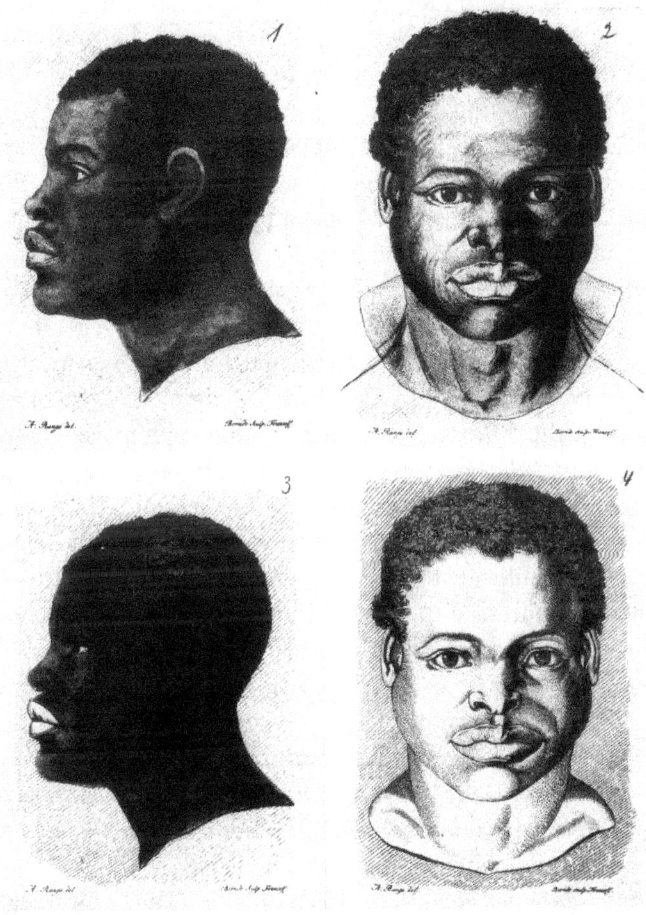

FIGURE 1.4
Four copper engravings in Samuel Thomas Soemmerring, *Über die körperliche Verschiedenheit des Mohren vom Europäer* (Frankfurt am Main 1785).

Andreas Range of Kassel. However, only a few copies of his work contained illustrations.[54] The dedication was addressed to his most trusted friend, a connoisseur of the world and of mankind, Georg Forster, the man most suited to judge it. Homeland, age, common zeal for the knowledge of nature, daily

54 See information about *Der anthropologische Nachlass Soemmerrings*, in Samuel Thomas Soemmerring, *Ueber die körperliche Verschiedenheit des Negers vom Europäer bearbeitet und herausgegeben von Sigrid Oehler-Klein* (Stuttgart-New York: Gustav Fischer, 1998) pp. 303-309. This anastatic edition contains also a long introduction by the editor (pp. 11-142), a detailed *Stellenkommentar* (pp. 253-296), and other *Materialen*. Interesting insights are given (pp. 115-142) on the reception of Soemmerring's findings, which ranged from the enthusiastic reception in popular racial-political writings to scientific approval or criticism of the possible consequences for the explosive problem of the slave trade and the hierarchical classification of races.

friendship bound them strongly and led them to share a perfect harmony of thought. The dedication has clear rhetorical tones but conveys the complicity that existed between those two brilliant young men.[55] Moderate rhetoric also reappears in the preface, beginning with a confession of guilt. Soemmerring complains that "Wir, Europäer" were and are not very fraternal in dealing with the unfortunate Africans, as if we considered them less perfect than the first animal on the planet. Only Europeans behave so badly, forced by greed and by practical prejudices that however, being so widespread, must have some real support. Human actions are often performed without the intellect (*Verstand*) understanding their motives. Significantly, Soemmerring resorts to a gendered example as a proof of what he means: a prince known to him had been brought up since his early childhood with his sister, roughly the same way both physically and morally, and yet the difference between feminine and masculine remained quite striking.

Investigating moral causes was not the anatomist's business, and his competence could only try to understand which definite and remarkable differences in the bodily structure might allocate the African to a lower step in the "Throne of Mankind". Although he addressed the issue, Montesquieu had not gone beyond the color of the skin or the flat nose; Soemmerring seems to be puzzled by the famous, ironic fifth chapter (Book 15) of *L'esprit des lois*, with its pretended defense of slavery. On the other hand, his great teacher and friend Camper had recalled the opinion of ancient authors on white peoples as more sublime and rational creatures (*erhabenere und vernünftigere Geschöpfe*); the popular philosopher Christoph Meiners, as well as Zimmermann the geographer and David Hume, had expressed similar thoughts.[56]

The main question Soemmerring's book tried to answer is: What if it were anatomically possible to show that the *Neger* is somewhat closer to a monkey than to the European? And what if the organ of the intellect – which distinguishes men from animals – penalizes Blacks? In theory, the undeniable similarity between monkeys and men suggests that there must be individual persons, if not whole nations, in which it is greater. So, for instance, according to Johann Reinhold Forster, among all the men he had seen during the Cook's

55 See Hans Querner, "Samuel Thomas Soemmerring und Johann Georg Forster – eine Freundschaft," in Mann, Dumont (Hrsg.), *Samuel Thomas Soemmerring* (cit. note 49), pp. 229-245. Fifty Forster's letters from 1779 to 1793 are in Rudolph Wagner, *Samuel Thomas Soemmerrings Leben und Verkehr mit seinen Zeitgenossen. Erste Abtheilung* (Leipzig: Voss, 1844), pp. 122-280, also reprinted and edited by Franz Dumont (Stuttgart-New York: Gustav Fischer, 1986).

56 Samuel Thomas Soemmerring, *Ueber die körperliche Verschiedenheit des Negers vom Europäer* (Frankfurt-Mainz: Barrentrapp Sohn und Wenner, 1785), pp. VII-XIII.

second Pacific voyage (1772-1775), the inhabitants of Malekula Island, in Melanesia, showed the closest kinship with the "tribe of monkeys".[57] Buffon himself had attributed an ape-like physiognomy to the inhabitants of Papua and New Guinea, without having seen them in person. Herder was right in noticing recently that it would be a miracle if man had remained unchanged in such different climates, and so was Zimmermann when he claimed that everything in creation is linked by nuances, from the human species to the lower monkey. Moreover, as the novelist Oliver Goldsmith had remarked in his *History of the Earth and Animated Nature*, of all animals the smallest physical differences occur in mankind.[58]

This was also Soemmerring's belief. At all latitudes, man keeps his upright gait and distinguished superiority, whereas the main morphological difference comes from the color of the skin, which gradually fades from light to dark. His concern, in the *Vorrede*, was also to declare that he had presented his results with the utmost caution, because of the judgmental consequences with which the unhappy slaves were everywhere treated; and yet he could not entirely escape the misunderstandings. So it had seemed necessary to condemn the inhumanity of the slavers from the beginning – "wie wenig brüderlich" – and even more emphatically to refute the doubts about there being one single ancestor for all mankind. In other words, Soemmerring claimed to be on the side of Buffon, Camper, Blumenbach, Forster, and Zimmermann against those who, like Voltaire and Lord Kames, had questioned the remote existence of a common *Stammvater*. All things considered, *Mohren* were human beings, "so gut wie wir", definitely above monkeys and distinct from quadrupeds: some of them stood very close to Whites or even, in particular cases, exceeded them. Soemmerring wished to avoid the accusation of having mixed black people and apes – contained in a letter to the *Journal des gens du monde* and transcribed by him – and also stated that the anatomical parts were kept in his collection, and therefore open to the observation of whoever might be interested.[59]

---

57 Soemmerring, *Ueber die körperliche Verschiedenheit* (cit. note 56), pp. XIV-XV. Here is the beginning of Forster's rather long description: "The natives of Mallicollo [Malekula] are a small, nimble, slender, black and ill-favoured set of beings, that of all men I ever saw, the nearest upon the tribe of monkies. Their skulls are of a very singular structure, being from the root of the nose more depressed backward, than in any of the other races of mankind". See John Reinold [*sic*] Forster, *Observations Made during a Voyage round the World on Physical Geography, Natural History, and Ethic Philosophy* (London: G. Robinson, 1778) p. 242.

58 Oliver Goldsmith, *A History of the Earth and Animated Nature in Four Volumes* (London: Henry Fisher, 1824), I, pp. 345-346. The first edition in eight volumes had been published in 1774.

59 Soemmerring, *Ueber die körperliche Verschiedenheit* (cit. note 56), pp. XVI-XXIV.

Flat nose, wool hair, skin color: these were commonly held to be the distinguishing features of Blacks, but they were not enough to Soemmerring the physiologist, who was searching for deeper differences, irrefutably rooted in the body. The first forty-eight paragraphs of his book reviewed a series of anatomic areas or organs from head to foot, partly resorting to what had been told by other authors. After listing numerous peculiarities in the structure, among others, of teeth and tongue, hand and foot, Soemmerring discussed Johann Gottlieb Walter's belief that the blood of blacks is very different from the European one: a volatile disgusting smell, a black color. Not to be understood literally – he corrected – if anything in the sense of a darker red. Also Barrère in his *Dissertation* of 1741 and, going back more in time, Jan Baptiste van Helmont, thought the same. Camper himself had seen darker blood in some Europeans.

About twenty other paragraphs focused on the most significant section of the body. Camper had proved, thanks to the principle of facial line (*Gesichtslinie*), that, in the ancient and finest Greek busts, the bones of the head proportionally exceed those of the face. The brow advances to a line with the nose, forming an ideal and gentle curve: a shape just fit to provide ample space for the brain. Nature seems instead to have reversed the proportions of her favorite model in the low and flat African skull. Comparatively, the frontal bone is always larger in the European. To the "deviations" of the *Mohren*'s skull, neither randomly nor artfully produced, Soemmerring paid attention, when craniology was at its dawn. Cranial differences in black people were persistent and customary, though individually present in varying degrees. In fact, by consulting the collections known to him – Camper's in Amsterdam, Walter's in Berlin, Blumenbach's in Göttingen – or the five specimens he owned, he could record a wide range of variations. After all, even the bones of Europeans vary a great deal, without being disfigured by any disease.

Soemmerring had measured the skulls of numerous Blacks and compared them with those of Whites. Differences in size were immediately evident to the eye: lower volume of the cranial cavity, smaller bones and a harder substance, almost similar to that of common animals. Evidence of variability was mentioned: from Herodotus, who had noticed the lighter and thin Persian skulls compared to the Greek ones, to Pieter Paaw – in 1616 the Dutch anatomist had noticed the hardest and thickest skulls of the Ethiopians, likened to an iron helmet – to Blumenbach, who had recently discovered a minor relationship between the brain-shell and the facial bones in the skull of Blacks. Designed in 1764 by Daubenton to show the position and inclination of the *foramen magnum*, the occipital angle clearly revealed its more backward position in monkeys, and it seemed to Soemmerring that something similar could be seen in black people. This position could also explain the fact that the skull of a Black, placed on a plane without its jaw, remained so steeply inclined that that the teeth did not touch the surface.[60]

As for the brain, two of the greatest German anatomists, Meckel and Walter, had divergent opinions on its color in Blacks: the former considering it darker (and also the ultimate cause of skin pigmentation), the latter denying the reliability of that observation. Curious to find out who was right, Soemmerring dissected three fresh brains of Blacks without seeing the slightest color difference either in gray or in white matter, and other colleagues had witnessed one of his autopsies. If anything, he had seemed to detect even a paler color, as it also happened to Camper. Instead, both Walter and his predecessor in Berlin, Meckel, had found a more solid brain substance in Blacks. On his part, Soemmerring added that nerves connected with the inferior cerebral surface were thicker and sturdier, at least the olfactory, the facial and the fifth nerves. If ears, eyes, and nasal organs of Blacks are larger – he argued – it is no wonder that their corresponding nerves have greater consistency.

The careful comparison between brains of various animal species had led Soemmerring to discover that man has the biggest brain with the smallest nerves, or rather that his brain is bigger in relation to the nerves. In the past, from Aristotle onwards, it was often repeated that the human species has the largest brain in relation to the body size. However, physiologists who wanted to determine this apparent truth more accurately by studying several animal species felt embarrassed when they found that the bird, the dolphin, the sea lion, but also small mammals like the mouse far exceed humans in the proportion of brain and body weights. From this unexpected finding Soemmerring drew the conviction that it was safer to establish a comparison between brain and nerves. The creature that, in addition to the cerebral portion strictly necessary for animal life, has a further increase in the brain will also be able to develop mental powers. This is what Herder had asserted, in the recent first volume of his *Ideen zur Philosophie der Geschichte der Menschheit*, and Soemmerring reported his statement approvingly. Not for nothing in all the monkeys he had been able to view, the sensory system was predominant.[61]

Having ascertained that the nerves are stronger in black people, the logical conclusion was that their brain should be smaller compared to that of Europeans. Given this inference, historical facts about their wild nature, indocility, and lower capacity for higher culture might perhaps be explained. In support of such a thesis, Soemmerring recruited some alleged authorities, among the many possible: the Jesuit Pierre-François-Xavier de Charlevoix, who had

60 Ibid., pp. 49-55.
61 Ibid., pp. 55-63.

spoken of the dullness of the inhabitants of Santo Domingo; Benjamin Franklin's opinion on the natural tendency of every slave for thieving; Buffon and Lord Kames, for both of whom much would contribute to the African's lack of mental exercise and miserable conditions. Even Europeans – his friend Blumenbach had written to him – easily slip into brutality and ferocity during a prolonged stay in the East and West Indies or in Africa. His teacher in Göttingen, Heinrich August Wrisberg, had associated a small brain and slender nerves with a weak sensorial reactivity, a phlegmatic-melancholic slowness, and a childish character. As his own experiences were partly in contrast with those of Wrisberg, Soemmerring asked him for an explanation of the sentence *cum parvo cerebro et exilibus nervis*, and also of the perception of a different cerebral weight in animals and men.[62]

Regarding the conformation of individual cerebral parts and the roots of the nerves, Soemmerring had not noticed any difference. Exposing in detail the results of the three dissections, he attributed the cause of death to a special disease of black people, originated in the lungs, not unlike that affecting the monkeys when carried to a temperate environment. Might this deadly disorder be caused, both in Africans and in monkeys, by inadequate food and a climate alien to them? All such assumptions, therefore, made a certain resemblance between the *Neger* and the *Affen* plausible. Nonetheless, the former remained entirely human, and even much higher than quadrupeds, as sufficiently demonstrated by Camper and Blumenbach. Undoubtedly subtle nuances of color and form divided humanity into a continuous series of characters.

On the one hand, a small group of naturalists and philosophers, albeit enumerating human varieties differently through the century, had always put Africans in a distinct class. On the other hand, although Linnaeus and the Swiss philosopher Johann Georg Sulzer had had difficulty distinguishing between man and the anthropoid, Soemmerring promised that he would elsewhere

62 Soemmerring, *Ueber die körperliche Verschiedenheit* (cit. note 56), pp. 67-71. Soemmerring will later provoke heated discussions with a small book *Über das Organ der Seele* (Königsberg: Nicolovius, 1796) dedicated to Kant. Its first part proved that the wall of the cerebral cavities receives nerve endings, but it was the second part that aroused controversy by proposing to localize the *sensorium commune* in the intra-ventricular cerebral fluid. An old, medieval idea seemed to be revived, just to provoke illustrious rejections. While appreciating Soemmerring's book, Goethe wrote him that its title was a mistake: better not to disturb the soul, restricting it to such tiny space. For his part Kant – "the pride of our age" as Soemmerring defined him – assessed that the question of the seat of the soul was insoluble, inherently contradictory: see his letter included in *Über das Organ der Seele*, pp. 80-86. Walther Riese judged that a figure attached to the text was "the first correct picture of the mesial aspect of the cerebral hemispheres [...] a master-piece of observation and reproduction": see his "The 150th Anniversary of S. T. Soemmerring's *Organ of the Soul*," *Bulletin of the History of Medicine*, 1946, 20: 310-321.

demonstrate the existence of at least fifteen anatomical differences in their respective brains, beyond size and weight.[63] Finally, he did not dare to decide whether the original man, if born in Asia or not, had become a European, or if he had degenerated into the African, his structural characters suited to the climate of the torrid continent, and from this point of view perhaps even more perfect than the European. In a reversed relationship Soemmerring thought that the mind often becomes the disadvantage of the body, and conversely, the body cultivates the disadvantage of the mind.[64]

Johann Friedrich Blumenbach was among those who discussed the two versions of Soemmerring's innovative work.[65] In both his comments, the reviewer aimed at averting possible interpretations that would dig a furrow between Africans and other human varieties. Undeniably, a more oblique facial line and a protruding chin were, already in black newborns, the most striking character, together with a smaller brain-shell (*Hirnschaale*) if compared to big *ossa faciei*. Blumenbach, a great collector of skulls, confirmed Soemmerring's craniological results, also referring to the principle, established by his colleague and friend, that smaller nerves would correspond to a larger brain. The main concern of the reviewer was in downplaying the morphological proximity of *Mohren* to the *Affengeschlecht*, that the author had affirmed. Such a statement could only be accepted as tantamount to saying that the Angora cat is closer to the Bolognese dog than to other feline races. Blumenbach was pleased that Soemmerring had nevertheless recognized Blacks as fully belonging to the human species: it was possible to shift from one variety to another through imperceptible transitions, subtle nuances of shape, color etc., essentially caused by the climate. And that's not all: there were also at least as many differences between single black individuals, as between the black and other races.

Since 1774, Soemmerring – at that time a student in Jena and Göttingen – had become a friend and admirer of Blumenbach, three years older than him, so that a lifelong fruitful relationship developed between them. It was 1775 when Blumenbach presented a dissertation *De generis humani varietate nativa* for his medical degree: the initial germ of a work that twenty years later would become a classic of physical anthropology in its early days.[66] The original

63 In 1790 the project of a *Pithecologie* was sketched – and never realized – by Forster and Soemmerring, about the relationship between ape and man: see letters to Gottlob Heyne, March 20 and 22, in Georg Forster, *Werke. Briefe 1790-1791* (Berlin: De Gruyter, 1980), Band 16, pp. 29 and 33.

64 Soemmerring, *Ueber die körperliche Verschiedenheit* (cit. note 56), pp. 71-80.

65 See Blumenbach's two reviews in "Göttingische Anzeigen von Gehlerten. Der erste Band auf das Jahr 1785", pp. 108-111; "Göttingische Anzeigen von Gehlerten. Der erste Band auf das Jahr 1786", pp. 302-303.

66 On his vast, long-lived, and extraordinary activity see Nicolaas Rupke and Gerhard Lauer (eds.), *Johann Friedrich Blumenbach. Race and Natural History, 1750-1850* (New

version consisted of one hundred pages with two plates, the first of which showed the skull and the cerebral base of a mandrill dissected by the author to compare it with the human. The classification of mankind in four varieties (European, Asiatic, African, American) was the same as the one given by Linnaeus a few decades before. Among other things, Blumenbach noted that eminent authors had formulated so many hypotheses about the seat and causes of the diversity of human pigmentation that only a large volume could contain them. Great disputes had taken place, above all, on the skin color of Blacks, from the belief in the alleged curse of Cain or Cham, and their posterity, to the search for natural causes, such as the role played by the bile. More recently, Meckel had certified a dark color of their blood, or even an ashy (*cinericea*) appearance of the brain and spinal marrow. Nevertheless Blumenbach thought that *clima, solis aerisque* and the way of life had the greatest influence, as the ancients had already supposed and as he intended to corroborate with appropriate examples and arguments. In actual fact, an almost insensible and indefinable transition (*insensilis fere et nullo modo definiendum transitus*) separated the candid skin of the German virgin from the deepest black of the Ethiopian, via intermediate shades.[67]

A new *Malaica* variety was already added to the second edition of 1781 and confirmed by the third one of 1795, greatly grown to nearly three hundred and fifty pages. Exploration voyages in the Pacific had revealed, in the meantime, the existence of other human groups, to be taken into account. Although the species formed a continuous whole, however Blumenbach, in arranging a large collection of human skulls (approximately three hundred), had managed to identify five varieties, described not only in their physical characters:

A) *Caucasiae*,
B) *Mongolicae*
C) *Aethiopicae*
D) *Americanae*
et E) *Malaicae*[68]

York: Routledge, 2019). Essential sources are provided by Frank Peter William Dougherty, *Bibliographie der Werke und Schriften von Johann Friedrich Blumenbach nebst ihren Übersetzungen und Digitalisierungen. Bibliography of Johann Friedrich Blumenbach. Durchgesehen, ergänzt und mit Anmerkungen und einer Einleitung herausgegeben von Norbert Klatt. Mit einem Geleitwort von Prof. Dr. Dr. Heinz Schott* (Göttingen: Norbert Klatt, 2009); Claudia Kroke, *Johann Friedrich Blumenbach. Bibliographie seiner Schriften* (Göttingen: Universitätsverlag, 2010).

67 Johann Friedrich Blumenbach, *De generis humani varietate nativa. Illustris facultatis medicae consensus pro gradu doctoris medicinae disputavit d. XVI Sept. MDCCLXXV* (Gottingae: typis Frid. Andr. Rosenbuschii, 1775), pp. 48-50.

68 Johann Friedrich Blumenbach, *De generis humani varietate nativa. Editio tertia: premissa est epistola ad virum perillustrem Iosephum Banks*...(Gottingae: Apud Vandenhoeck et

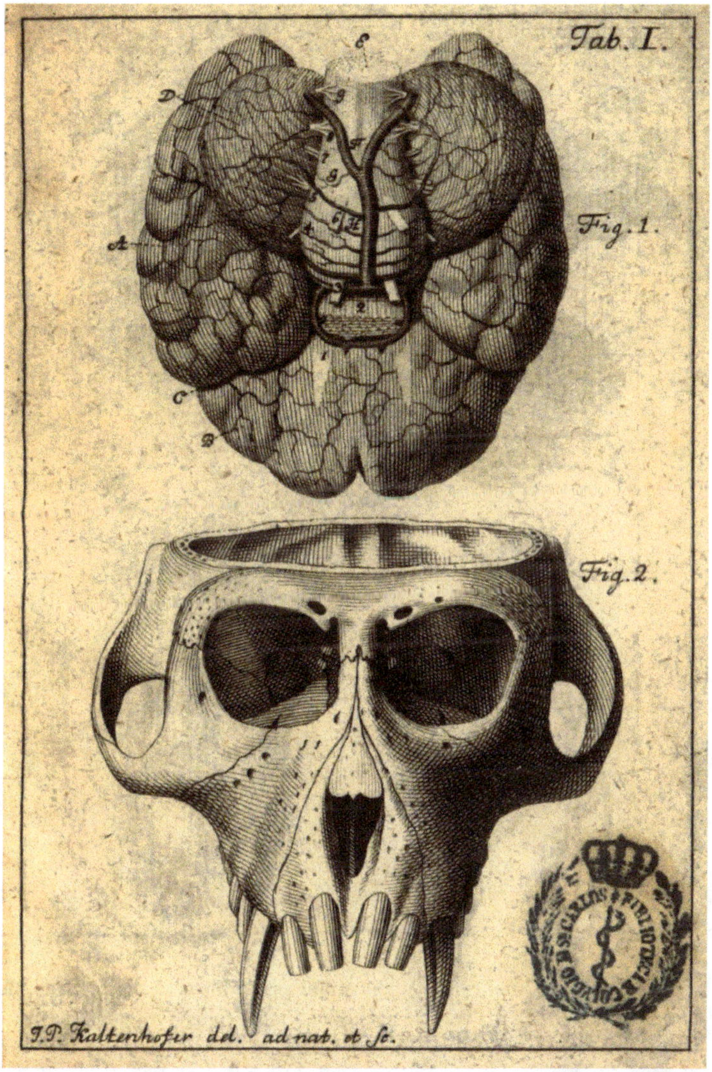

FIGURE 1.5
"Basis cerebri papionis Mandril" (brain base of the Mandril). Engraving in Johann Friedrich Blumenbach, *De generis humani varietate nativa* (Gottingae 1775).

At the end of the book, two plates exhibit drawings of the five essential skulls, compared by means of the so-called *norma verticalis*, while the anthropological

---

Ruprecht, 1795), p. 286. It is anecdotal that Blumenbach took up comparative osteology already at ten years of age, when he would spend hours studying the skeleton owned by a doctor who practiced in his hometown of Gotha, or visiting cemeteries in search of bones: see Pierre Flourens, "Éloge historique de Jean-Frédéric Blumenbach", in *Mémoires de l'Académie royale des sciences de l'Institut de France*, 1847:*21*, pp. i-xxi, p. iii. For detailed information on his collecting skulls see Wolfgang Böker, "Zur Geschichte der Schädelsammlung Johann Friedrich Blumenbach," *Annals of the History and Philosophy of Biology* (2018), *23*:3-29.

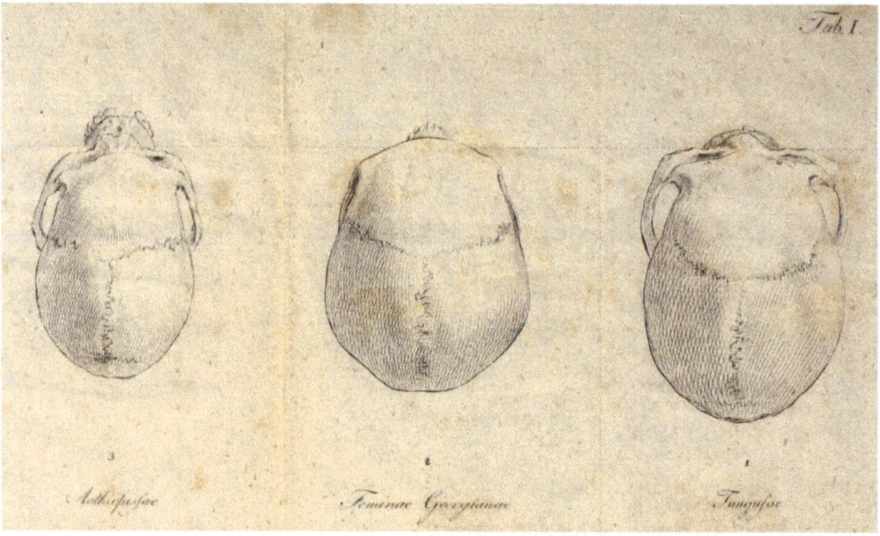

a

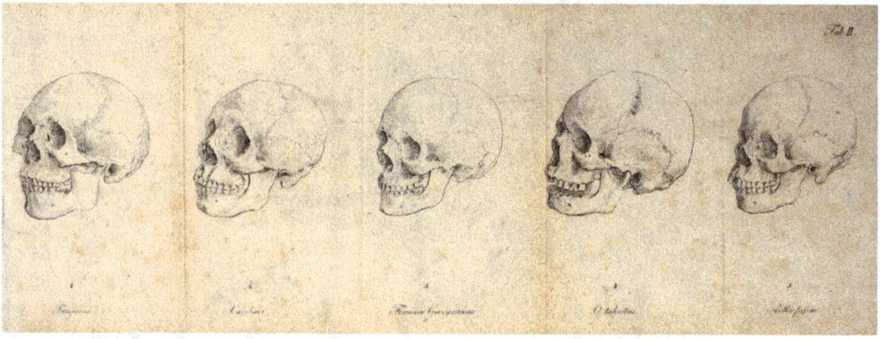

b

FIGURE 1.6A-B a. "Schema synopticum exhibit ad illustrandam norma verticalem" (Synoptic drawings illustrating the vertical norm); b. "Crania collectionis meae quina selectissima adumbrat, ad totidem generis humani varietatum principalium diversitatem demonstrandam" (Drawings of five selected skulls in my collection, to show the diversity of as many varieties of mankind). Two plates in Johann Friedrich Blumenbach, *De generis humani varietate native. Editio tertia* (Gottingae 1795).

reliability of Camper's facial angle was questioned: in fact, different human groups often presented a similar gradation. Blumenbach revealed that he had taken the name of the first human variety from Mount Caucasus, both because its surroundings produced the "pulcherrima hominum stirps" and because all physiological reasons concurred to place the likely birthplace of the human species in that region. The Caucasian stock displayed the most beautiful form of skull, a sort of primeval type from which the others "simplicissima gradatione fluunt", to the opposite extremes of the Mongolian and the Ethiopian. Besides, he assumed that precisely the Caucasians possessed the original color of mankind, for the ease with which white could "degenerate" into brown, while the opposite could not occur once the secretion of the carbonaceous pigment had once taken deep roots in the organisms.

For its part, *Varietas Aethiopica* was so differently colored that it led many to think of it as a separate species: among them was Voltaire, whom he defined *facetus* but badly instructed in physiology. It was superfluous, according to Blumenbach, to devote too many words to the refutation of that unmotivated belief. Suffice it to say, on the one hand, that no character was so peculiar to the Ethiopians that it could not be found in other human varieties; on the other hand, that many *Nigritae* were found to lack some of them. Finally, Blumenbach rejected the meaningless thesis – also proposed by his friend Soemmerring – of a greater proximity between Blacks and apes: likewise, it could be said that the solid-hoofed (*solidungula*) variety of the domestic pig was nearer to the horse than other varieties. Essentially, other non-European populations had also suffered the same debasing treatment.[69]

As far as the brain is concerned, in 1795 Blumenbach simply pointed out that its mass is greater in man than in any other animals, though not in proportion to the whole body – as from Aristotle onwards many had repeated – but to the size of the nerves that branch off. Therefore, the results of Soemmerring's anatomical work were acceptable. With the consequence that man, in the animal kingdom, would have a greater amount of the noblest part of the brain. On the same basis, a paragraph from the sixth edition of Blumenbach's *Handbuch der Naturgeschichte* (1799) still confirmed the inverse proportion of the cranial nerves with the *Geisteskräften*, thus implying the cerebral superiority of the human species[70]

In an article of 1787 Blumenbach had already written *Von den Negern*, as he had been given the opportunity to see and talk to a few of them. He pointed out that the portrait of four *Negerköpfe*, at that time owned by the Pommersfelder Bildergalerie and attributed to Antoon van Dyck (but now to Rubens's school), showed no particularly oblique facial line, with the effect of deviating very little from the typical European expression. In any event, he stressed the fact that, for such physical features as color and *Gesichtslinie*, there were

69 Blumenbach, *De generis humani* (cit. note 68), pp. 303-304, 307-310. Blumenbach's library – Pierre Flourens recalled in his *Éloge* of 1847 – had a section composed of books written by black authors.

70 Ibid., pp. 42-43; Id., *Handbuch der Naturgeschichte. Sechste Auflage* (Göttingen: Johann Christian Dieterich, 1799), p. 37.

FIGURE 1.7 *Negerköpfe*, Four Studies of a Male Head, Workshop of Peter Paul Rubens (about 1617-1620), Oil on panel, The J. Paul Getty Museum, Los Angeles.

inter-individual differences in Blacks as well as in the other human varieties. In three of their skulls that he examined, the first one had a jaw so prominent as to suggest that it might descend from another Adam; the third one showed traits such as to make his belonging to the Ethiopian variety doubtful, if it were not certain; the second one represented an intermediate form between the previous two. It was generally thought that, in physical structure and mental abilities, Blacks were not up to the rest of the human species. The second part of Blumenbach's article collected a series of opinions aimed at falsifying this assumption. In particular, some emphasis was placed on the case of the philosopher Anton Wilhelm Amo, brought from Ghana to Amsterdam as a child by the Dutch and given as a present to the Duke of Brunswick-Wolfenbüttel, who had raised him as a family member: the first African man to have attended and then taught in the Universities of Halle and Jena. Its conclusion insisted on black people being fairly equal (*ziemlich Gleichheit*) to the rest of Adam's children.[71]

71 Johann Friedrich Blumenbach, "Einige naturhistorische Bemerkungen bey Gelegenheit einer Schweizerreise," *Magazin für das Neueste aus der Physik und Naturgeschichte*, 1787, 4/3: 1-12. While Blumenbach had mostly hesitated to establish precise relationships between the physical and the moral elements, a few years later a shift would have started "from a dynamic model of differentiated adaptation among all humans in contextual conditions, to a moralizing, static, and deterministic model that would dominate naturalist and anthropological thinking throughout the nineteenth century": an example of this shift can be found in the *Discours préliminaire du traducteur* which opens the French version of Blumenbach's book, made in 1804 by Frédéric Charles Chardel (*De l'unité du genre humain et de ses variétés*). About this shift, see Francesco Panese, "The creation of the 'Negro' at the Turn of the Nineteenth Century. Petrus Camper, Johann Friedrich Blumenbach, and Julien-Joseph Virey," in Bancel et al. (eds.), *The Invention of Race* (cit. note 44), pp. 52-55.

A further fruit of that fertile corner of *spätaufklärischer* Germany must be mentioned. Mainz, 1788: Jacob Fidelis Ackermann discussed a medical thesis focused upon distinctions between the two sexes *praeter genitalia*, dedicated to his *praeceptor* Soemmerring. Believing that anatomists, in the past, had mostly neglected the numerous differences which would have allowed to really distinguish the female body from the male one, he presented them analytically: so skin, hair, bones, skull, mouth, eyes, voice, blood vessels, sweat, brain and nerves were put through his descriptive sieve. The four last chapters concerned the brain, nerves and sensory skills. A rather constant structure was detected in the brain, with no peculiarity distinguishing the masculine from the feminine, except for smaller median lobes in the latter, due to the narrower frontal part of the skull. Soemmerring took the leading edge in the practice of weighing brains, which Ackermann had learned and continued, finding that the female brain was generally lighter than the male one. Only as an absolute value, though, because the opposite was true if one considered the weight of the brain weight in relation to body weight. Besides, he reported that women had thinner nerves than men, this particular conformation resulting in greater sensitivity, which predisposed them to intellectual activities (*in universum ingeniis operibus*). Finally, the pineal gland was found by him to be larger, on average, in the female brain than in the male one.[72]

3 Speculations and Objections

The third edition of Blumenbach's *De generis humani varietate nativa* was dedicated to Sir Joseph Banks, who had participated as a young botanist in Cook's first voyage on HMS Endeavour (1768-1771). Since 1778 he had been President of the Royal Society, a position he held for over four decades. Blumenbach owed him much, having been offered in London, three years before, free use of his library and collections of treasures that had been fundamental both "ad studium anthropologicum" and in transforming his 1775 medical thesis into a

72 Jacob Fidelis Ackermann, *Dissertatio inauguralis anatomica de discrimine sexuum praeter genitalia* (Moguntiae: Hered. Haeffner, 1788), pp. 89-95; also in German: *Ueber die körperliche Verschiedenheit des Mannes vom Weibe außer den Geschlechtstheilen. Uebersezt nebst einer Vorrede und einigen Bemerkungen von Joseph Wenzel* (Mainz: in der Winkoppischen Buchhandlung, 1788), pp. 140-150. See also Frank Stahnisch, "Über die Natur des weiblichen Gehirns. Geschlechterpolitik im Werk der Mainzer Anatomen Jacob Fidelis Ackermann (1765-1815), in Rüdiger Schultka, Josef N. Neumann (Hrsg.), *Anatomie und anatomische Sammlungen in 18. Jahrhundert. Anlässlich der 250. Wiederkehr des Geburtstages von Philipp Friedrich Theodor Meckel (1755-1803)* (Berlin: LIT, 2007), pp. 421-435.

full-bodied book. The University of Göttingen, where Blumenbach spent his entire life of research and teaching, was rather young, having been founded in 1737 by George III, Duke of Brunswick-Lüneburg and King of Great Britain and Ireland, a prince-elector of the Holy Roman Empire. The Hanoverian connection allowed the University established in Lower Saxony – a kind of continental outpost of English culture – to enjoy a cosmopolitan dimension, especially in natural history, because the British maritime hegemony could provide specimens from all over the world. Significantly, more than 350 objects from the voyages from Cook's voyages enriched Blumenbach's collection, by royal gift. Incidentally, on visiting Göttingen in 1779, Camper had demonstrated the metamorphoses of living beings up to the best *Gesichtslinie* to an audience composed of Lichtenberg, Wrisberg, Soemmerring and Blumenbach.[73]

For the second of those explorations, which ranged across the Pacific and Antarctic in search of a southern land, King George III had recruited the Forsters, father and son, who became familiar with a series of local human groups and cultures. Even before the *Observations Made during a Voyage round the World* by his father Johann Reinhold Forster, Georg published a longer report in two volumes, where his desire to establish a new practice of broad spectrum anthropology found a precocious expression. All was recorded in those hundreds of pages: seascape and landscape, weather and oceanic conditions, plants, animals, and especially human groups, interactions between sailors and natives, whose living conditions, food and clothing, tools and material objects were of particular interest to Forster. It has often been noted that the three years of his youth spent sailing the ocean marked forever his worldview and conception of mankind. Eight years after Cook's violent death at Kealakekua Bay, Hawaii, he praised him in a long portrait as a charismatic leader and exemplary man, a genius of practical reason and learning ability.[74]

Forster loved to make fun of bookworms and of those conceited philosophers who believed they could abstractly know men. Nothing more alien to him than his father-in-law's invitation to enjoy life behind a desk: quite comprehensible, since it came from the classical scholar Gottlob Heyne, a

73 Blumenbach, *De generis humani* (cit. note 68), p. vi. See also John Gascoigne, "Blumenbach, Banks, and the Beginnings of Anthropology in Germany," in Nicolaas Rupke (ed.), *Göttingen and the Development of the Natural Sciences* (Göttingen: Wallstein, 2002), pp. 86-98; Reinhard Hildebrand, "Petrus Camper in his relationship to Samuel Thomas Soemmerring and other German scientists of the *Goethezeit*," in van Berkel, Ramakers (eds.), *Petrus Camper in Context* (cit. note 44), p. 134.

74 Georg Forster, *James Cook, der Entdecker* (1787), in Id., *Kleine Schriften. Ein Beytrag zur Völker- und Länderkunde, Naturgeschichte und Philosophie des Lebens* (Leipzig: Paul Gotthelf Kummer, 1789), pp. 1-232.

long-time director of the Göttingen State and University Library. Paradoxically, in 1788, he had to accept a post of librarian in Mainz and, converted to Jacobinism, was one of the founders of the Mainzer Republik, which lasted only five months. The Prussians soon conquered the city, and Forster fled to Paris to die there in January 1794, not yet forty.[75]

When Herder, in the first volume of his *Ideen zur Philosophie der Geschichte der Menschheit* (1784), tried to question the validity of the concept of race, Kant wrote an article to reinforce this and to criticize the explorers, whose investigation of exotic peoples had left him dissatisfied. In natural history too, any concept – he prescribed – has to be defined clearly and in advance before resorting to a dubious "experience".[76] Then, defending the Herderian perspective, Forster essentially accused the prince of German philosophers – who had never encountered and personally studied other cultures – of speculating in vain and of meddling in things he did not understand. Who would not prefer the survey of a discerning and reliable empiricist – Forster asked rhetorically – to the makeup of a partisan *Systematiker*? Attention, judgment and impartiality are the requirements on which everything depends: the business of the philosopher is to deduce general concepts from single true statements, and in this kind of mental activity delusion is as possible as at the moment of observation. Forster wished to evaluate the contributions that travellers were gradually adding to the knowledge of the human species, for among them were credible people, capable of providing accurate, useful experiences.[77]

---

75 George Forster, *A Voyage round the World in His Britannic Majesty's Sloop, Resolution, commanded by Capt. James Cook, during the Years 1772, 3, 4, and 5* (London: B. White, J. Robson, P. Elmsly, and G. Robinson, 1777). A 19th century portrait was written by Jacob Moleschott, *Georg Forster, der Naturforscher des Volks* (Frankfurt am Main: Eidinger Sohn & Cie., 1857). A more recent and well-known interpretation is in Wolf Lepenies, *Autoren und Wissenschaftler im 18. Jahrhundert. Buffon, Linné, Winckelmann, Georg Forster, Erasmus Darwin* (München-Wien: Carl Hanser, 1988), pp. 121-154; see also Hans Erich Bödeker, "Aufklärerische ethnologische Praxis: Johann Reinhold Forster und Georg Forster," in Id., Peter Hanns Reill und Jürgen Schlumbohm (Hrsg.), *Wissenschaft als kulturelle Praxis, 1750-1900* (Göttingen: Vandenhoeck & Ruprecht, 1999), pp. 227-253. For a wider and more complete reconstruction, see Tanja van Hoorn, *Dem Leibe abgelesen. Georg Forster im Kontext der physischen Anthropologie des 18. Jahrhunderts* (Tübingen: Niemeyer, 2004); Jürgen Goldstein, *Georg Forster. Zwischen Freiheit und Naturgewalt* (Berlin: Matthes & Seitz, 2015).

76 Immanuel Kant, "Bestimmung des Begriffs einer Menschenrace," *Berlinische Monatsschrift*, 1785, *16*: 390-417; Forster to Soemmerring, 19/21 Januar 1787, in *Georg Forsters Werke. Briefe 1784-Juni 1787* (Berlin: De Gruyter, 1978), Band 14, p. 617-618.

77 Georg Forster, "Noch etwas über die Menschenraßen," *Der Teutsche Merkur*, 1786, *4 Viertel-jahr*: 57-86, 150-166. Forster's article aroused in Kant irritated reaction. See Mario Marino, "Noch etwas über die Menschenrassen. Eine Lektüre der Kant-Herder-Forster Kontroverse," in Simone De Angelis, Florian Gelzer, Lucas Marco Gisi (Hrsg.), *Natur, Naturrecht*

However, there was something more in Forster's attack. He disliked Kant's thesis that all humans came from the original couple, as narrated by the Book of Genesis, because there was no proof of its truth. Moreover, the various populations encountered during his Pacific voyage were so different from each other that they seemed to belong to distinct stocks. Particularly the inhabitants of Malekula insinuated doubts in him about the unity of the human species, and also on the real power of the environment in shaping morphology and human characters: so many physiognomies and attitudes under the same sky. Though registering their ugliness, he had nevertheless seen them as curious and sociable, waiting for some impulse to rise to a higher degree of civilization.[78] When in 1784/85 Forster read his friend Soemmerring's *Ueber die körperliche Verschiedenheit*, he admired its anatomical-comparative arguments – which finally did not use only the color of the skin as a distinctive criterion – and found reasons to momentarily lean towards a heterodox hypothesis that would be called "polygenist" in the following century. However, the belief that the human races were varieties of a single and vast *Gattung* finally prevailed in him, the future Jacobin, and also ruled out the possibility of establishing a precise hierarchy. Soemmerring himself had been advised by Camper not to disclose his suspicion that Europeans and Africans differed less by variety than by species, and that there could have been, so to speak, "two Adams".[79]

In 1789, Forster was inspired by the reading of Matthew Prior's humorous poem *Alma; or, The Progress of Mind* (1718), where ancient and modern theories were blamed for their confusion. Matthew, the principal interlocutor of a dialogue with Richard, brings forward his own theory: during life, at different ages, the individual mind would progress stage by stage through the body, "from the Feet upward to the Head". A droll but intriguing idea – Forster commented – as the infant, for example, can only live in his feet, bumping and wriggling for a long time before crawling and learning to move other parts of his body. Even later, the mind dwells in the extremities: sticking and jumping, the child cannot stand still for a moment and his legs determine his will. Afterwards the mind rises higher; other organs dominate the body of the flowering

und Geschichte. Aspekte eines fundamentalen Begründungsdiskurses der Neuzeit (1600-1900) (Heidelberg: Universitätsverlag Winter, 2010), pp. 393-413.

78 See the Chapter I of Book III, *An account of our stay at Mallicollo, and discovery of the New Hebrides*, in Forster, *A Voyage round the World* (cit. note 75), vol. II, pp. 199-261.

79 Sigrid Oehler-Klein, "Der «Mohr» auf der «niedrigeren Staffel am Throne der Menschheit»? Georg Forsters Rezeption der Anthropologie Soemmerrings," in Dippel, Horst und Scheuer, Helmut (Hrsg.), *Georg-Forster-Studien III* (Kassel: Kassel University Press, 1999), pp. 119-166; Raphaël Lagier, "Un outsider de la fondation de l'anthropologie: Georg Forster," *Revue d'Histoire des Sciences humaines*, 2006, *14*: 137-152.

youth; a new impulse fills the whole being, fixes all the activity on one point, tied to the Belt of Love (*Gürtel der Liebe*).

As soon as Forster read Prior's poem, a whole system of the so-called history of humanity took shape in his head. The connecting link was the usual – and misused – comparison between different phases of individual life with the stages of the historical development of mankind, and consequently he had a "dream" about four stages of the so-called muscular, spermatic, heroic, and sensitive cultures. He promised himself to work out that system in a big book; for the moment, it was enough for him to say that the wettest, softest, most impressive organ, the brain – performer of sensation, memory, and consciousness – receives and collects, from childhood onwards, the influences of the external objects, by means of the senses and of the whole nervous system. Its mass remains soft and acquires a certain, but always very low, strength at later ages. No wonder, then, that only in the period of maturity (*in der Periode des Stillstands*) do the vital forces of the brain express their highest activity, and increase the clarity of consciousness. If at that point of life the bones are already brittle, the muscles stiff, the senses dull, and the nerves less susceptible, the efficacy of an admirable brain might still be preserved. Retreated from its greater sphere of activity, man then stays within himself, and in the delicate cerebral tissue he finds the universe again, when outside it scarcely exists for him. This enjoyment, too, and perhaps the most peculiar of all, is the heightened awareness of man, who thus attains the last heights of his education.

It goes without saying that such scheme could be possibly applied to the history of mankind. Dance and war are the first abilities of the savage, which rises just by a single step over the needs of animality. He feels his power in annihilation; in the tumult of victory he irresistibly stomps the earth with his feet; all about him is irrepressible boyish will, and inner striving without direction. From a sort of proto-evolutionary perspective, Forster continued to illustrate the next steps. Abundance – whether from hunting or cattle raising or agriculture – increased the sexual instinct, by the softer charm of proliferating juices. A mild climate, a fertile land, a quiet neighborhood, hastened growth among the Chinese and Indians, as well as the Africans, making them the most populous nations on earth. In the hearts and minds of these nations slept the invigorating force, or it convulsively twitched. Born into slavery, they needed it, and still need the wisdom of a despot who leads them to the arts of peace and awakens mechanical craft in them.

By another combination of circumstances, and over time, the quiet enjoyment of property brought with it a greater population, favoring the development of a germ into great and sublime passions, already lying in the brutal, destructive barbarian. Only those peoples, who had grown into the arms of liberty, could quickly ascend to the highest peak of education, where all the energy is contained most actively in the finer tools of sensation and intellect. Only three times, only in Europe, and each time in a different guise did history witness the spectacle of this last stage of education. Athens was the first, as its blossoming imagination and sense of beauty created the first fruits of art and science. Rome was no longer free, and the spoils of the classic world had already set fire to the most unrestrained moral corruption when it absorbed the ruins of Attic culture, and shinier by opulence than by high spirits of genius, preserved for its future conquerors. Last came the offspring of a fortunate barbarian tribe, after whom the romantic fire blazed so beautifully into an autumn with its ripe fruits; the harvest was for an unpredictable winter to come.[80]

Intertwining the history of the body and the history of mankind, Forster's poetic and *weltgeschichtlich* imagination pointed to the likely existence of a stage of development both incapable of freedom and primitive, as far as the brain was concerned. This did not prevent him from becoming a public prosecutor of the explicitly reactionary and racist view elaborated by the polygraph Christoph Meiners, professor of *Weltweisheit* in Göttingen for almost forty years, who was part of that synergy of forces that for a while turned the Lower Saxony University into a hot bed working around historical and anthropological themes. Between 1789 and 1791, with his own field expertise contributing to the *Allgemeine Literatur-Zeitung*, Forster fustigated the excess of erudition, the lack of rigor and the prejudices that filled Meiners' overabundant and eclectic production. It completely lacked what it would have been necessary, that is, a study of every single people in itself, a description of all its living conditions, and an understanding of its adaptation to the place it occupied on earth. Even his correspondence contains harsh criticism: in writing to Herder in January 1787, he blamed Meiners' habit of indiscriminately exploiting travel compilations for the purpose of supporting particular theses. He was also enraged by the pro-slavery stance of the Göttingen professor – prone to consider most of mankind as *Halbmenschen*– as proven by other private letters.[81]

80 [Georg Forster], "Leitfaden zu einer künftigen Geschichte der Menschheit," *Neues Deutsches Museum*, 1789, 1/3: 269-283

81 Georg Forster to Herder, 21 Januar 1787, in *Werke in vier Bänden* (Leipzig: Insel Verlag, 1968), 4, pp. 448-449. On Meiners' controversial stances see Friedrich Lotter, "Christoph Meiners und die Lehre von der unterschiedlichen Wertigkeit der Menschenrassen," in Hartmut Boockmann, Hermann Wellenreuther (Hrsg.), *Geschichtswissenschaft in Göttingen. Eine Vorlesungsreihe* (Göttingen: Vandenhoeck & Ruprecht, 1987), pp. 30-75; Britta Rupp-Eisenreich, "La science nouvelle de Christoph Meiners (1747-1810)," in *L'invention de la race* (cit. note 44), pp. 83-99.

In 1785, when the first edition of Meiners' *Grundriss der Geschichte der Menschheit* came out, he announced an alleged "new science", of which only contradictory fragments seemed to be around: a very ambitious *Wissenschaft* aimed at discovering the 'nature' of man through the study of the conditions and customs of savage and barbarous peoples up to the highest stages of civilization. The history of mankind should have started with its corporeity, to discover how physical and moral causes could transform it with repercussions on the invisible and higher faculties. Meiners was convinced that until then no comparison had ever been seriously established among the various nations from that particular point of view. Of course the material to work on, in view of general ideas and laws, was largely extracted from the amount of travelogues that had accumulated over the previous two centuries. It is worth noting that a few years later Blumenbach – Meiners' colleague in Göttingen but also his severe opponent – in the preface to a collection of *Reisegeschichten* warned against excessive confidence in their use and urged to read them with the utmost caution and critical attitude.[82] The Association for Promoting the Discovery of the Interior Parts of Africa, led by Sir Joseph Banks, had just been founded in London, and Blumenbach gave his contribution to their efforts to fill the blanks in the map of the continent, beyond the coastline and Egypt. Four expeditions were organized by Banks, and provided with special instructions, with the support of the Association, whose members were antislavery. All of them failed, however, and no participant survived.[83]

In 1790, after the start of a chain of rebellion in the French colony of Saint-Domingue, a long article by Meiners had dealt with the nature of the Africans, and formulated a clear judgment on the legitimacy of their enslaved condition. At that time – he complained – an increasing love for freedom was escalating into a rebellion against princes and nobility, and the hatred of oppression had turned into a feverish enthusiasm for the impossible and unjust equality of all *Stände* and peoples. The same rights and freedoms could never be granted to subjects and sovereigns, children and adults, women and men, servants and masters; likewise, Jews and Negroes could not demand to have the same privileges enjoyed by Christians and Whites among whom they lived or whom they obeyed. Iniquitous was the will to equalize those whom nature itself, or other

82 Christoph Meiners, *Grundriß der Geschichte der Menschheit* (Lemgo: im Verlage der Meyerschen Buchhandlung, 1785); Johann Friedrich Blumenbach, "Vorrede," in *Sammlung seltener und merkwürdiger Reisegeschichten* (Memmingen: Seyler, 1789), I, pp. 4-8.

83 See Hans Plischke, *Johann Friedrich Blumenbachs Einfluss aud die Entdeckungsreisenden seiner Zeit* (Göttingen: Vandenhoeck & Ruprecht, 1937).

insurmountable causes, had made unequal.[84] Meiners' entire writing consisted of a repertoire of innumerable, irreducible differential traits between Whites and Blacks: the former being more beautiful, larger, stronger, more courageous, richer in spirit, more noble; the latter uglier, smaller, weaker, more vile and sick by nature. Nothing new, really, compared to the stereotypes he himself and other authors of the late eighteenth century had repeatedly advanced, were it not that in his opinion no more than two essential human stocks (*Hauptstämme*) coexisted on earth, the dark-skinned and lower-valued Mongolian (which subsumed Blacks and Amerindians), opposed to the *hellhäutig* Caucasian nations, in turn divided into two races (Celtic and Slavic), with such hierarchy culminating in Germany. The latter represented the realization of the ideal nation, not subjected to a single sovereign or committed to plundering other peoples: prosperous, moral, wise. In the following years Meiners often returned to extolling the marvelous virtues of the Celtic race.[85]

Color and exterior shape distinguished races, implying a correspondence with spiritual characters from a physiognomic point of view, but Meiners was also sure that the relative proportions of the nerves and brain represented distinctive, hereditary and unbreakable features equally well, according to what Soemmerring's dissections had demonstrated a few years earlier.[86] The heads of the *Neger* were undeniably different from those destined by nature for Europeans, just as much by their size as by their shape. The skull, and especially the forehead, the occiput, the cranial cavity itself looked many degrees smaller in Blacks, and the brain firmer (*fester*) but also more fragile: two peculiarities that in Europe were perceived as typical of the feeble-minded and insane. But, even regardless of their narrower skulls, smaller and less pliant brains, coarser nerves, one would still be bound to conclude that they were irremediably inferior, based on the way they lived and behaved.[87] After all, in 1788, Meiners had already stigmatized the great irritability and fragility of the weaker

84 Christoph Meiners, "Ueber die Natur des Afrikanischen Neger, und die davon abhangende Befreyung, oder Einschränkung der Schwarzen," *Göttingisches Historisches Magazin*, 1790, *6*: 385-456, pp. 386-387.

85 Meiners, *Grundriß der Geschichte der Menschheit* (cit. note 82), pp. 16-80. See also Id., "Ueber die grosse Verschiedenheit der Biegsamkeit und Unbiegsamkeit, der Härte und Weichheit der verschiedenen Stämme und Racen der Menschen", *Göttingisches Historisches Magazin*, 1787, *1*: 210-246 and "Ueber die Natur der Germanischen und übrigen Celtischen Völker", *Göttingisches Historisches Magazin*, 1791, *8*: 1-48, 67-124.

86 On the conflict between Forster and Meiner, see Siegrid Oehler-Klein, "Der «Mohr" (cit. n. 79), p. 122.

87 Meiners, "Ueber die Natur des Afrikanischen Neger," (cit. note 84), pp. 402-404. See also the *Nachwort* of Frank Schäfer to the re-edition of Meiners' text (Hannover: Wehrhahn Verlag, 2000), pp. 67-78.

(*schwächer*) peoples, which would often fall prey to fits of rage and convulsions, or epileptic syncopes, due to the structural flimsiness of their nervous system.[88]

Among Meiners' countless readings, there was certainly the *History of Jamaica*, three volumes anonymously published in 1774 by the Jamaican-born British colonial administrator, planter, and slave-owner Edward Long. An economic, social, and political account with a survey of the island, parish by parish, which also contained an assessment of the striking physical resemblance between the orangutan and man, such as to make of the former a human being "*quoad* his form and organs, but of an inferior species, *quoad* his intellect". From a morphological point of view, the orangutan would be closer to the *Negroe* race than the latter to the white man. Therefore, the supposition was well founded, that the brain and intellectual organs, though similar in texture and modification to those of other men, may be so constituted in Blacks, "as not to result to the same effect". If it was his pleasure, the Deity might diversify his work in such a manner, and either withhold the "superior principle" entirely or in part only, or infuse it into the different classes and races of human creatures. Consisting of varieties, the *Negroe* race will then seem to rise in the scale of intellect by as much as it progressively rises above orangutans and brutes.[89]

In writing about the recently founded State of Virginia, at the request of a French diplomat interested in the new country, the enlightened slave-owner Thomas Jefferson pointed out that the color difference between Whites and Blacks – whatever its cause – was "fixed in nature" from the Caribbeans to North America. It represented the foundation of "a greater or less share of beauty in the two races", with the monotony of the dark veil which covers all the emotions, while the fine mixtures of red and white would allow one to express every passion by a greater or lesser suffusion of color. Jefferson's physical

88 Christoph Meiners, "Ueber die sympathetische Reizbarkeit, und einige daraus zu erklärende Erscheinungen in den schwächern Völkern," *Göttingisches Historisches Magazin*, 1788, 2:40-56. Meiner's aversion and fear for the degradation of non-European human groups were satirized by the emigrated Huguenot Auguste Lafontaine – a popular author at the time – in a four-volume pseudonym novel: see [Gustav Freier], *Leben und Thaten des Freiherrn Quinctius Heymeran von Flaming* (Berlin: In der Vossischen Buchhandlung, 1795-1796). Its protagonist is a professor of Roman and Germanic descent, who conceives an attempt to improve the non-Celts, black and ugly, by the blood of blond and blue-eyed subjects, following the *Züchtung* model applied by animal breeders. It will be paradoxically an intelligent African woman, who had been a slave, to heal the *ante-litteram* eugenicist from his obsession, by loving and marrying him.

89 [Edward Long], *The History of Jamaica. Or, General Survey of the Antient and Modern State of that Island: with Reflections on its Situation, Settlements, Inhabitants, Climate, Products, Commerce, Laws, and Government* (London: T. Lowndes, 1774), II, pp. 369-371.

and moral portrait of Blacks, which echoed the received view of his times, is well known: in addition to ugliness, their lack of true hair, the strong and disagreeable odor due to a transpiration that renders them more tolerant to heat, their endurance in labor, sexual ardor, an existence that appears to participate more of sensation than reflection. An animal whose body is at rest, and does not reflect, must inevitably yield to sleep: equal to Whites in memory, in reason much inferior, in imagination "dull, tasteless, and anomalous".

Although not as advantageously familiar with Whites, Indians often carve figures on their pipes not destitute of design and merit, or give proof of an astonishing oratory; instead, Jefferson could never find Blacks uttering a thought above the level of narration. However he recognized their musical superiority, while the few examples of their poetry left much to be desired, and the slave condition was not to blame for this, since in Roman antiquity, despite the poorest condition of slaves, some of them often excelled in art and science: "but they were of the race of Whites". Jefferson did not know whether further scientific observation might verify the conjecture that nature had been less bountiful to Blacks in the endowments of their heads, but he believed that in those of the heart she had done them justice. Justifying a general conclusion about their physical and mental inferiority still required much study, in a subject submitted to the anatomical knife, to optical glasses, or to analysis by fire or solvents. Thus the future president of the United States hinted at a sort of sketchy program for future research, which, by the way, would soon begin to be realized.[90]

In the same year, 1787, an oration delivered by the Reverend Samuel Stanhope Smith – professor of Moral Philosophy at the College of New Jersey – before the Philosophical Society in Philadelphia faced similar questions, which were stirring the newly founded nation. He criticized the extreme tenets held by Lord Kames in his *Sketches on the History of Man*, where the dilemma had been clearly formulated from the first lines:

> Whether there be different races of men, or whether all men are of one race, without any difference but what proceeds from climate or other accident, is a profound question of natural history, which remains still

90 Thomas Jefferson, *On the State of Virginia. Illustrated with a Map, including the States of Virginia, Maryland, Delaware and Pennsylvania* (London: John Stockdale, 1787), pp. 229-240. On Jefferson's "racism" see Ari Helo, Peter S. Onuf, "Jefferson, Morality, and the Problem of Slavery," *The William and Mary Quarterly*, 2003, *60*: 583-614; Peter S. Onuf, *The Mind of Thomas Jefferson* (Charlottesville: University of Virginia Press, 2007), pp. 205-270.

> undetermined after all that has been said upon it philosophers differ widely about.[91]

While in 1774 the Scottish philosopher had insisted on the unbridgeable distance between human races, as if they were distinct and separate stocks, Smith countered this "arbitrary hypothesis" by remarking, first and foremost, that every permanent and characteristic variety in human nature is performed only through flow and almost imperceptible gradations. In his opinion, climate was the powerful, supreme factor, as countless examples could prove. One for all: Negroes born in the United States were gradually losing the strong smell of their African brothers; their hair was growing less curled, and becoming denser and longer. The other effective factor of change consisted in the state of society. For instance, the field slaves were badly fed, clothed and lodged: living by themselves on the plantations, they looked ill shaped and retained many of the customs and manners of their African ancestors: dull genius, sleepy and stupid expressions. On the other hand, the domestic servants, employed in the families of their masters, had advanced far in acquiring the agreeable, regular features, and the expressive countenance of civilized society.[92]

In the wake of Lord Kames, whose theses were gaining consensus at the turn of the century, and in contrast to Smith, the eminent surgeon and obstetrician Charles White held some lectures at the Literary and Philosophical Society of Manchester in 1795, published in 1799. To begin with, readers were warned that the author did not intend "to give the smallest countenance to the pernicious practice of enslaving mankind", which he wished instead to see abolished throughout the world. Neither was he desirous of assigning to anyone superiority over another, "except that which naturally arises from superior bodily strength, mental powers, and industry". Also in his conclusions he declared the slave trade indefensible "on any hypothesis": laws ought not to allow less freedom to men of inferior capacity or most unshapely or ill formed. What interested him was to investigate the truth and to discover the laws of nature in that respect. White also recognized his debt to John Hunter, who in Manchester had recently read some *Remarks on the Gradation of Skulls*, ranking these from

91 [Henry Home, Lord Kames], *Sketches of the History of Man. In two Volumes* (Edinburgh: W. Creech, London: W. Strahan and T. Cadell, 1774), I, p. 1. The work, "intended for men, who, equally removed from the corruption of opulence, and from the depression of bodily labour, are ben on useful knowledge" (p. v), had numerous editions, increasing over time from two to four volumes.

92 Samuel Stanhope Smith, *An Essay on the Causes of the Variety of Complexion and Figure in the Human Species. To Which are Added Strictures on Lord Kaims's* Discourse on the Original Diversity of Mankind (Philadelphia: Robert Aitken, 1787), pp. 34, 57-58.

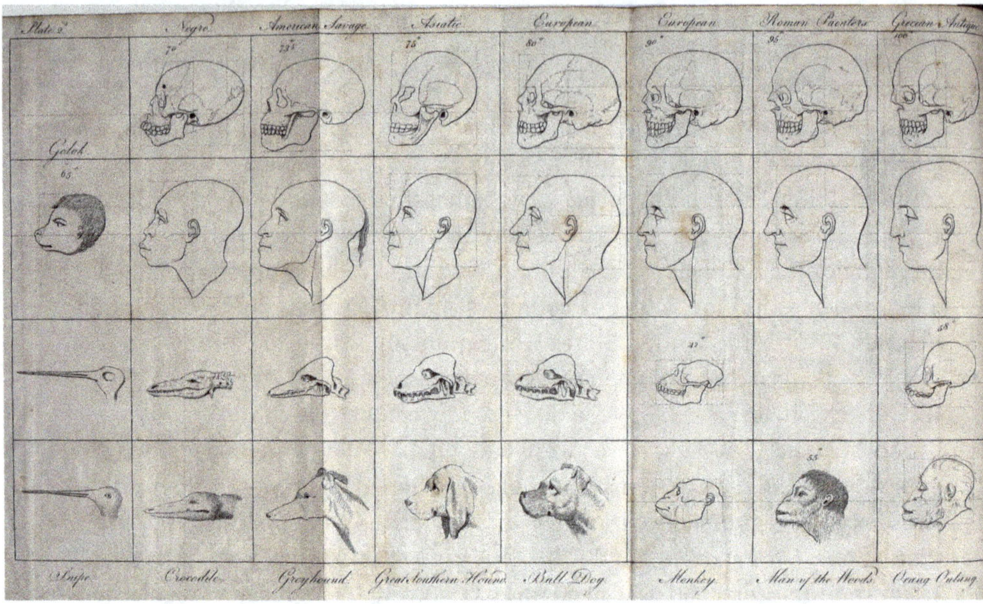

a

FIGURE 1.8A-B a. "This Plate is intended to shew the facial line in Man, and in different Animals, from the perpendicular line in the European Man, to the horizontal one in the Snipe or Woodcock"; b. " This Plate exhibits copies of the best authenticated engravings that have yet been published of four different kind of Apes, which approach nearest to Man: likewise the skull of Dr. Tyson Pigmy – the skull of a Monkey from Lavater – the profiles of a native of Botany Bay and an European – and the profiles of an African and a European". Two plates in Charles White, *An Account of the Regular Gradation in Man, and in Different Animals and Vegetables; and from the Former to the Latter. Illustrated with Engravings Adapted to the Subject* (London 1799).

European to Asiatic, American, African, monkey, and so on, and he attached a plate to show the gradual changing of the facial angle in various human and animal forms, and another one that compared the skull of Tyson's Pigmy with that of monkey from Lavater, the profiles of a native of Botany Bay with that of a European, that of an African with that of a European.[93]

93 Charles White, *An Account of the Regular Gradation in Man, and in Different Animals and Vegetables; and from the Former to the Latter. Illustrated with Engravings Adapted to the Subject* (London: C. Dilly, 1799), pp. iii, 137-138. Hunter's manuscript on gradation was among those burned in 1823 by his brother-in-law, assistant, and executor of will Everard Home: see Jane M. Oppenheimer, *New Aspects of John and William Hunter. I. Everard Home and the Destruction of John Hunter Manuscripts. II. William Hunter and his* Contemporaries (New York: Henry Schuman, 1946). A famous portrait made in 1789 by Joshua

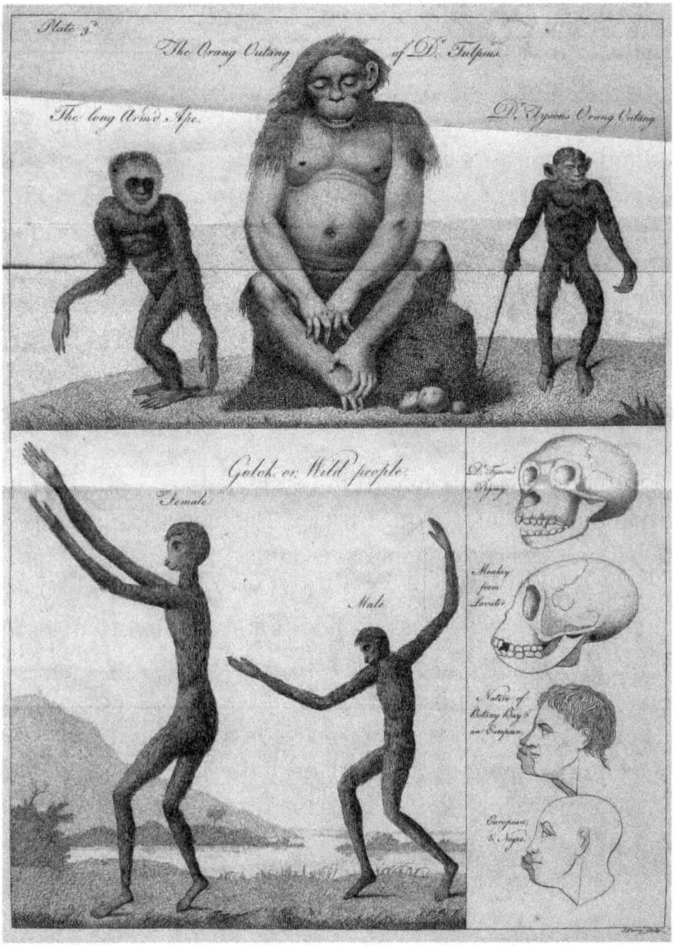

b

Any student of natural history – White maintained, quoting various sources of inspiration, including Charles Bonnet's *échelle des êtres*, Camper's facial angle and Lavater's physiognomy – could contemplate the beautiful gradation among the created beings, from the highest to the lowest, "from man down to the smallest reptile", an immense chain endued with various degrees of intelligence and powers. In its wide phenomenology, that gradation might be also traced "from the brain of the human European [...] to the polypus, who has none at all", so that White endeavored to prove it in man, and especially with

Reynolds (Royal College of Surgeon, London) shows Hunter next to an open sketchbook with the profiles of six skulls in descending order: from the European to the crocodile passing through intermediate stages.

reference to the skull. He had found frontal and occipital bones to be narrower in the Negro than in the European, the foramen magnum more backward, and the occipital bone itself pointing upwards, forming a more obtuse angle with the spine. Furthermore, the internal capacity of the skull was less in the former, and the lower and upper jaws exhibited remarkable, peculiar features. The chin, instead of projecting, receded, and the *meatus auditorius* was wider: "in all these points it [the Negro] differed from the European, and approached the ape". No organ or aspect seems to be neglected by White in his search for proofs of the alleged gradation in nature. As for the cavity of the skull, it was clear that man has the largest brain of any animal, and that in turn of all man the European possesses the largest one. Yet it was fair to admit that some animals have a larger brain in proportion to their body, as do mice, squirrels and some birds.[94]

So little was known about the physiology of the brain and nerves that White abstained from stating much about them. However, the observations already available indicated that animals with a greater amount of brain also had more reason or sagacity. The original capacity of the different tribes of mankind might as well depend on their brain mass. In addition to Thomas Jefferson's evidence, White relied on that given by the Swedish naturalist and traveller Carl Peter Thunberg, about the dull torpid brain of the inhabitants of the warmer climates. Their power of thinking would be less profound and their conversation rather trifling. Without meaning to insult the dark brown peoples of the East Indies, Thunberg too had noticed a greater proximity between them and the monkeys.[95] Up the line of gradation, White at last came to the white European,

> who being most removed from the brute creation, may be considered as the most beautiful of the human race. No one will doubt his superiority in intellectual powers; and I believe it will be found that his capacity is naturally superior also to that of every white man. Where shall we find, unless in the European, that nobly arched head, containing such a quantity of brain, and supported by a hollow conical pillar, entering its centre?[96]

---

94 White, *An Account* (cit. note 93), pp. 14, 41-63.

95 Ibid., pp. 64-68. One of the Linnean "apostles", Thunberg had published between 1788 and 1793 an account in four volumes of his travels and stays on three continents (*Resa uti Europa, Africa, Asia, Förrättad Åren 1770 – 1779*), translated in French, English, and German: see Marie-Christine Skuncke, "Suède, Europe, Japon: le botaniste Carl Peter Thunberg sur le marché international", *La Révolution française*, 2018:13, URL : <http://journals.openedition.org/lrf/1928 ; DOI : 10.4000/lrf.1928>.

96 White, *An Account of the Regular Gradation in Man* (cit. note 93), pp. 134-135. Significantly, an appendix to the book (pp. CXXXIX-CLXVI) contained "detached passages selected

As already noticed, at the end of the eighteenth century complaints about deficiencies in the science of the central nervous system were quite frequent. Richard Saumarez, surgeon at the Magdalen Hospital of Streatham and a prolific medical author, reiterated it and on that occasion quoted Busick Harwood, professor of anatomy at Cambridge and a fellow of the Royal Society:

> The skill of the anatomist has demonstrated every process, explored every cavity, and would, if possible, have traced every filament of this inexplicable mass of that wonderful and anomalous organ, placed on the doubtful confines of the material and spiritual worlds! Nor have the physiologist and the metaphysician been less eager to discover, or to assign to each part its peculiar office. Whatever may be due to the former for accuracy, and to the latter for ingenuity and zeal, we must lament that little knowledge has resulted from their labour.[97]

Although knowledge was poor, and the field open to more specific enquiries, still Saumarez was sure that science had acquired sufficient facts to uphold at least that

> the most rational systems are endued with a greater proportion of brain than the less rational, and that the brain of rational beings in general is proportionably larger than the most instinctive. The skull of a white is considerably larger than that of a black, of an African than of a monkey, of a monkey than of a dog, of a dog than of a sheep, or of an ox. [...] As we ascend in the general scale of rational beings, the magnitude of the brain bears an increased and strongly marked proportion to the size of the system in general. In the African, it is as one to 54. In a European, as one of the 50th part of the system together!!![98]

In a note (p. 163), Saumarez reported that a friend of him – whose anatomical accuracy he depended upon and valued more than his own – had made a number of experiments. The skull of an African was compared with thirty-six different European skulls, and they were all found to be uniformly and considerable

from Soemmerring's Essay on the Comparative Anatomy of the Negro and the European" (1785).

97 Richard Saumarez, *A New System of Physiology, Comprehending the Laws by which Animated Beings in General, and the Human Species in Particular, are Governed, in their Several States of Health and Disease* (London: T. Cox, 1798), I, pp. 159-160. The quotation was drawn from Busick Harwood, *A System of Comparative Anatomy and Physiology* (Cambridge: J. Burges, 1796), I, pp. 1-2.

98 Saumarez, *A New System of Physiology* (cit. note 97), pp. 160-161.

larger: "and on filling them with water, the latter were found capable of containing a much larger quantity than the former."

While Saumarez, White and others deplored the dearth of safe notions about cerebral and nervous physiology, someone in Vienna wrote an open letter to Baron Joseph von Retzer – head of the Austrian imperial censorship – explaining a research program to fill in the gaps. To put it very briefly, Franz Joseph Gall – a physician of German birth – asserted: 1) that moral qualities and intellectual faculties are innate; 2) that their exercise depends on the brain's morphology; 3) that the brain acts as an organ of all qualities and faculties; 4) that it is composed of many particular organs as the original and primitive faculties. At the time, before beginning his scientific and proselytizing journey, he was not yet able to exactly explain methods and objects. However he confessed to a wish: if all kinds of genius had appointed him as heir to their skulls, he predicted he would erect a beautiful building in ten years, while at the moment the materials were still scarce. Surely, it would be quite dangerous for Kästner, the mathematician, Kant, Wieland, or similarly celebrated figures, if he had had David's destroying angel at his command.[99] Adding more information about the work to be done, Gall also promised he would write a section on "national heads" and on the difference between the heads of men and women. As for the first issue, one might come to understand why "some of our brothers" (*einige von unsern Brüdern*) cannot count over three, why others do not have a notion of private property, or even why eternal peace among mankind would always remain a dream. On the second topic, he solicited the complicity of his interlocutor: both of them knew very well how difficult it was to decipher female heads.[100]

A magazine directed in Weimar by Christoph Martin Wieland, who had gathered around him the leading exponents of the Goethezeit, published Gall's letter. In a short answer, von Retzer encouraged Gall to continue and to ignore the risk of being misunderstood: a common fate – according to the high court official – for all those who, since Aristotle down to Bacon, Newton and Kant, had discovered some truth. Mentioning those big names was certainly an exaggeration, but Gall had flattered him by writing no less hyperbolically that

99 "Des Herrn Dr. F. J. Gall Schreiben über seinen bereits geendigten Prodromus über die Verrichtungen des Gehirns der Menschen und der Thiere an Herrn Jos. Fr. von Retzer," *Der Neue Teutscher Merkur*, 1798:3, pp. 310-335, p. 326. Gall and phrenology have inspired an impressive mass of secondary literature in the last half-century. See his recent biography, which also contains a comprehensive bibliography: Stanley Finger, Paul Eling, *Franz Joseph Gall. Naturalist of the Mind. Visionary of the Brain* (Oxford: Oxford University Press, 2019).

100 "Des Herrn Dr. F. J. Gall Schreiben" (cit. note 99), pp. 329-330.

such a man – his protector in Vienna – had more than twice the brain of a silly bigoted woman (*dumme Betschwester*) and at least two twelfths more than the wisest elephant.[101]

101 Jos. Fr. von Retzer, "Antwort an Hernn Doktor Gall", *Der Neue Teutscher Merkur*, 1798:3, pp. 332-335; "Des Herrn Dr. F. J. Gall Schreiben" (cit. note 99), p. 318.

CHAPTER 2

Rising Tide

1 The "Phrenological Wedge"

In 1801, Louis-Joseph-Marie Robert was thirty when he dedicated to the members of the Institut National de France a weird book on the art of making *enfans d'esprit* bound to become great men. He had studied medicine in Paris, while his professional life took place in Marseilles, as a professor of Naval Hygiene and as a doctor of the local Lazaret. The distinguished *Citoyens* of the Institut were offered – so he wrote without any false modesty – the most useful book ever. By adopting the plan proposed by him, the Republican government could soon obtain the frequent birth and the multiplication of excellent men. Robert assured them that it was no less easy to get *enfans d'esprit* than special animals like an Arabian horse, a dachshund with crooked legs or a pure canary. The *facies propria* of human excellence consisted of a series of outer features, combined with a peculiar form of brain, which were inheritable. Modern physiologists considered the cerebral pulp as the secretory organ of that thought, which seemed more or less rapidly conceived and expressed depending on whether the organ-filter was elastic, vibrating and expansive. Extremely sensitive were "le jeu et les ressorts de ce mécanisme", but the motor principle was still unknown. Nevertheless, would the brain of a Lapp be able to produce the *Théodicée*? Or how distant was the brain of an Eskimo, who could only count up to six, from that of the extraordinary mind who had weighed the planets and analyzed the solar radiation?[1]

Robert claimed that his system was not a dream of unbridled imagination or a sign of delirium. Knowing the action of nature in creating brilliant individuals could lead to imitate it, and to have soon a "pépinière toujours vivante" of little geniuses. The *Mégalanthropogénésie*, a practice based on evidence of the inheritance of physiognomic traits – Lavater as the main authority – and entrusted to the care of a wise political power, would have a favorable effect on civilization.[2] The perspective outlined by the French doctor may seem rather bizarre, and yet the cultural context of the early nineteenth century was such

1 Robert le jeune [Louis-Joseph-Marie Robert], *Mégalanthropogénésie, ou l'Art de faire des Enfans d'esprit, qui deviennent de grands-hommes; Suivie des traits physiognomoniques propres à les faire reconnaître, décrits par Lavater, et du meilleur mode de génération* (Paris: Debray-Bailleul, 1801), pp. iii-vi, 7-8, 31-33, 97-98.

2 Ibid., pp. 222-236.

© KONINKLIJKE BRILL NV, LEIDEN, 2020 DOI:10.1163/9789004431881_004

as to favor the beginning of a long succession of studies, data collection, theories that marked the pervasive discourse on brains belonging to men of genius.[3]

On February 12th, 1804, Kant died after a few years of mental decline. A young anatomist at the University of Königsberg, Wilhelm Gottlieb Kelch examined his head on the deathbed, carefully applying the criteria of Gall's organology – already controversial – and disclosed the results through the publisher who had printed Kant's works in the 1790s. Kelch had detected remarkable prominences on the philosopher's forehead, associated to a poor development in the occipital part: his sagacity, wit, and memory, as well as his total shortage of sexual drive, were expected to have a proper cranial representation.[4] Soemmerring approvingly reviewed Kelch's booklet – sixty pages in all – commenting on how extremely advantageous, for the propagation of Gall's wonderful (*herrliche*) discoveries, was their application to a great man, whose brain organization, because of its brilliant achievements, must have been a privileged one. If the organological doctrine had continued to receive such support, it would have gained more and more consensus, as every truth does.[5]

As a matter of fact, Franz Joseph Gall's open letter to von Retzer, published in 1798, did not protect him for as long as it should have. On December 24, 1801, Francis II, the last Roman emperor, decided to forbid his lessons, on suspicion of materialism and atheism. Despite numerous and authoritative voices in his favor, the interdict against his theory persuaded him to leave Vienna, an increasingly hostile residence. In March 1805 he set off on a long journey, with his pupil Johann Gaspar Spurzheim. For two and a half years Gall visited Germany, Denmark, Sweden, Holland, Switzerland, stopping at major cultural centers and courts, often as a guest of distinguished people, holding conferences and demonstrations, visiting scientific cabinets, hospitals and prisons. Although someone sniffed out the charlatan in him, a few universities respectfully received him; in Berlin, two medals were coined in his honor; among others, Goethe himself attended Gall's courses in Halle and Weimar and let a plaster cast of his face be made. Harsh criticism was about to fall on

3 The reference text is by Michael Hagner, *Geniale Gehirne. Zur Geschichte der Elitegehirnforschung* (Göttingen: Wallstein, 2004).

4 Wilhelm Gottlieb Kelch, *Ueber den Schädel Kants. Ein Beytrag zu Galls Hirn- und Schädellehre* (Königsberg: Friedrich Nicolovius, 1804). No dissection of Kant's head was performed and its plaster cast became part of the Gall collection.

5 u.s.f. [Samuel Thomas Soemmerring], review of Kelch's *Ueber den Schädel Kants*, *Göttingische gelehrte Anzeigen*, 1804, 3:1457-1460. Soemmerring offered a summary of the content, also reporting some meaningful correspondences with an anonymous biography of Kant just published in Leipzig.

Schädellehre – the term phrenology was introduced only later – as in 1807 Hegel's *Phänomenologie des Geistes* mocked his reduction of spiritual life to a bone. And in a short article of that same year, Schelling noticed how Gall understood the anatomy of the brain much better than the faculties that it supposedly contained.[6]

In November 1807, on his way to Paris, Gall did not know that he would remain there until his death, except for a brief stay in London. Quite a lot of colleagues welcomed him and helped him put together a rich clientele. What interested him was to perfect and spread his theory: since February 1808, he could do it at the Athénée des arts and, shortly afterwards, at the Société de Médecine. Along with Spurzheim, in March 1808, he took a big step forward: a *Mémoire* about their research in the nervous system was submitted to the Institut de France for a review. Napoleon's hostility towards the heterodox couple of doctors influenced the judgment of the panel, whose *Rapport* – signed by Cuvier, Pinel, Portal, Sabatier and Tenon – though recognizing the novelty and importance of the anatomy practiced by Gall, rejected his physiological ambitions and claims. In other words, the official scientific community turned down precisely his attempts at reconnecting the organic and the mental, from which the whole operation had started. In the aftermath of the response, Gall and Spurzheim published their *Mémoire* with defensive remarks.[7]

The academic dismissal had the effect of intensifying their activity: between 1810 and 1819, the four large volumes of the *Anatomie et physiologie du système nerveux* – accompanied by an atlas of a hundred tables – finally described the analytical elements of organology. Only the first two were written in collaboration, since in 1813 Gall and Spurzheim parted for various reasons – pride, rivalry, and money – and went their separate ways.[8] In the first volume, they maintained that, given the distribution of accidental differences, any brain organ could be more developed and determine the peculiar character of individuals, races, and nations:

6 Friedrich Wilhelm Joseph Schelling, "Einiges über die Schädellehre," *Morgenblatt für gebildete Stände*, 1807, 1/74:293-294.

7 Franz Joseph Gall, Johann Gaspar Spurzheim, *Recherches sur le système nerveux en général et sur celui du cerveau en particulier. Mémoire présenté à l'Institut de France, le 14 mars 1808; suivi d'Observations sur le rapport qui en a été fait à cette compagnie par ses commissaires* (Paris: Schoell-Nicole, 1809).

8 Harry Whitaker, Gonia Jarema, "The Split between Gall and Spurzheim," *Journal of the History of the Neurosciences*, 2017, 26/2:216-223; John van Wyhe, *Phrenology and the Origins of Victorian Scientific Naturalism* (Aldershot-Burlington: Ashgate, 2004), pp. 27-29.

> Il y a longtemps que l'on espère pouvoir expliquer le caractère national de tel ou tel peuple, en découvrant le caractère particulier de l'organisation de la tête, qui seroit commune à tous les individus de cette nation. Cela seroit sans doute possible, si l'on cherchoit dans un très grand nombre d'individus l'organisation propre, commune et uniforme, au lieu de déduire, comme on l'a fait jusqu'ici, une consequence générale d'un très-petit nombre d'observations. [...] Au reste, que la conformité d'organisation ait lieu par hérédité ou par toute autre cause accidentelle, elle est accompagné d'une conformité de dispositions, de penchants et de facultés de l'esprit.[9]

Word by word, the same perspective of investigation was repeated in a more 'philosophical' book of 1811, as it often happened to the two apostles, in their attempt to reach an ever-wider readership.[10] And again in 1819, Gall – now as a single author – commented on the practice of some *savants* to collect skulls of different nations and races to look for typical distinctive features of the head instead of trying to discover the organic causes of their characters. Different qualities and faculties were associated with certain countries and certain races of men as well. Quite predictably, Gall claimed that in a temperate climate, man had always attained the highest degree of perfection. The geographical extremes of heat and cold had the effect of slowing down his activity: in extreme heat he was violent in affections, of weak judgment, prone to animal pleasures; in extreme cold, "lourd et stupide", peaceful, and with moderate drives. In a long quote from Adam Ferguson's *Essay on the History of Civil Society* (1767), Gall gave a better explanation of what he meant and added that the history of man was an inexhaustible source of further reflections similarly to those put forward by the Scottish philosopher and historian.

In addition, he predicted that research of that kind would grow and become more important, insofar as the viewpoint that intellectual operations depend on the state of the physical organization was well received. A frequent error had to be averted in particular, that is, believing that the general character of a nation could be deciphered by observing a small number of skulls. Actually, because of the existence of too many inter-individual differences, nothing

9 Franz Joseph Gall, Johann Gaspar Spurzheim, *Anatomie et physiologie du système nerveux en général et du cerveau en particulier, avec des observations sur la possibilité de reconnoître plusieurs dispositions intellectuelles et morales de l'homme et des animaux, par la configuration de leurs tête* (Paris: Schoell, 1810), I, pp. 24-25.

10 Franz Joseph Gall, Johann Gaspar Spurzheim, *Des dispositions innées de l'âme et de l'esprit, du materialisme, du fatalisme et de la liberté morale, avec une réflexion sur l'éducation et sur la legislation criminelle* (Paris: Schoell, 1811), pp. 38-39.

certain could be inferred from few data, and only large numbers would enable organology to discover a racial type in the structure of a head. It seemed generally true that the *Nègre* was inferior to the European in his intellectual faculties, and that he had a smaller head and brain mass. Likewise, it was a common notion that the English and French people were less talented at music than the Italians and Germans, and so on, about various national characters. However, Gall was wary of any generalization and he warned that historical circumstances – often surprisingly bizarre – called for much prudence in judging.[11]

Two young French naturalists, Jean René Constant Quoy and Joseph Paul Gaimard, participated as surgeons in the around-the-world expedition of the *Uranie* and *Physicienne*, two corvettes, between 1817 and 1820. Of the seven volumes containing the whole report of the long journey, they were the authors of the third one on zoology that, among so many other things, began with a description of the inhabitants of the Papua island:

> Les observations que nous avons faites sur les Papous sont favorable à la doctrine du docteur Gall: leur justesse nous ayant paru confirmée, jusqu'à un certain point, per l'étude des moeurs des individus qui en font le sujet, nous semblent contredire les paradoxes de ces philosophes chagrins qui, s'indignant des vices de l'homme en société, ont inventé l'homme de la nature tel qu'il n'existe pas […]. Nous devons ajouter que les Papous seroient susceptibles d'éducation, que leurs facultés intellectuelles ne demanderoient qu'à être exercées et développées pour leur faire tenir un rang distingués parmi les nombreuses variétés de l'espèce humain.[12]

Despite Papuans' supposed educability, the examination of some cranial specimens – also submitted to Gall's judgment – seemed to reveal aspects associated with salient traits of their behavior. In addition to an inclination for thieving, what struck the two observers was their pronounced *instinct carnassier*: a phrenological term that meant an innate tendency to kill, "affreux

11 Franz Joseph Gall, *Anatomie et physiologie du système nerveux en général et du cerveau en particulier, avec des observations sur la possibilité de reconnoître plusieurs dispositions intellectuelles et morales de l'homme et des animaux, par la configuration de leurs têtes* (Paris: N. Maze, 1819), IV, pp. 279-284.

12 *Voyage autour du monde, entrepris par ordre du Roi* […] *exécuté sur les corvettes l'Uranie et la Physicienne pendant les années 1817, 1818, 1819 et 1820* […] *Zoologie, par MM. Quoy et Gaimard, médecins de l'expédition* (Paris: Pillet aîné, 1824), pp. 10-11. As the *dessinateur de l'expédition*, it was Jacques Arago who had also written in an epistolary form the two volumes of a vivid *Promenade autour du monde, pendant les années 1817, 1818, 1819 et 1820, sur les corvettes du Roi l'Uranie et la Physicienne, commandés par M. Freycinet* (Paris: Leblanc, 1822).

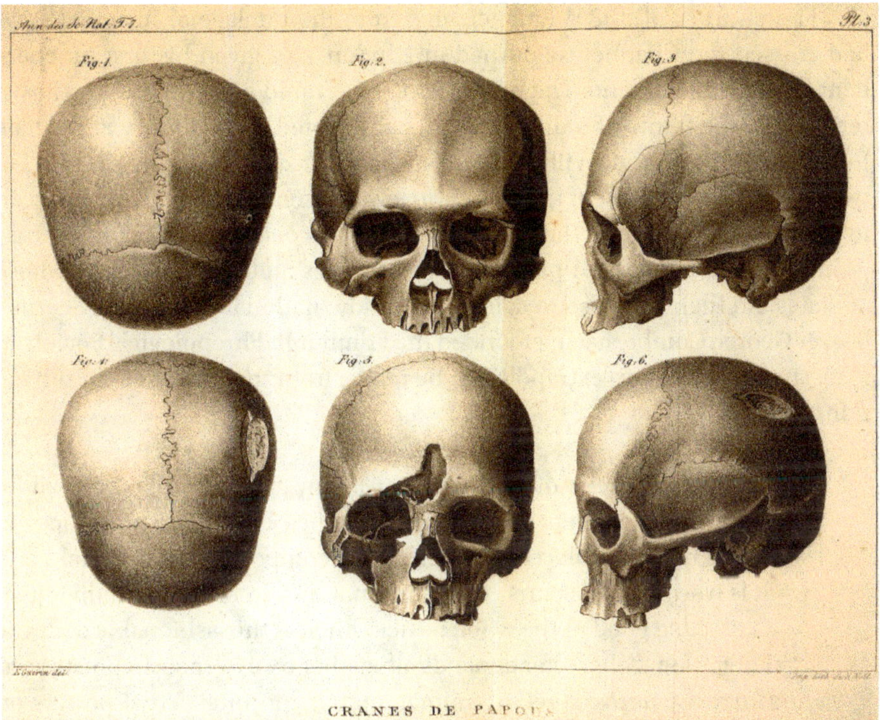

FIGURE 2.1 *Crânes de Papous de l'île de Rawack*, plate in MM. Quoy et Gaimard, "Observations sur la Constitution physique des Papous qui habitant les îles Rawak et Vaigiou," *Annales des sciences naturelles*, 1826, 7:27-38.

penchant auquel ces insulaires s'abandonnent avec fureur, et dont les ossemens qui nous occupent sont probablement des témoignages." A plate showed two skulls of Papuans seen from above, from the front and in profile, also included in an article published by Quoy and Guimard on the same topic.[13]

A young astronomer and naturalist, who at that time was a medical student in Cambridge, coined the term *phrenology* in 1815. Thomas Forster had met Spurzheim in London, and became passionate about that "wonderful and perfect system of anthropology" capable of improving the knowledge and treatment of mental illnesses, making educational methods more effective, distinguishing those predisposed to crime from those who committed it only occasionally. Unlike Gall, Spurzheim gladly adopted the new term in the title

13 MM. Quoy et Gaimard, "Observations sur la Constitution physique des Papous qui habitant les îles Rawak et Vaigiou," *Annales des sciences naturelles*, 1826, 7:27-38.

of a French work of 1818.[14] Emancipated from the tutelage of his mentor, he had crossed the Channel, sojourned in London, and given his first independent lectures in Scotland. The new discipline of mind and conduct was easily received in the frame of a divine plan locally rooted for a century and a half. Just think of the fame of William Paley's *Natural Theology* – twenty reprints between 1802 and 1820 – or of the eight Bridgewater Treatises, published between 1833 and 1836, illustrating the power, wisdom and goodness of God in Nature.

In Britain, a network of phrenological societies quickly took shape, whose affiliates did their best also to collect and study "national" skulls. The Scottish lawyer George Combe, having founded the Edinburgh Phrenological Society in 1820, suggested that an extrapolation be made, from the individual to the nation:

> The mental character of an individual, at any given time, is the result of his natural endowment of faculties, modified by the circumstances in which he has been placed [...] A nation is composed of individuals, and what is true of all the parts (which in a nation preserve their individuality), must hold good of the whole; – nevertheless the fashionable doctrine is, that national character depends altogether on external circumstances; and that the *native* stock of animal, moral, and intellectual powers on which these operate, is the same in New Holland and in England, in Hindostan and in France.[15]

Such supposed uniformity is nowhere to be found in reality. Combe was struck with the extreme dissimilarity in the attainments of the human varieties, with their distinct and permanent features of character and mental constitution across the continents. Due to their strong tendency towards moral and intellectual improvement, Europeans surpassed Asians, Africans and Native Americans; notable differences also existed within the white race: one only had to look at the constitutional opposition between the Scots and the Irish.

In 1828 Combe produced a manual of popular science aimed at magnifying the order of creation, by highlighting the inheritance of dispositions and

14 Thomas Forster, "Observations on a new system of phrenology, or the anatomy and the physiology of the brain, of Drs. Gall and Spurzheim," *Philosophical Magazine & Journal*, 1815: *45*:44-50; Johann Gaspar Spurzheim, *Observations sur la phrénologie, ou la connaissance de l'homme moral et intellectuel, fondée sur les fonctions du système nerveux* (Paris: Treuttel & Würtz, 1818).

15 George Combe, "On the Coincidence between the Natural Talents and Dispositions of Nations and the Development of their Brains," *The Phrenological Journal and Miscellany*, 1824, 2/*1*:1-19, pp. 1-2.

talents, the innate and the perfectible of man, drawing the table of his duties, in full observance of the laws of nature. His *Constitution of Man* flooded the market: with an increasing number of pages, edition by edition, one hundred thousand copies were sold in Great Britain and two hundred thousand in the United States before 1860. Almost a sort of "serial production", it has been written.[16] Every phrenologist knows – according to Combe's bestseller – that the New Hollanders, Charibs and other savage human groups are distinguished by great deficiencies in their moral and intellectual brain organs. Inasmuch as their substantial development is indispensable to the comprehension of science and the practice of virtue, and if the law of hereditary descent were hypothetically not valid, then it would be impossible to raise the New Hollanders, as a people, one step higher in their capacity. One might cultivate each generation up to the limits of its powers, but then the possible improvement would stop; no principle of increasing amelioration would exist for the next generation, born with brains equally deficient. The same remarks were applicable to every human tribe.

Taking Europeans "as the standard", and simulating the abrogation of the law of hereditary descent, every deficiency attributable to an imperfect development of the brain would be irremediable, and continue as long as the race existed. Understandably, Combe's 'meliorism' made him lean towards a more optimistic conclusion. It appeared to him that savages could not, with their brains, adopt European civilization: in any phrenological collection, specimens of their skulls showed organs of reflecting intellect, ideality, consciousness, and benevolence greatly inferior in size to those of Europeans. However, if these organs could be enlarged, the savages would certainly long for civilization, and would adopt it when offered, as long as the improvement efforts were performed during childhood, when the brain is more malleable.[17]

For a while, *The Phrenological Journal and Miscellany* published a series of articles centered on the description and deciphering of "national skulls" belonging to some population of the British colonies, and in its vastness the Empire facilitated the circulation of phrenological goods.[18] Just to make an

16 George Combe, *The Constitution of Man Considered in Relation to External Objects* (Edinburgh: John Anderson Jun., 1828). About Combe see the second part of van Wyhe, *Phrenology* (cit. note 8); Bill Jenkins, "Phrenology, Heredity and Progress in George Combe's *Constitution of Man*", *British Journal for the History of Science*, 2015, *48*: 455-473; James A. Secord, *Visions of Science: Books and Readers at the Dawn of the Victorian Age* (Oxford: Oxford University Press, 2014), p. 194.

17 Combe, *The Constitution of Man* (cit. note 16), pp. 164-168.

18 See Paul Turnbull, "«Rare Work amongst the Professors»: the Capture of Indigenous Skulls within Phrenological Knowledge in early colonial Australia," in Barbara Creed and Jeanette Horn (eds.), *Body Trade: Captivity, Cannibalism and Colonialism in the Pacific*

example: seven of the many hundreds of skulls that are in the University Medical School of Edinburgh came from India, dating back to 1833 and qualified as "Thug", a religious sect of bandits and killers who had become a nightmare for the colonial officials. The culprits were tried and executed, their heads were removed before the cremation or burial, then sent to the Edinburgh Phrenological Society, where a perfect correspondence with the wickedness shown in life was noticed: strong protuberances in the areas of animal instincts and a poor development in those of moral feelings.[19]

In the motherland, itinerant phrenologists, besides lecturing and exhibiting skulls and casts, deduced the character of bystanders from their bone structure: it has been calculated that in 1836 about five thousand of those proselytizers traveled far and wide in the British territory.[20] Beside individual examinations and counseling, they often proposed comparisons between different races. Thus arguments in support of the superiority of the Whites circulated and penetrated the common sense, while the fear grew that the interchange with the colonies could, in the long run, alter the pre-eminence of a race that was supposedly bound to rule the world. With the complicity of phrenological naturalism, the improvement of the species seemed to have become a mission entrusted to a selected segment of it. A new constellation of ideas and intentions hoped to graft the restraint of animal appetites and passions into a fairly pliable biological material.

Having helped export phrenology beyond the Channel, in June 1832 Spurzheim decided to embark for New York at Le Havre. Almost twenty years had passed since his break with Gall – who died in 1828 – and meanwhile he

---

(New York: Routledge, 2001), pp. 3-23. According to Turnbull, in the concrescence of racialised perceptions of Indigenous Australians, "phrenology reinforced what, by the early 1820s, had become a staple of mainstream scientific discourse", and favored the finding of indigenous skulls: see his *Science, Museums and Collecting the Indigenous Dead in Colonial Australia* (London: Palgrave Macmillan, 2012), p. 152. More generally, an original perspective on collecting skulls – which played a crucial role in accrediting anthropology as a science – is offered by Ricardo Roque, "Race and the Mobility of Human as Things," *Science, Technology, & Human Values*, 2014, *39*:607-617; Id., "Introduction. Anthropologie et materialité de la race," in *Revue d'histoire des sciences humaines*, 2015, *n. 27*:7-27;

19 Henry H. Spry, "Some Account of the Gang-Murderers of Central India, Commonly Called Thugs; Accompanying the Seven Skulls of Them," *The Phrenological Journal and Miscellany*, 1834, *8*:511-524; Robert Cox, "Remarks on the Skulls and Character of the Thugs", ibid., pp. 524-530; Kim A. Wagner, "Confessions of a Skull: Phrenology and Colonial Knowledge in Early Nineteenth-Century India," *History Workshop Journal*, 2010, *69*:27-51.

20 John van Wyhe, "The Diffusion of Phrenology through Public Lecturing", in Aileen Fyfe, Bernard Lightman (eds.), *Science in the Marketplace. Nineteenth-Century Sites and Experiences* (Chicago: University of Chicago Press, 2007), pp. 60-96.

had changed the nomenclature, increased the number of organs and faculties, and emphasized the practical attitude of phrenology, which became a resource for an even more ambitious program of reform. Then Spurzheim thought it was time at last to study the living conditions, the physiognomy, skull and brain of Blacks and Amerindians. From New York he headed for New England, warmly welcomed everywhere. In a few days, his message aroused the devotion and energies of Boston's educated ruling class, proud of its intellectual and moral primacy. And it was just about education that Spurzheim held his first speech, in front of a selected audience that crowded the hall of the Massachusetts State House. In mid-September, he started two cycles of conferences for a wider audience, while a more esoteric course of lectures on cerebral anatomy was addressed to the teaching staff of the medical faculty and to the best professionals. Too many engagements and the burden of his charisma wore out Spurzheim's strength in a couple of months. As all the treatments for his fever turned out to be useless, he died in November, assisted by powerless doctors, and his death was announced as an irreparable loss.

On the evening of the solemn funeral ceremony held for Spurzheim, the young publisher and bookseller Nahum Capen founded the Boston Phrenological Society, which instantly attracted lots of distinguished members. In subsequent years, with his business partners, he edited and published some of Spurzheim's works. In the first of them, phrenology was said to admit the existence of just one species of man, endowed with a certain number of fundamental powers, each of which are more or less active in different races or in different individuals of the same race. Their activity might be increased or diminished, and to this operation the perfectibility of man was confined.[21] In connection with the talents and characters of whole nations, however, a second work – containing also Capen's biography of its author – specified that no general inference could be drawn from observations made on a few individuals[22] One of the purposes of Spurzheim's overseas travel was revealed in the fourth American edition of an earlier *View of the Elementary Principles of Education* printed by a Scottish publisher in 1821:

21 Johann Gaspar Spurzheim, *Outlines of Phrenology, Being also a Manual of Reference for the Marked Bust* (Boston: Marsh, Capen and Lyon, 1832), p. 94.

22 Johann Gaspar Spurzheim, *Phrenology, in Connexion with the Study of Physiognomy. Illustration of Characters. First American Edition. To which is Prefixed a Biography of the Author, by Nahum Capen* (Boston: Marsh, Capen and Lyon, 1833), p. 44. An anonymous note (supposedly by Capen) informed that "This principle the author strictly adhered to, while in the United States, as he invariably refused to give an opinion upon our national character." (p. 44 n.)

> If there be several species of Man, there can be no universal principles of human conduct; human nature cannot be included in any one system; and the rules which are suitable for one nation will not be fit for another [...] Moreover, if there were several species, and one superior to the others, the White to the Negro, for example, slavery might be contended for as an institution of Nature; but if the species be only one, neither the primitive moral character, nor Christianity, can excuse this most selfish of all barbarities. [...]
>
> Whoever could demonstrate, that one part of the brain in Europeans is wanting in Negroes, would prove that there is a natural difference between them; I hope, during my stay in the United States, to be able to examine the brains of Negroes, and to ascertain that the same essential parts exist in them, subject, merely, to modifications, as it is the case in different individuals of the white race.[23]

Of course his "hope" – which deserves to be pointed out – could not be fulfilled because of his premature death. Another one of Spurzheim's texts remarked that several anatomists and physiologists had until then endeavored to set out specific national head shapes. Their observations, "although very defective", seemed to him "still rather in favor, than in opposition to, our physiology of the brain". Nevertheless, that important kind of study was quite in its infancy, and many valuable data ought to be collected by travellers and visitors of distant countries: to assist them, Spurzheim compiled a short list of instructions on how to proceed.[24]

Actually, the diffusion of the organological doctrine in the new American nation had preceded Spurzheim's arrival. During a stay in Europe, the young Nicholas Biddle – a future banker and journalist – witnessed the coronation of Napoleon and had also the opportunity to attend one of the lecture cycles held by Gall and Spurzheim in Karlsruhe. Returning to Philadelphia in 1807, he brought, with other travel souvenirs, a skull with the contours of the cerebral regions engraved on it. It took a few years for some doctors to open a Central Phrenological Society there, and to solicit the sympathy of the local public by divulging second-hand materials. Charles Caldwell had graduated from the medical school of the University of Pennsylvania in 1796, and had then practiced the profession in Philadelphia. In charge of acquiring books and

23 Johann Gaspar Spurzheim, *A View of the Elementary Principles of Education, Founded on the Study of the Nature of Man* (Boston: Marsh, Capen and Lyon, 1835), pp. 19, 28.

24 Johann Gaspar Spurzheim, *Phrenology, or the Doctrine of Mental Phenomena. Vol. I Physiological Part. Second American Edition, greatly Improved by the Author, from the Third London Edition* (Boston: Marsh, Capen and Lyon, 1833), pp. 99-101 and note.

instruments in Paris for the Transylvania University of Lexington (Kentucky), Caldwell followed some of Gall's lessons in 1821. It took three years for him to write the hundred pages of his *Elements of Phrenology*, the first American manual on the subject, specifically addressed to medical students and making sure the new doctrine met the principles of Christianity. He stressed that Revelation itself would confirm the material organs as essential for practicing the mental processes, thus teaching that, after death, the spirit must rejoin the matter in order to receive a reward or a punishment.[25]

So religiously oriented, and challenging the thesis advanced by James Cowles Prichard (or even by Spurzheim, whose name was left off), in 1830 Caldwell tried also to prove that a disbelief in the hypothesis of the original unity of mankind would by no means be unfriendly to the Christian faith. Whether it sprang from the same primitive root was a question pertaining solely to natural science. On the other hand, faith entered into the constitution of man as an essential ingredient: deprive him of it, and he would be rendered monstrous. Every new scientific truth might be made to minister to religious feeling, by the light it throws on the beautiful and beneficent arrangement of nature. The theory about the existence of different species of men, some inferior to others, did not imply injustice towards the former; rather, inferior beings should become objects of kindness. Men, in their higher species, protect and cherish women, because of her feebleness. Similarly, a hierarchical vision would give no endorsement to the cruelty and tyranny imposed on Africans or Indians, and no justification to Caucasians in either enslaving the former or to destroying the latter. Caldwell supported that kind of view, because it was true – resting on facts – and beneficial in its tendency.[26]

One of these overt facts was the general diversity, corporeal and mental, between the Caucasian and the African races: color, texture and figure on one side, intellect and moral feeling on the other. In particular, the bones of the head were thicker, more compact, stronger and heavier in the African, and the cavity of the cranium much smaller: after having described them in every detail, Caldwell focused on its content, persuaded by Soemmerring's finding:

> To this may be added, as a further diversity, in an important organ, that by far the greatest portion of his brain lies behind a perpendicular line drawn from the external opening of the ear to the top of the head, while, in the Caucasian, the portions on each side of such a line are much more

25 Charles Caldwell, *Elements of Phrenology* (Lexington, KY: Thomas T. Skillman, 1824).

26 Charles Caldwell, *Thoughts on the Original Unity of the Human Race* (New York: E. Bliss, 1830), pp. iii-x.

> nearly equal. [...] The foramen magnum, in the occipital bone, is larger in the African than in the Caucasian race. The necessary consequence of this is, that the medulla oblongata, which passes through it and fills it, is also larger, as is indeed the whole of the spinal cord, in common with many of the nerves. We may here remark, that the nerves of the African generally are larger in proportion to his brain, than those of the Caucasian. In this he resembles the inferior animals, occupying a station between them and the individuals of the race with which we are contrasting him.[27]

In his pamphlet Caldwell wrote a good number of pages to challenge the idea that differences only came from the climate, enhancing instead the power of the "constitutional" factor. Consequently, the Caucasian was superior to the other races in its "native intellectual faculties", so endowed with a higher and better organization that he can read and interpret the book of Nature, the true source of knowledge and wisdom. White man was to be credited for all the progress made in knowledge since the origins, while the African and Indian races were "motionless", fixed to a spot, like the rocks and trees, each generation always pursuing the same time-beaten track. Let some of their tribes be instructed in science and arts by Caucasian teachers, and then be abandoned to themselves: their course would be definitely retrograde to their original ignorance. Nor were the moral deficiencies of Africans and Indians less striking. Cannibalism, for instance, appeared to belong exclusively to them. Caldwell was especially concerned with the Indians, assuming that they could not be civilized and that any effort to that effect would only contribute to their predictable extinction.[28]

Six years after Spurzheim's death, George Combe undertook a journey of proselytism, sailing for New York to follow up on his predecessor's work overseas, and commenced a course of lectures in Boston in October 1838 that was followed by others in the main northeastern centers, going all the way to Philadelphia. He kept a notebook, eventually published in two volumes, which reveal his curiosity toward that world so different from his own, and did not refrain from condemning Negro slavery as "a canker in the moral constitution of the country, that must produce evil continually until it is removed."[29] Although a foreigner had no right to interfere, however he could express an

27 Ibid., pp. 80-81.
28 Ibid., pp. 137-145.
29 George Combe, *Notes on the United States of North America during a Phrenological Visit in 1838-9-40* (Edinburgh: MacLachlan, Stewart, & Co., 1841), I, pp. 254-255.

opinion on the institutions of the visited country: so, Combe felt the duty to blame the most cruel injustice perpetrated on a subdued race, and did not fear that emancipation would entail risks anyway. On the other hand, though, by applying his phrenological tools, he found that

> the existing races of native American Indians shew skulls inferior in their moral and intellectual development to those of the Anglo-Saxon race, and that, morally and intellectually, these Indians are inferior to their Anglo-Saxon invaders, and have receded before them. These facts are tangible and authentic, they all harmonize, and go to support one conclusion, namely, that diminutive size in the anterior lobe of the brain is concomitant with feeble, and large size with great, intellectual powers.[30]

Besides, when the "practical phrenologist" Orson Squire Fowler – an astute entrepreneur, who had opened a first workshop in Philadelphia in 1835, soon moved to New York – wrote profusely about *Hereditary Descent*, he also dealt with races (colored, Indian) and nations (Jews mainly). His characterizations obviously made use of a specialist terminology:

> The African head is longer from the root of the nose to both Philoprogenitiveness and to Self Esteem, than the European, longer and higher in the crown, but not as wide [...] they are proverbially polite and urbane, and hence make excellent waiters; are fond of ornament and show; love to swell, and are noted for feeling large and swaggering. In harmony with their greater development of Philoprogenitiveness, they make our best nurses, as far as fondness and patience with children are concerned, and evince a most passionate attachment to their children, and the strongest attachment to friends. [...]
>
> Make a slave of an Indian! Who ever heard of such a thing? [...] the love of freedom, and the fierce spirit that dies sooner than submit to servitude are *born* in the American race, as is also gratitude for favors, and revenge for wrongs. [...] By civilizing and educating Indian parents, you will without doubt be able to make additional improvements in the children, and, in a series of generations, to civilize and adorn the race, but his still predominant Destructiveness will render him revengeful and vindictive, his

30 Ibid., pp. 141-142.

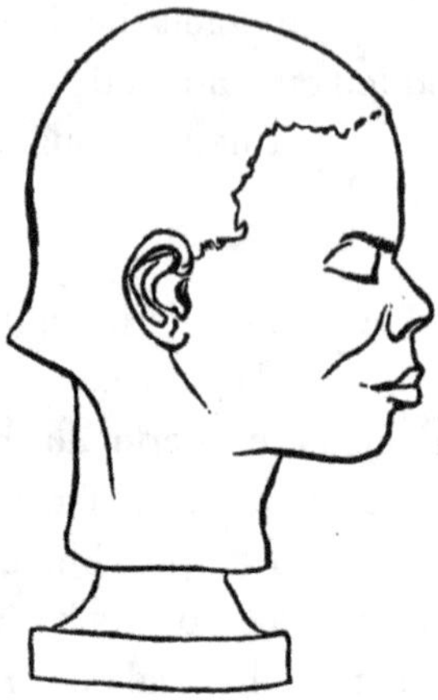

FIGURE 2.2
Silhouette of a plaster cast of Cinquez' head, in Lorenzo Niles Fowler, "Phrenological Development of Joseph Cinquez, alias Ginqua," *The American Phrenological Journal and* Miscellany, 1840, 2:136-138, p. 136.

> powerful Secretiveness and Cautiousness, crafty and cunning, and his great Perceptive organs, knowing and intelligent.[31]

While he was lecturing in Hartford, Combe happened to deal also with the famous case of the Amistad captives, a pack of fifty-three young people that had been enslaved in Sierra Leone and transported to Cuba. Two Spaniards bought them in Havana, and transferred them into the schooner bearing the ironic name of Amistad, heading for another Caribbean destination. But the slaves rebelled, killed Captain Ferrer and forced the crew to change course and bring them back to Africa, without noticing that the sailors turned the route to northwest. In Long Island Sound, the U.S. Coast Guard apprehended the ship and a long judicial process began against the Africans who, imprisoned in New Haven jail, became an ideal lever for the abolitionist campaign.[32]

Lorenzo Niles Fowler, brother and business partner of the aforementioned Orson Squire, visited the captives in September 1839. His detailed report of their leader Cinquez' "phrenological development" was all in all positive – also supported by a plaster cast of his head, copies of which were made for sale:

31 Orson Squire Fowler, "Hereditary Descent: its Laws and Facts, Illustrated and Applied to the Improvement of Mankind; with Hints to Woman etc.," *The American Phrenological Journal and Miscellany*, 1843, 5:385-650, pp. 419-422.

32 See Marcus Rediker, *The Amistad Rebellion. An Atlantic Odissey of Slavery and Freedom* (New York: Viking, 2012); Susan Branson, "Phrenology and the Science of Race in Antebellum America," *Early American Studies: An Interdisciplinary Journal*, 2017, 15:164-193.

> The base of his brain is inferior in size; consequently the lower animal propensities do not constitute the leading elements of his character. [...] His head measures most in the region of those faculties giving a love of liberty, independence, determination, ambition, regard for his country, and for what he thinks is sacred and right; also, good practical talents and powers of observation, shrewdness, tact, and management, joined with an uncommon degree of moral courage and pride of character. [...] His intellect is generally well-balanced, and better developed than most persons' belonging to his race. Still he is quite deficient in those faculties giving natural refinement, delicacy of feeling, imagination, powers of adaptation, and construction. [...] His cerebral organization, as a whole, I should think, was also superior to the majority of negroes' in our own country.[33]

Twenty days later, George Combe repeated Fowler's experience, recording in his diary how the public opinion was split: were the prisoners heroes who had nobly risen against their oppressors, or just bandits to be tried and executed? And he too looked at the same subject with phrenological sympathy:

> Their leader Cinquez or Jinquez, who killed the captain of the schooner, is a well-made man of 24 or 25 years of age. His head is long from the front to the back, and rises high above the ear, particularly in the regions of Self-Esteem, and Firmness. The breadth is moderate, and Destructiveness is large, but moderate. Benevolence and Veneration are well marked, and rise above the lateral organs; but the coronal region altogether is narrow. The anterior lobe also is narrow; but it is long from front to back. The middle perpendicular portion, including Comparison and Eventuality, is decidedly large. Individuality is full. The temperament seems to be nervous-bilious. This size and form of brain indicate considerable mental

33 Lorenzo Niles Fowler, "Phrenological Development of Joseph Cinquez, alias Ginqua," *The American Phrenological Journal and* Miscellany, 1840, 2:136-138. Thirty-six descriptions and silhouettes of the captives were instead provided by John Warner Barber, *A History of the Amistad Captives: Being a Circumstantial Account of the Capture of the Spanish Schooner Amistad, by the Africans on Board; their Voyage, and Capture near Long Island, New York; with Biographical Sketches of Each of the Surviving Africans. Also, an Account of the Trials had on their Case etc.* (New Haven, CT: E.L. & J.W. Barber, 1840). The abolitionist painter and engraver Nathaniel Jocelyn made an ennobling portrait of Cinquez – dressed in a kind of toga against the backdrop of an African landscape – held by the New Haven Colony Historical Society.

> power, decision, self-reliance, prompt perception, and readiness of action.[34]

After the release of the Africans from jail in March 1841, an Amistad committee of abolitionists organized a tour through some northern cities, where they attracted an audience that was eager to hear the story of their experiences and Christian conversions. Most of them returned to Sierra Leone in November 1841.

In an article on Robert Chambers' *Vestiges of the Natural History of Creation* – the anonymous first edition of which, in 1844, adopted Gall's authority in discussing man's mental endowments – James Secord points to a "phrenological wedge" as the major agency for the introduction of naturalism into Victorian Britain.[35] More generally, *wedge* is a good metaphorical term for designating the effects of the mission undertaken by Gall and his numerous followers during the early decades of the century: in Western culture principles and notions concerning the centrality of the brain and the need to study it in depth spread and took hold with great force and determination, promising to shed new light on the most diverse fields.

2 Shrunken Brains

Most probably the young pharmacist Julien-Joseph Virey had not yet heard much about the construction of Gall's system when, in 1803, he identified the best *zoomètre*, an efficient method for measuring and classifying the various degrees of the animal kingdom, in the nervous system. What constituted the animal as such – "la racine de l'animalité" – was *le sentiment*, the faculty to have sensations, decreasing from the upper forms down to more ambiguous beings such as the sea urchin, the starfish, the sea anemone, the coral, or the organisms that can be observed under a microscope in aqueous infusions. In other words, the more perfect is the nervous system, the more sensitive it is, the higher it rises in the scale of animal nature. About ten years later, acknowledging Virey as the author of the idea, Cuvier adopted the same taxonomic criterion.[36]

---

34 George Combe, *Notes on the United States of North America during a Phrenological Visit in 1838-9-40* (Philadelphia: Carey & Hart, 1841), II, pp. 138-139.

35 James A. Secord, "Behind the Veil: Robert Chambers and *Vestiges*" in James R. Moore (ed.), *History, Humanity and Evolution. Essays for John C. Greene* (Cambridge: Cambridge University Press, 1989), p. 172.

36 v. [Julien-Joseph Virey], "Animal," in *Nouveau Dictionnaire d'Histoire naturelle appliquée aux Arts, principalement à l'Agriculture et à l'Economie rurale et domestique par une Société*

Virey was born near Langres in 1775, from a middle class small-town family, and attended the local college before taking an apprenticeship with a pharmacist uncle. He spent his youth and early maturity in the period of the political and military events of the Revolution, and almost twenty years in the medical corps. Thanks to his patron, the famous pharmacist and agronomist Antoine-Augustin Parmentier, he took up service at the military hospital of Val-de-Grâce and could benefit from the Parisian intellectual climate of the time, so as to get somehow close to the Idéologues – protagonists of the *secondes Lumières,* mediators between the two centuries – and to have contacts with, but not formally belonging to, the short experience of the Société des Observateurs de l'Homme, disliked by Napoleon, the Emperor.[37] Rather, Virey's own and peculiar orientation was emerging in an alternative way, fully exposed in 1800-1801 by the two volumes of his *Histoire naturelle du genre humain*, dedicated to Buffon, "le plus eloquent des naturalists; philosophe profond", "premicr de ses maîtres", although he had never met him as he was too young when the great naturalist died.

A bit of an understatement, Virey presented it as a "small book" – actually, more than eight hundred pages – on an immense subject, and not fearing to move away from the path traced by his predecessors; who implicitly included Buffon, in whose *Oeuvres*, updated and edited by Sonnini de Manoncourt, he was to participate. No battle flag of any sect to wave: Virey claimed he detested polemics and was only moved only by his love for the science of nature. Even though his work generally did not show such peaceful intentions, as Armand de Quatrefages remarked a few decades later:

> Rappelons ici un fait important dans l'histoire des sciences et qui signale le début des discussions sur lesquelles nous reviendrons plus tard. Virey

des Naturalistes et d'Agriculteurs (Paris: Crapelet-Deterville, 1803), tome I, pp. 419-466. According to Cuvier, "Le système nerveux est au fond tout l'animal; les autres systèmes ne sont là que pour le servir ou pour l'entretenir; il n'est donc pas étonnant que ce soit d'après lui qui'ils se règlent": see his "Sur un nouveau rapprochement à établir entre les classes qui composent le Règne Animal," in *Annales du Muséum d'Histoire naturelle*, 1812, *19*:73-84, pp. 76-77.

37 See Jean-Luc Chappey, *La Société des Observateurs de l'homme. Des anthropologues au temps de Bonaparte* (Paris: Société des études robespierristes, 2002), p. 349. Virey participated in the debates on the *enfant sauvage de l'Aveyron* – whom the *Observateur* Jean Itard was studying and trying to re-educate – with a few articles then resumed in his first work of 1800-1801: see Julien-Joseph Virey, *Histoire naturelle du genre humain, ou Recherches sur ses principaux Fondemens physiques et moraux; precedées d'un Discours sur la nature des êtres organiques, et sur l'ensemble de leurs physiologie. On y a joint une dissertation sur le sauvage de l'Aveyron* (Paris: F. Dufart, an IX [1800-1801]), II, pp. 287-350.

> fut le premier à combattre, au nom des sciences naturelles, l'unité spécifique des hommes, admise par Linné, Buffon et Blumenbach.[38]

For a long time, with successive variations, the dispute over that original unity permeated the developments of the natural science of man. On the one hand, people loyal to Buffon's perspective, inclined to see fortuitous deviations in the human races, caused by climate and ways of life, and for whom intraspecific fertility was a quite effective proof. On the other hand, from the anatomical and physiological data some naturalists and philosophers drew the opposite conclusion, namely that the races were veritable biological *species*, with a different origin, in time or space, and tried to demolish the arguments of the mainstream theory, one by one. As is known, statements that were eventually called *polygenist* had widely circulated during the previous century (Benoît de Maillet, Voltaire, Hume, Lord Kames, Meiners ecc.), so that the "rupture doctrinale" sometimes attributed to Virey had a number of antecedents, which he quoted in some places. Yet later on he passed for being the first *savant* who had truly denied the common origin of mankind.[39]

Soon, the cultural and political phase in which his *Histoire naturelle du genre humain* appeared was to dictate the *décret consulaire* of 30 floréal an X (20 mai 1802), repealing the laws that had abolished slavery throughout the territory of the French Republic in 1792 and 1794. Regardless of its different enforcement in each colony, the legal event had something highly symbolic in it. So, in a still republican period, Virey could assume that man's intelligence was due to the mass of his brain, which would shrink in the lower races like the black one, to the extent that the jaws are prolonged and form a sharper angle: with Soemmerring and Camper recruited as the best authorities on the subject. Moreover, regardless of the influences of the sunlight that colors the body, each race would have a certain natural and hereditary idiosyncrasy.

In addition to color, which Malpighi found to be located in the *rete mucosum* and which from then on was repeatedly investigated, a peculiar and constant body shape was undeniable. Finally and importantly, races differed in the more fixed and radical parts of their organization – blood, bones, nerves, and brain: Virey randomly described a host of racial characters, relying on the numerous sources cited in the notes. What he called "espèce nègre ou ethiopienne" was divided into two varieties, the first one inhabiting a part of Africa,

38 Armand de Quatrefages, *Rapport sur les progrès de l'anthropologie* (Paris: A l'Imprimerie imperial, 1867), p. 20.

39 See Claude Blanckaert, "J.-J. Virey, Observateur de l'Homme," in Claude Benichou et Claude Blanckaert (sous la direction de), *Julien-Joseph Virey naturaliste et anthropologue* (Paris: Vrin, 1988), pp. 102-104.

Papua and New Guinea and having a facial angle of about 75 degrees: "leur cerveau est rétréci par la Nature elle-même." The geometric measuring instrument invented by Camper in his search for beauty was consistently used by Virey, who took it as a "règle fondamentale invariable".[40] From various sources, including the Dutch anatomist, Virey borrowed the illustrations distributed on a series of *planches*, with the *caveat* that, due to their high variability, it was not easy to correctly represent the national characters. On the other hand, their authors, mostly travelers, were not always skilled artists. It would have been better – he noted – to reproduce heads in their natural size, in their original colors and by excellent designers, "d'après l'individu".[41]

Because man, and not just an ordinary animal, was at stake, the second volume of the *Histoire naturelle du genre humain* dealt with culture: customs, moral aspects, rituals, religions, music, education, philosophy. Nonetheless, even here, Virey did enter information about "l'homme physique", in particular, his cerebral peculiarity. It seemed probable that the continuous use of thought had developed the civilized brain, more voluminous than the wild one, and that, in this respect, some races seemed less favored by nature. Such were Hottentots and *nègres*, whose cervical nerves would be proportionally larger than the European ones, with a smaller brain, as Soemmerring had shown. Different were also the Mongolian heads, whose parietals would be closer to each other, as in carnivorous animals, in order to leave more space to the temporal muscles for better chewing. A further example: the depressed forehead of the Caribbean or Malay could not fail to reduce their brain mass. However, Virey recommended that no immediate conclusions be drawn from the outward appearance of the head. For instance, the circumpolar people were macrocephalic without being civilized, just because of the extraordinary thickness of their skull.

In order to learn more about the brain, one should also study the genital organs. An excess of sexual pleasures would tend to destroy the forces of the intellect and to diminish the cerebral volume, while the occupations of the highest minds would render one impotent, as attested by the example of Newton's virginity. For Virey, civilized Europeans live mainly in their organs of thought; on the contrary, savages concentrate on the organs of appetite, such as the stomach and the reproductive parts. There would be no other activity more suitable for their simple nature than eating and having sex.[42] In 1808, the

40 Virey, *Histoire naturelle du genre humain* (cit. note 37), I, pp. 120-126, 136, 197-202, 297-299, 418-419.

41 Ibid., II, pp. 595-596.

42 Ibid., pp. 20-22, 182-183n, 256n., 327.

a

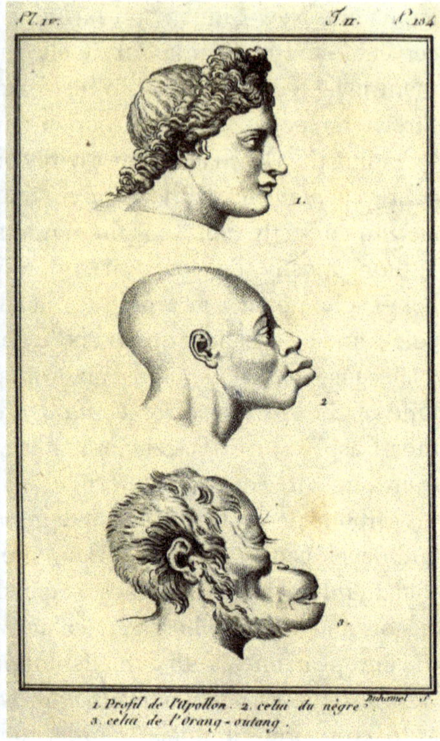

b

FIGURE 2.3A-C *Planches* in Julien-Joseph Virey, *Histoire naturelle du genre humain* (Paris an IX [1800-1801]), I, p. 149; II, pp. 134, 136.

same theme was taken up by his *L'Art de perfectionner l'homme*: "comme le cerveau est le foyer de l'esprit, l'organe sexuel est le foyer du sentiment le plus vif"[43]; again, those two volumes, which reflect a deeply felt concern at the beginning of the century, clearly represented that opposition between the races:

> Une vie à la manière des brutes, continuée pendant beaucoup de siècles sur une terre brûlée des feux du soleil, a donné aux nègres une physionomie animal, un teint noir et des cheveux crêpus, avec des moeurs grossières et sans esprit. Mais la vie civilisée, sous un ciel tempéré, donne aux abitans de l'Europe un visage droit et régulier, un teint blanc, des cheveux longs, et surtout des moeurs polies avec un esprit susceptible d'être développé par la culture. Plus la face est droite, plus elle présente de noblesse

43 Julien-Joseph Virey, *L'Art de perfectionner l'homme, ou de la médecine spirituelle et morale* (Paris: Deterville, 1808), I, p. 329.

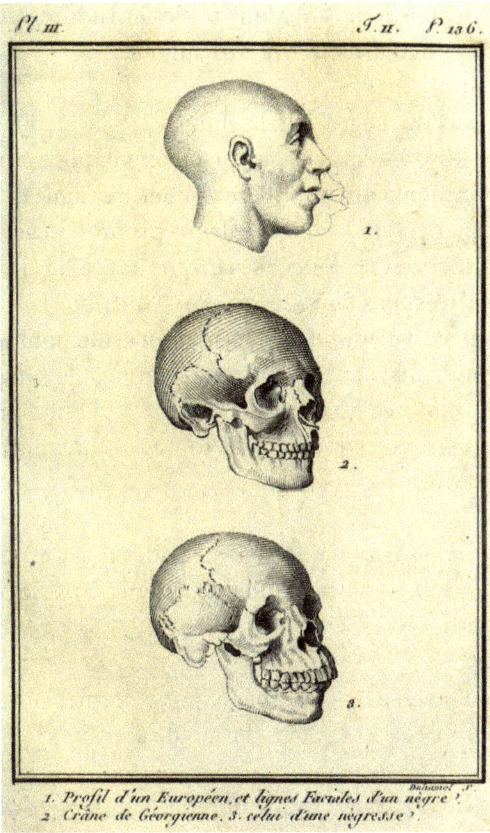

c

> et d'excellence; et, au contraire, si elle s'allonge en museau comme chex les Hottentots et d'autres nations, elle annonce davantage un caractère de brute, et une dégradation vers le genre des singes. A mesure que les mâchoires s'agrandissent, que la bouche et les organes des sens sont dominans, l'organe de la pensée, ou le cerveau se rétrécit à proportion.[44]

Only by reading two long dictionary entries can one understand the subsequent development of Virey's racial view, which ended up singling out just two human "species". In 1817, he wrote a Buffonian hymn to man, "le premier des êtres sur notre globe; placé à la tête du règne animal", "roi et dominateur", by

44 Ibid., I, pp. 143-144.

virtue of his superb brain and industrious hands.[45] Many topics are addressed here, and among others the extreme singularity of the African race:

> L'on a cru expliquer le museau du nègre par la moue que la chaleur semble produire, en gonflant les traits du visage, lorsqu'on s'expose aux ardeurs du soleil. Mais [...] comment cette moue, continuée pendant des siècles, a-t-elle pu reculer le trou occipital, enfoncer les os propres du nez, rétrécir la cavité cérébrale du nègre, et prolonger ses mâchoires? Si la chaleur a noirci le nègre (comme il n'est pas douteux qu'elle y contribue), il la faudrait supposer terriblement active pour charbonner non seulement le réseau muqueux sous l'épiderme, mais encore noircir davantage le sang, les chairs, la partie corticale ou cendrée du cerveau du nègre [...] Enfin on trouve dans le nègre une diathèse noire radicale en toute son organization.[46]

An entry in the *Nouveau Dictionnaire d'Histoire naturelle* provides a variation on the subject of *Homme*, with a short but trenchant definition and a *planche* that Virey's visual rhetoric used in other works:

> Le nègre est en quelque sorte l'inverse de l'Européen, par la forme, la capacité de son crâne, et par la foiblesse et la dégradation de son âme.[47]

Blacks and Whites would therefore be separated by a difference similar to that existing between the hare and the rabbit, i.e. between two species. Significantly, another distinctive trait of the *nègre* is not only the habit of crushing parasitic insects with the teeth, in the manner of monkeys, but also his louse, *pediculus nigritarum*, feeding on a black humor that gives it a darker color, besides having a particular shape: triangular head, wrinkled abdomen.

In another volume of the *Dictionnaire des science médicales*, Virey filled over fifty pages with an entry, *Nègre*, now definitely presented, right from the first lines, as one of the two main and distinct species. Not only for his skin color, but also for other deeper, lasting and indelible characters, which imply a precise anatomical and physiological diversity, unexplainable by traditional

45 Julien-Joseph Virey, "Homme," in *Dictionnaire des sciences médicales, par une société de médecins et de chirurgiens* (Paris: Panckoucke, 1817), vol. 21, pp. 191-344.

46 Ibid., pp. 263-264.

47 Julien-Joseph Virey, "Homme," in *Nouveau Dictionnaire d'Histoire naturelle, appliquée aux Arts, à l'Agriculture, à l'Économie rurale et domestique, à la Médecine, etc. par une Société de Naturalistes et d'Agriculteurs* (Paris: Deterville, 1817), vol. 15, pp. 1-268 (quote p. 168).

climatic determinism.[48] Following the scale of the creatures, the apes seemed to be the original root of humanity, so much so that it was possible to shift through shades, from the orangutan to the Hottentot, to the Black and finally to the white man, "l'homme par excellence". Providentially, the latter – "ce roi du globe par son génie" – did not stand alongside the orangutan, thanks to the presence of intermediate human links. What Virey's entry does, by means of a verbose prose, is an effective, merciless animalization of those "nomades qui parcourent les solitudes africaines", gloomily vegetating in the most complete stupidity.[49]

How could the sole influence of the sun and heat make the brain and nerves of Blacks darker, an innate and radical quality observed by a few anatomists in their dissections? It was not just a question of color: they had also noticed a comparatively narrower brain and more voluminous nerves. Together with the naturalist Palisot de Beauvois – a great connoisseur of Africa and the Antilles – Virey himself had weighed the amount of liquid that skulls of Whites and Blacks could respectively contain, and found nine ounces less in the latter. The comparison led to make meaningful considerations based on a quantitative approach, which was also endorsed by the phrenological doctrine that was thriving at that time:

> En effet, plus un organe se développe, plus il obtient d'activité et de puissance; de meme, à mesure qu'il perd de son étendue, cette puissance est diminuée. On voit donc que si le cerveau se rapetisse, et si les cordons médullaires qui en sortent grandissent, le nègre sera moins porté a faire usage de la pensée, qu'à se livrer à ses appetits physiques, tandis qu'il en sera tout autrement dans le blanc. [...] Chez nous le fronte avance, et la bouche semble se rapetisser, se reculer, comme si nous étions destinés à penser plutôt qu'à manger. [...] De même, les membres et les sens ne se perfectionnent beaucoup à l'extérieur qu'aux dépenses des facultés intellectuelles chez le nègre: il semble que son cerveau se soit écoulé en grande partie dans ses nerfs, tant qu'il a les sens actifs et les fibres mobiles; il est tout sensations.[50]

As a result of this organic endowment, Blacks would get an incredibly fine sense of smell, acute hearing, a sensual and vigorous taste; a kind of agility,

---

48 Julien-Joseph Virey, "Nègre," in *Dictionnaire des sciences médicales, par une société de médecins et de chirurgiens* (Paris: Panckoucke, 1819), vol. 35, pp. 378-432.

49 Ibid., pp. 385-387.

50 Ibid., pp. 389-390.

flexibility, and imitative powers that surpass those of any other human variety. Only by capturing their senses by means of pleasure or by causing fear could one act upon them. Some had been raised with care and the same education as Whites, and yet they proved unable to appropriate the received knowledge: "ils sont donc de grands enfans". Not to mention the Hottentots, whose distance from the monkeys and animals seemed to Virey further reduced, due to a number of impressive somatic similarities; *inter alia*, their footprint on sand, unique among humans because of the morphology of their feet and close to that of the climbing monkey; the limited articulation of their vocal sounds, reminiscent of the noise made by turkeys.[51]

Nevertheless, Virey blamed "cet odieux et avilissant esclavage" that degraded people who were naturally equipped with a generous and sensitive heart, with all the virtues of simple souls. Heroic examples of attachment had been reported, of slaves who died to save their masters. Such were the men oppressed by Europeans for a very long time, and even when the progress of philanthropy had abolished the slave trade in some places. The more opulent and corrupt the nation, the more slaves it had: an eloquent and frightful relationship. And indeed it was easy to list names of a large group of opponents of slavery, more or less illustrious, both in France and in Great Britain, who had chosen the job of stretching our a protective hand, of elevating that unfortunate human species to an acceptable rank.[52]

Virey's defense of Blacks against the inflicted harassment was fostered by a detailed historical analysis of the phenomenon and did not contradict his conviction that nature itself had predisposed the existence of both the strongest and the weakest. In 1824-1826, the second edition of his *Histoire naturelle du genre humain* – "augmentée et entièrement refondue" – gave center stage to the "espèce nègre" and reaffirmed the previous antithetical scheme, whereby White and Black were on opposite poles and the otherness of the latter could also serve as a lower step in the scale of beings.[53] In the first volume Virey reviewed the value of the physiological reasons given by those who – like Blumenbach – supported the unity of the human species, and patiently countered them point by point.[54]

---

51 Ibid., pp. 405-406, 410.

52 Ibid., pp. 406-426.

53 Julien-Joseph Virey, *Histoire naturelle du genre humain, nouvelle édition augmentée et entièrement refondue. Tome premier* (Paris: Crochard, 1824); *Tome second et troisième* (Bruxelles: Aug. Wahlen, 1826). In 1834 the four volumes of a third edition came out (Bruxelles: Louis Hauman et Comp.).

54 Virey, *Histoire naturelle du genre humain* (cit. note 53), I, pp. 429-435.

Although Virey elevated the European to the top of the human hierarchy, his judgment of the present state of Western civilization was not at all flattering: it had seemed to him, in 1801, that the old continent was suffering from a pathological weakness and afflicted by corruption of the customs: "on pourrait appeler notre siècle si vanté, un *siècle-femme*, avec autant de raison qu'il a pris le titre de philosophique."[55] His expertise on women as a *médecin-philosophe* was widely explained in a book of over four hundred pages that followed those by other authors on the subject, nothing new. In Virey's case, it is significant that, after some *Vues générales*, the second chapter addresses the varieties of the female sex in the five races identified by Blumenbach. As the first character of the black female, he points to a greater tendency to lasciviousness, along with a special conformation of her sexual organs, which are minutely described.[56]

In general, differences between man and woman are not limited – according to Virey – to the organs of generation, but all the parts of their bodies reflect some peculiarity: the hair, skin and flesh, the hips and thighs, the extremities. The upper parts of the male body – the chest, shoulders and head – are stout and powerful, the capacity of the brain is conspicuous and contains three to four ounces more, "suivant nos expériences", than that of the female. Her organs of affection are more developed than those in charge of intellectual activity:

> Chez l'homme, en revanche, ces derniers sont plus fortement caractérisé. C'est parce qu'en effet l'homme pense plus qu'il ne sent, tandis que la femme sent toujours plus qu'elle ne réfléchit. Cette différence de conformation est analogue aux fonctions de chaque sexe; l'homme est destiné per la nature au travail, à l'emploi des forces physiques, à l'usage de la pensée, à se servir de la raison et du génie pour soutenir la famille don't il doit être le chef; la femme, à qui le dépôt de la génération devait être confié, avait besoin d'un basin spacieux [...]. Le mot de famille vient aussi de *femina* [*sic*], car la femme ne fait qu'un avec ses enfants. En effet la femme se rapporte à l'enfance en beaucoup de choses: ses os restent plus petits, plus minces que ceux de l'homme adulte [...].[57]

55 Virey, *Histoire naturelle du genre humain* (cit. note 37), II, p. 65.

56 Julien-Joseph Virey, *De la femme sous ses rapports physiologique, moral et littéraire* (Paris: Crochard, 1823), pp. 29-44.

57 Ibid., pp. 186-194.

A lower nature, therefore, but perfect for reproduction purposes: an old stereotype repeated here without much modification. In the last of his works, Virey took up a point that had already been raised before, and stressed the antagonism between *foyer cérébral* and *foyer génital*, two ends of a nervous chain always susceptible to tension: the abuse of the former might determine the destruction of the other. Consequently, for the mission entrusted to them by nature, it would be better if women developed their own brain only to a certain extent.[58]

Those who wanted to repudiate (*dishéritér*) Blacks, had often invoked anatomy. Noting this circumstance, in 1808 the abbé Grégoire, founder of the Conservatoire National des Arts et Métiers – who had sided, as a member of the Assemblée National, with *de gens de couleur ou sang-mêlés* since 1789 – pointed out that, for example, almost all the anatomists disagreed with Meckel, about the existence of a dark-colored brain. Neither did the widespread thesis of a quantitative difference in the volume of the brains of Whites and Blacks sound reliable to him. Georges Cuvier thought that intelligence was measurable not by the overall volume of the brain, but by that of the hemispheres alone: however, the naturalist had correctly warned that it was a simple conjecture, and in Grégoire's opinion it would have taken centuries before science could penetrate the mystery of the physical conditions of the intellect and morality. As an admirer of Blumenbach – with whom he shared the belief in the original unity of mankind – he mentioned Camper's faith in the facial angle, adopted by several anatomists, and the claim, attributed to Gall, to root the alleged inferiority of *Nègres* in the structure of the skull and brain. Faced with the objection that many of them showed undeniable talents, the founder of *Organologie* had replied that their cranial shapes approached the proper structure of the white people, and likewise the stupid White must have a conformation similar to that of Blacks. Even the most eminent of men – Grégoire commented – can draw questionable conclusions from valuable observations.[59]

58 Julien-Joseph Virey, *De la physiologie dans ses rapport avec la philosophie* (Paris: Baillière, 1844), pp. 84-96.

59 Henri Grégoire, *De la Littérature des Nègres, ou Recherches sur leurs facultés intellectuelles, leurs qualités morales et leur littérature; suivies de Notices sur les ouvrages des Nègres qui se sont distingués dans les Sciences, les Lettres et les Arts* (Paris: Maradan, 1808), pp. 14-28. Regarding the route that led to the publication of the book, see Bernard Gainot, "L'abbé Grégoire et la place des Noirs dans l'histoire universelle," *Gradhiva. Revue d'anthropologie et d'histoire des arts*, 2009, *10*: 22-39. More generally, see Jeremy D. Popkin and Richard H. Popkin (eds.), *The Abbé Grégoire and his World* (Dordrecht: Kluwer, 2000) and Alyssa Goldstein Sepinwall, *The Abbé Grégoire and the French Revolution. The Making of Modern Universalism* (Berkeley: University of California Press, 2005).

Grégoire's *nigrophile* book – as his detractors called it – came out in a particular conjuncture of the Napoleonic era, when an underground opposition challenged the authoritarian tendencies of the French Empire. Since the restoration of slavery in 1802, the legislation of the Consulate had introduced discriminatory rules both in the colonies and in the motherland. With a good dose of courage, *De la Littérature des Nègres* was dedicated to all those "brave men" who had supported the cause of Blacks and mongrels with their works, or had been active supporters of associations for the abolition of the slave trade and slavery. Two hundred and sixty-one names of them were listed: half English, a quarter French, and then Americans, *Nègres et sang-mêlés,* Germans, Danish, Swedes, Dutch, Italians, only one Spaniard. For the rest, the three hundred pages of the book – also translated into English and German – contained a vast collection of examples, aimed at showing the many qualities of Blacks through historical records and had a mixed response in France. The author's mission rhetorically set egalitarianism and common sense on a par:

> Des hommes qui ne consultent que leur bon sens, et qui n'ont pas suivi les discussions relatives aux colonies, douteront peut-être qu'on ait pu ravaler les Nègres au rang des Brutes, et mettre en problème leur capacité intellectuelle et morale. Cependant cette doctrine, aussi absurde qu'abominable, est insinuée ou professée dans une foule d'écrits. [...] Depuis trois siècle, l'Europe, qui se dit chrétienne et civilisées, torture sans pitié, sans rélache, en Amérique et en Afrique, des peuples qu'elle appelle sauvages et barbares. [...] Je m'étois imposé le devoir de prouver que les Nègres sont capable de vertus et des talens; je l'ai établi par le raisonnement, plus encore par les faits; ces faits n'annoncent pas des découvertes sublimes; ces ouvrages ne sont pas des chef-d'oeuvres; mais ils sont des argumens sans réplique contre les détracteurs des Nègres.[60]

In the same year in which the Catholic abbé Grégoire published his major work, a young British Quaker graduated in medicine at the University of Edinburgh, where his religious affiliation was no bar, with a *Disputatio inauguralis de generis humani varietate* – suggestive of an undeniably Blumenbachian title and orientation.[61] From his family, James Cowles Prichard had inherited a

---

60 Grégoire, *De la Littérature des Nègres* (cit. note 59), pp. 275, 278-279. Grégoire's campaign against prejudice continued after the restoration of monarchy: see his *De la noblesse de la peau, ou du préjudice des blancs contre la couleur des Africains et celle de leurs descendans noirs et sang-mêlés* (Paris: Baudouin Frères, 1826).

61 James Cowles Prichard, *Disputatio inauguralis de generis humani varietate* (Edinburgi: Abernethy & Walker), 1808. The thesis shows an excellent knowledge of both classic and

clear religious commitment to the doctrine of human unity, which pervaded the general approach of his thesis in 1808, as well as the rest of his work. After the Scottish period, he went back to England to complete his studies at the University of Cambridge, which was then involved in an Evangelical Revival that led him to convert to a form of Anglicanism. At the Trinity College in Oxford he quickly managed to learn Arabic, Celtic, Hebrew, and Sanskrit. It was 1810 when he finally moved to Bristol to set up a private practice – also hired by St. Peter's Hospital to work in the lunatic asylum – while still dealing with the doctoral topic up to the publication, three years later, of his *Researches into the Physical History of Man*. From their first pages, Prichard wanted to clarify that he by no means wished to exploit any religious dogma in favor of his opinion "that all mankind constitute but one race or proceed from a single family". He denied, in other words, that the truth of the Scriptures were involved in the solution of the question. Thus, only scientific knowledge was at stake, and the contributions of those who had previously dealt with the topic were presented and discussed accordingly.[62]

Prichard took a double perspective to find out the ratio of brain power to function, and to determine to what comparative degrees individuals or groups "are intellectual or merely sensitive". In the past, several different schemes – which he discussed – had been proposed for this purpose. One of the first and "rudest" attempts had consisted in comparing the capacity of the cranium with the bulk of the body, just to discover that "the ratio of the former to the latter is in general greater in Europeans than in Negroes". Still, for Prichard this notion was formed on very imperfect grounds, and fully refuted by the infinite diversity observable in the dimension of the head without any corresponding difference of mind. He judged "a more specious [plausible] method, and one which approaches much nearer to the attainment of the object" that of comparing – as Soemmerring had done – the relative magnitude of the brain with the brain-derived nerves. Even on such grounds, the African would still be inferior to the European. A similar principle had inspired Camper: the greater the facial

---

eighteenth-century sources, and is divided in the following chapters: *De coloris varietate, De sede coloris et causa varietatis; De crispa nigritarum coma; De formae varietate; De varia gentium statura; De causis mutationes in animalium specie praecipueque hominum producentibus; De sententia quae varietates hominum de coelo pendere facit; De aliis discriminibus coeli vulgo adscriptis; Conclusiones generales*. In Prichard's description of the human varieties both the skin pigmentation and the shape and size of the skull played the major role.

62 James Cowles Prichard, *Researches into the Physical History of Man* (London: John and Arthur Arch, 1813), p. iii. A reprint of the book (Chicago-London: The University of Chicago Press, 1973) is introduced by its editor: see George W. Stocking, Jr., "From Chronology to Ethnology. James Cowles Prichard and British Anthropology", pp. ix-cx.

angle, the higher the forehead and the cranial capacity. Although his basic principle seemed reasonable, at the same time Prichard maintained that it was liable to objection, because the facial angle matched the size of the skull in just one direction, while its capacity might also vary laterally or at the back. So, that kind of measurement delivered a totally incorrect estimate of the comparative magnitude of the brain.

Cuvier was credited by Prichard for having remedied the limit inherent in Camper's method. Two sections of the cranium and facial bones had to be made: one of them vertical, the other longitudinal. By measuring these sections, one could compare the area of the head as occupied by the "intellectual" brain with that of the "sensitive" face, the lower jaw being excluded from the calculation. The ratio of cranium to face calculated by Cuvier was as 4:1 in European heads, while the ratio of the facial area increased by 1/10 in the skull of Kalmuk and by 1/5 in that of the African. In these two varieties "a greater provision is made in the conformation of the head for the perfection of the senses, and less proportionably for the intellectual organ, than in Europeans". An even higher value was measurable in the orangutan, and then gradually descending through the scale of animated beings. Highly olfactory were the Africans, indeed, and it was generally known that the sensitive powers were greater in non-European races, but Prichard did not reckon it equally evident that their intellect was weaker, only on the basis of a comparison with other separate species.[63]

Blumenbach had found many defects in Camper's method, such as to make him reject it. If on the one side the facial angle displayed an intra-racial variation, on the other side skulls very different in many particulars might have similar angles. So, his own mode of examination had afforded a much more ample and accurate view of cranial diversities: Prichard explained its features

63 Prichard, *Researches* (cit. note 62), pp. 49-55. Through physical and moral criteria, a few years later Cuvier thus identified three basic races: "La caucasique, à laquelle nous appartenons, se distingue par la beauté de l'ovale que forme sa tête; et c'est elle qui a donné naissance aux peuples les plus civilisés, à ceux qui ont le plus généralement dominé les autres: elle varie par le teinte et par la couleur des cheveux. Le mongolique se reconnaît à ses pommettes saillantes, à son visage plat, à ses yeux étroits et obliques, à ses cheveux droit et noirs, à sa barbe grêle, à sa teint olivastre. Elle à formé de grands empires à la Chine et au Japon, et elle a quelquefois étendue ses conquêtes en deçà du grand desert; mais sa civilization est toujours resté stationnaire. La race nègre est confinée au midi de l'Atlas; son teint est noire, ses cheveux crépus, son crane comprimé, et son nez ecrasé; son museau saillant et ses grosses lèvres, la rapprochent manifestement des singes: les peuplades qui la composent sont toujours restées barbares.": George Cuvier, *Le Régne animal distribué d'après son organization, pour servir de base à l'histoire naturelle des animaux et d'introduction à l'anatomie comparée* (Paris: Deterville, 1817), 1, pp. 94-95.

and reproduced Blumenbach's description of skull varieties in the European, Mongol, and Negro.[64] He thought the dark races best adapted by their organization to a rude and uncivilized state, "which we must conceive to have been the primitive state of mankind": their dense and firm fibers render them much more able to endure fatigue and the inclemency of the seasons. Their more perfect senses are of most importance to the savages and less necessary to the civilized man, whose more capacious skull leaves space for an ample conformation of the brain, "on which an increase of intellectual power is probably dependent". Prichard suggested that moving away from a savage state was also correlated with a shift from dark to light pigmentation:

> Wherever we find the people naked, destitute barbarians running wild in the woods, there we also observe them to be black, and to partake considerably of the Negro form and character. Wherever we see any progress towards civilization, there we also find deviation towards a lighter colour and a different form."[65]

Prichard's *Researches* were a sort of work in progress: the two volumes of a second edition came out in 1826, with a slightly different title (*Mankind* instead of *Man*) and more than doubled the pages. The portrait of a black man, Abbas Gregorius, opened the first volume, with a dedication "to the venerable and justly celebrated Professor Blumenbach".[66] In a note to the preface, Prichard recognized his debt also towards certain researches and lectures by William Lawrence, from whom, however, he differed widely on some particular aspects. Again, an objection concerned the validity of the facial angle: some had supposed that a greater breadth or development of the brain – implied by a large and elevated forehead – was a presumption of great intellectual vigor, just as great muscles denoted strength of body and limbs:

---

64 Prichard, *Researches* (cit. note 62), pp. 55-62.

65 Ibid., p. 236. The idea of an original blackness was not new: for instance, Camper and John Hunter had suggested a similar hypothesis: see Hannah Franziska Augstein, *James Cowles Prichard's Anthropology: Remaking the Science of Man in Early Nineteenth Century Britain* (Amsterdam-Atlanta: Rodopi, 1999), pp. 131-133. In later editions of the *Researches* Prichard corrected his view, perhaps having realized that it had been improperly exploited, and to avoid theological objection to a possible "black Adam". Anyway he did not dismiss the notion of civilization as a "whitening agent".

66 Abbas Gregorius, an Abyssinian clergyman living in Rome at the end of the 17th century, had taught the Ethiopian language to the orientalist Hiob Ludolf. His portrait was redesigned for Prichard – who had discovered it through Blumenbach – by the Maltese lithographer Maxim (or Massimo) Gauci, on a print made in 1681. I owe this information to my friend Renato G. Mazzolini, whom I thank.

FIGURE 2.4
Abbas Gregorius as an "Ethiopian" specimen, in James Cowles Prichard, *Researches into the Physical History of Man* (London 1826).

> But this analogy is a very weak argument, and we require some other proof, that an ample size of nervous and cerebral parts, is in general associated with great power of the proper functions of those parts, with which functions the exercise of the mental faculties is allowed to be nearly connected. This cannot be gratuitously assumed: but it is obvious that this is a position capable of being proved or disproved by an appeal to facts.[67]

Those who were devoting themselves to the study of craniology with much zeal might supply, "at no very distant period", a clear conclusion to the general inquiry. Nevertheless for Prichard the question whether in Blacks mental faculties were less perfect than in Whites was still not fully resolved; information collected on the subject from the most judicious observers assured that "Negroes are not by any means inferior in intellect to Europeans", at least in their

67 James Cowles Prichard, *Researches into the Physical History of Mankind. Second edition in two volumes* (London: John and Arthur Arch, 1826), I, pp. 161-162.

sphere of action. The almost uniform testimony of many intelligent planters and medical practitioners from the West Indies, with whom he had conversed, was in favor of a "natural equality". What's more, instances were not rare of Blacks whose superiority of endowment had elevated them above their degraded conditions and despite so many political and social disadvantages: "distinguished as men of science and literature, and poetical genius".[68]

The third edition of his *Researches* – which grew up to five illustrated volumes with thousands of pages between 1836 and 1847 – had no scruples in expressing Prichard's uncertainty on the subject:

> There is nothing more probable than the supposition that the average degree of perfection in the development of the brain as of other parts of the system, differs in different nations with the diversities of climate and other elements of the external condition, and with the degrees of social culture. It is probable that the condition of man in civilized society produces some modification in the intellectual capabilities of the race. But without going into any of these conjectural discussions, it will be quite sufficient for my present argument, if it is allowed, that there are *some* Negroes whose mental faculties fully attain the standard of European intellect.
>
> I am not prepared to deny that any connexion exists between perfection and the development and structure of the brain, and vigour in the exercise of mental functions; but it appears to me that those who dogmatize on this subject, overlook many opposing facts, which they are bound to explain.[69]

In other words, it was by no means irrefutably established that intelligence bears some proportion to the development of the brain, and that, consequently, its mass gives an objective measure of the extent of mental power. Not surprisingly, therefore, in 1836 the skeptical Prichard welcomed a recent study by the German anatomist Friedrich Tiedemann – which will be discussed later – where several common tenets on the relationship between brain and race were falsified.[70]

68 Ibid., I, pp. 177-178.

69 James Cowles Prichard, *Researches into the Physical History of Mankind. Third Edition* (London: Sherwood, Gilbert, and Piper, 1836), I, pp. 216, 304.

70 Prichard, *Researches* (cit. note 67), II, pp. 350-355. Tiedemann's study "On the Brain of the Negro, compared with that of the European and the Orang-Outang" was published in the *Philosophical Transactions of the Royal Society of London*, 1836, *126*: 497-527.

3 Materialism and the Recapitulation Theory

William Lawrence's lectures, mentioned by Prichard, had been delivered in 1817-1818 at the Royal College of Surgeons and dedicated to Blumenbach, whose example "of combining together anatomical, physiological, and zoological pursuits, and advancing them by reciprocal illustration" was explicitly imitated. The contents of these lectures had been severely criticized for being inclined to materialism, so that, after their publication, the surgeon was accused of "the very unworthy design of propagating opinions detrimental to society, and of endeavoring to enforce them for the purpose of loosening those restraints, on which the welfare of mankind depends".[71] Young and bold, Lawrence did not fear to be ironic, in a country where (and when) philosophical orthodoxy was forceful:

> If the mental processes be not the function of the brain, what is its office? In animals which possess only a small part of the human cerebral structure, sensation exists, and in many cases is more acute than in man. What employment shall we find for all that man possesses over and above this portion, – for the large and prodigiously-developed human hemispheres? Are we to believe that these serve only to round the figure of the organ, or to fill the cranium? [...] the various forms of insanity [...] are only evidences of cerebral affections, disordered manifestations of those organs whose healthy action produces the phenomena called mental; in short, symptoms of diseased brain. The symptoms have the same relation to the brain, as vomiting, indigestion, heartburn, to the stomach; cough, asthma, to the lungs; or any other deranged functions to their corresponding organs.[72]

Prompted by these assumptions, most of Lawrence's book – that is, over four hundred pages – focused on the natural history of man. Craniology could not be lacking, as it increasingly provided an essential support to the growth of the anthropological discourse. Relying on Blumenbach's *Decades* – which were periodically exposing his collection of craniums *Diversarum Gentium* – and on Soemmerring's work, the British surgeon registered the lateral compression and the reduced capacity of the skull in the so-called Ethiopian variety.

---

71 William Lawrence, *Lectures on Physiology, Zoology, and the Natural History of Man, delivered at the Royal College of Surgeons* (London: J. Smith, 1822), p. 1. His main opponent was the anatomist and surgeon John Abernethy under whose guidance Lawrence had studied for many years.

72 Ibid., p. 104.

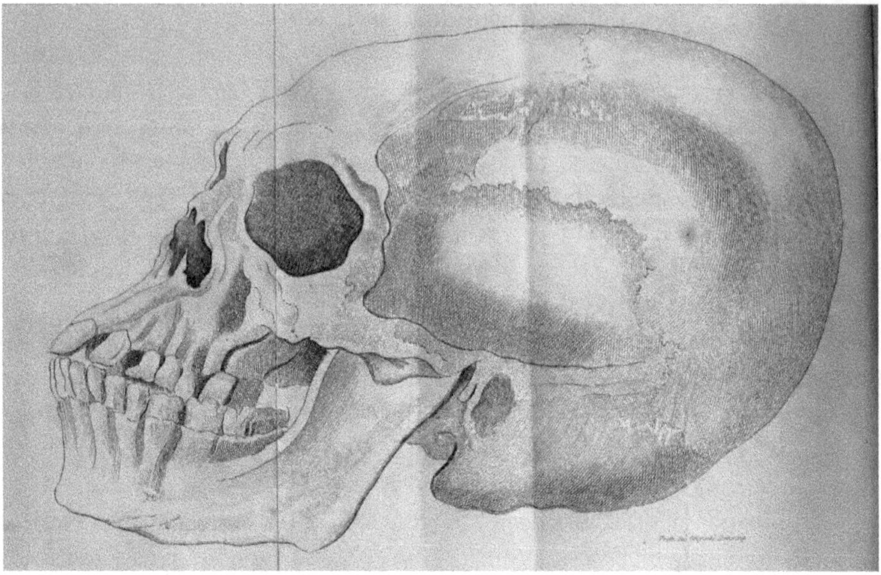

FIGURE 2.5 *The skull of a Negro*, plate in William Lawrence, *Lectures on Physiology, Zoology, and the Natural History of Man, delivered at the Royal College of Surgeons* (London 1822), after p. 330.

Lawrence himself warned that the description of its form and traits – illustrated by a plate – had to be taken only "in a general sense", being drawn from strongly marked specimens, and therefore not universally and strictly applicable.

Such diversity was abundantly proved, after all, by delineations of Sub-Saharan Africans executed by the best artists. No knowledgeable person, however, could fail to recognize their inferiority of organization and corresponding inferiority of faculties. Lawrence stated those plain results of his observation and experience, but ruled out that they could imply either an apology or an excuse for Negro slavery. In the heated disputes on the subject, both parties were in the wrong: the abolitionists erred in denying such an evident natural inferiority; their opponents committed the more serious moral mistake of perverting what should constitute a reason for kindness and indulgence into a justification of the "revolting and antichristian practice of traffic in human flesh". Conversely, European superiority should be employed to extend the blessings of civilization, and multiply the enjoyments of social life.[73]

After surveying the bodily diversities exhibited by the various races of men, Lawrence proceeded to review their moral and intellectual characters, to

73 Ibid., pp. 330-336.

understand whether these have such peculiarities as the numerous modifications of physical structure would suggest; whether propensities, moral feelings and the capabilities of knowledge were as different as the cerebral organs of which they seemed to be functions. All this persuaded him that that the races of mankind were no less characterized by diversity of moral endowments, than by differences of organization. As an example, Lawrence thought the distinction of color between white and black people no more striking than the pre-eminence of the former in moral feeling and mental endowments. The hideous savages of Van Diemen's Land, of New Holland, New Guinea, the Negroes of Congo etc. displayed the most disgusting moral as well as physical shape. The literature on this subject was broad and easily quotable, nullifying Rousseau's "delusive declamations" on the prerogatives and presumed advantages of the natural man. Once again, the negative judgment was balanced by acknowledging a natural goodness of heart and warmth of affection in Blacks. Several qualities were to be attributed to the original inhabitants of the New World, or to the Mongolian people. On the other hand, the superiority of the white races might be equally distinguished in the least-advanced state of civilization: just compare the ancient Germans, as delineated by Caesar and Tacitus, with a horde of Hottentots or a tribe of American Indians. The question was: why have the white races invariably raised themselves to a considerable height in the scale of cultivation, whereas the dark had almost as universally continued in the savage or barbarous state?[74]

All that Lawrence thought about the differential value among the races did not prevent him from condemning the responsibilities of the slave system: Negroes used to behave according to the treatment received, as confirmed by lots of evidence. Moreover, he listed the classic cases of excellent Blacks, as others had recently done. All the external influences – climate, way of life, degree of civilizations, form of government, education etc. – were nonetheless inadequate to account for the very marked differences which at all times and in all countries had distinguished the white from the dark races. Hence it was necessary to look deeper for their causes, and seek them in some circumstances inseparably interwoven in the human constitution, though Lawrence – incidentally – refuted the representation of a darker brain in Blacks:

> If the nobler attributes of man reside in the cerebral hemispheres; if the prerogatives which lift him so much above the brute are satisfactorily accounted for by the superior development of these important parts; the various degrees and kinds of moral feeling and of intellectual power may

74 Ibid., pp. 437-451.

> be consistently explained by the numerous and obvious differences of size in the various cerebral parts, besides which there may be peculiarities of internal organization, not appreciable by our means of inquiry.[75]

Because of the limitations of the available means, at that moment only craniology was able to provide a sufficient explanation of the natural capabilities of the subject. Benevolence and philanthropy had to be respected, but the retreating forehead and the depressed vertex of the dark varieties of mankind made Lawrence resolutely doubt whether they were susceptible of high destinies. These obstacles were too powerful for the best efforts of Missionaries and Bible Societies. Bulldogs might never equal the greyhound in speed, as the mastiff might never rival the sagacious and docile poodle in talents.

A curious criticism of Lawrence's treatment of race came from Thomas De Quincey, in an article written for the *London Magazine* in June 1824, which lamented – along Kantian or Coleridgean lines – the languishing condition of English physiology "for the want of first principles and a more philosophic spirit of study". On studying certain African skulls, Lawrence and many other physiologists had found clues of inferior intellectual faculties in the bony structures as compared with that of the Caucasian skull. De Quincey was disposed to agree with that, for nothing "unphilosophic" resided in supposing a scale of intellectual gradations among different races of men, any more than in seeing such a gradation among the different individuals of the same nation. Quite the contrary, guessing that a fair Nature would ever vary her workmanship "for the sake of absolute degradation" would be unphilosophic:

> If, therefore, the negro intellect be in some of the higher qualities inferior to that of the European, we may reasonably presume that this inferiority exists for the purpose of obtaining some compensatory excellence in lower qualities that could not else have existed. [...] By thus creating no absolute and entire superiority in any quarter, but distributing her gifts in parts, and making the several divisions of men the complements, as it were, of each other, she would point to that same intermixture of all the races with each other which on other grounds, *a priori* as well as empirical, we have reason to suppose one of her final purposes, and which the course of human events is manifestly preparing.[76]

---

75 Ibid., p. 460.

76 X.Y.Z. [Thomas De Quincey], "Notes from the Pocket Book of a late Opium Eater. English Physiology," *The London Magazine*, June 1824, p. 646. In a note to a later article he repeated something similar: "The crossing of races almost uniformly terminates in producing splendor, at any rate energy, of intellect. If the role of great men, or at least of energetic

Romantic writers seem to have been rather alert to what naturalists and physicians were practicing and divulging. In 1798-99, a young Samuel T. Coleridge had embarked on a ten-months journey to Germany and spent four of them studying at the University of Göttingen, where he assiduously attended Blumenbach's lectures and also made arrangements for private seminars with him.[77] There, he thought he could find an alternative option to France's scientific French materialism and welcomed Blumenbach's notion of *Bildungstrieb*, i.e. the vital principle grafted into living forms and developing in response to external stimuli. For his surgeon friend – and afterwards his literary executor – Joseph H. Green, Coleridge prepared some notes to be used in 1828 during his annual course of lectures. These attest to his adherence to the principle of a single human species, the summit of the organic creation, clearly separated from the animal kingdom, as well as to his opposition to the theses put forward in 1826 by the physiologist Louis-Antoine Desmoulins, whose work he defined "quintessential" French for having banished God and Providence from the system. The differences between humans were so slight in comparison those existing between animals that they could not account for a classification of different species. Above all, he intended to defend the "necessary idea" of the dispersal of Noah's sons, for its narrative and symbolic value. At the same time, Coleridge saw a precise hierarchy within the single species: from the white "Historic Race" all the others had degenerated down to "the wretched state of the "Boschesman in the wilds of Caffari or the New Hollanders". With an astronomical metaphor, he imagined a superior race that would act as a central Sun relative to the Planets, communicating light and heat, ameliorating them even by leaving them in their peculiar orbits.[78]

---

men, in Christendom, were carefully examined, it would astonish us to observe how many have been the children of mixed marriages; *i.e.* of alliances between two bloods as to nation." See [Thomas De Quincey], "Style No. IV," *Blackwoods Edinburgh Magazine*, February 1841, p. 217.

77 Kathleen Coburn, Anthony John Harding, *The Notebooks of Samuel Taylor Coleridge. Volume 5 1827-1834. Notes* (Princeton: Princeton University Press, 2002), Entry 5506 and notes.

78 Samuel Taylor Coleridge, *Marginalia. Edited by George Whalley* (London-Princeton: Routledge and Princeton University Press, 1980), II, pp. 176-177; Id., *Shorter Works and Fragments. Edited by H. J. Jackson and J. R. De J. Jackson* (London-Princeton: Routledge and Princeton University Press, 1995), pp. 1401-1405, 1457. See C.U.M. Smith, "Coleridge's «Theory of Life»," *Journal of the History of Biology*, 1999, *32*: 31-50; Tim Fulford, Debbie Lee and Peter J. Kitson, *Literature, Science and Exploration in the Romantic Era. Bodies of Knowledge* (Cambridge: Cambridge University Press, 2004), pp. 142-145; Peter J. Kitson, *Romantic Literature, Race, and Colonial Encounter* (New York: Palgrave Macmillan, 2007), pp. 39-42;

At the Muséum d'Histoire naturelle Desmoulins – whose materialism Coleridge feared – had worked with the assistance of Cuvier, after his medical studies in Paris. From the beginning, neuroanatomy had been his field of interest: in May 1820 he read a *Mémoire* to the first class of the Institut about the condition of the nervous system in cases of *marasma*, and again in December he presented to the same audience the results of a cerebral dissection carried out on a demented epileptic, who had died at the Hospice de Bicêtre. The young Desmoulins had ascertained that, as the cerebral volume decreases in the *vieillard*, the specific weight of the organ lessens, while its fibers take on greater hardness and cohesion. The same laws of the organization generally also applied to the brain: "ici, l'on voit décroître l'activité du cerveau et ses fonctions, à mesure que son volume, sa masse, la fluidité de ses molecules et partant l'intensité de ses fonctions diminuent." Non-senile forms of *marasma* were characterized by different phenomena. Howsoever, the quantitative approach resembled the style that phrenologists were advocating: a proportional relationship would bind the mass of the nervous matter with the intensity of nerve functions.[79]

In collaboration with François Magendie, in 1825 Desmoulins published an anatomical work on the nervous system. Three conventional beliefs were to be dispelled: that the material composition of the cerebro-spinal system in animals is constantly identical; that the same holds for the mechanism of its union with the nerves; that the number of its parts is always uniform. Twenty years earlier – Desmoulins recalled – Xavier Bichat had started to unravel that jumble of knowledge, and afterwards comparative and pathological anatomy offered precious physiological inductions. One of these tended to assert that the number and perfection of intellectual faculties – within the animal species and in individuals of the same species – are in proportion to the extent of the cerebral surfaces: no other measure of such faculties was possible, except for the relative quantity of cortical *plissement*. Desmoulins claimed to have been the first, in 1822, to suggest that kind of proportion. There was however no relationship between the amount of those folds and the extension of the *boîte cérébrale*, because a very large brain can have five or six times less surface than one two thirds smaller. Phrenological localizations seemed plausible to him, but Gall's and Spurzheim's demonstrations were far from being persuasive, resting

79 Louis-Antoine Desmoulins, "De l'état du système nerveux sous ses rapports de volume et de masse dans le marasme non sénile, et de l'influence de cet état sur les fonctions nerveuses," *Journal de physique, de chimie, d'histoire naturelle et des arts*, 1820, 90:442-452, p. 443; Id., "Suite des recherches sur l'état du volume et de la masse du système nerveux, et l'influence de cet état sur les fonctions nerveuses," *Journal de physique, de chimie, d'histoire naturelle et des arts*, 1821, 92:165-179.

only on the external configuration of the skull. So, he preferred to avoid the question, waiting for better investigations.[80]

Just one year later, the eclectic Desmoulins put forward his opinion on the natural history of human races, accompanied by a long complaining letter addressed in apparently humble tones to Cuvier ("Je ne suis pas de la moindre société littéraire ou savante, et vous êtes de toutes les académie du monde [...] je ai à peine trente ans, et votre âge est double du mien"), with whom relations had bitterly cooled, after an initial phase of closeness. Desmoulins grudgingly rebuked his former teacher for ignoring and mistreating him.[81] While Virey had divided the human genus into two species, he multiplied the species and races. What matters most here is not so much his particular classification – one of the many that sprouted in the nineteenth century – but some of his suggestions about the brain. For instance, Desmoulins commented about the political revolutions that left their mark on the history of China, without changing the rites of worship and administration:

> Où trouver les causes suffisantes d'une pareille stabilité, sinon dans une organisation particulière du cerveau dont les différences encore indéterminées anatomiquement n'en sont pas moins physiologiquement démontrée? [...] C'est surtout dans le langage et l'écriture, ce double moyen des communications intellectuelles et de la perfectibilité de l'humanité entière, par le perfectionnement et la multiplication des idées, que se montrent cette uniformité, cette fixité de l'organisation intellectuelle de la race indo-sinique. [...] Nous avons prouvé ailleurs (*Anat. des syst. nerv.*, *t.* 2, *liv.* IV, *chap.* V) qu'il y a dans le cerveau un organ particulier pour cet ordre de phénomènes intellectuelles; [...] Cela posé, on conçoit que cet organe puisse varier d'une race, à plus forte raison d'une espèce humaine à l'autre.[82]

The organ of language therefore coordinated with the cerebral organization of each race or species, and language could not be reformed by the race in which it is innate, but only by a race organized for a progressive tendency.

80 Louis-Antoine Desmoulins, *Anatomie du système nerveux des animaux à vertèbres, appliquée à la physiologie et à la zoologie. Ouvrage dont la partie physiologique est faite conjointement avec F. Magendie* (Paris: Méquignon-Marvis, 1825), II partie, pp. 606-609.

81 Louis-Antoine Desmoulins, *Histoire naturelle des races humaines du nord-est de l'Europe, de l'Asie boréale et orientale, et de l'Afrique australe, d'après des recherches spéciales d'antiquités, de physiologie, d'anatomie et de zoologie, appliquée à la recherche des origines des anciens peuples, à la science éthymologique, à la critique de l'histoire etc.* (Paris: Méquignon-Marvis, 1826), pp. vii-xxxiv.

82 Ibid., pp. 219-220.

Another cue comes from Desmoulins' treatment of the Hottentot race, whose cerebellum and posterior lobes appeared more developed than in any other. Contrariwise, the most perfect monkeys to whom Hottentots and Bushmen had been compared had a decidedly reduced occipital chamber. Moreover, the forehead of those African races would show a development of the anterior lobes that was virtually equal to that of the Europeans, while only the cerebral parts of their vertex seemed depressed and restricted. On the other hand – Desmoulins noticed – "ce n'est pas au seul volume du cerveau que tient la supériorité de l'entendement", otherwise the enormous volume of the head of the inhabitants of the far north (Eskimoes, Samoyeds, Lapps) would have contradicted their sterility in ideas and inventions.[83]

No animal could approach man in the number of folds of the cerebral hemispheres: in 1825, the naturalist and traveler Jean-Baptiste Bory de Saint-Vincent echoed Desmoulins' statement of 1822: intellectual superiority always seemed to be associated with a "cerveau profondément plissé".[84] A little over twenty years of age, in 1800 he had joined Captain Nicolas Baudin's expedition to New Holland: sailors and scientists united for the purpose of mapping new lands or maritime routes, and discovering new natural products. Once landed on the Île de France (Mauritius), under the pretext of a disease, Bory left the vessel and explored the Mascarene Islands, in the Indian Ocean, for a couple of years, fascinated by the tropics; back home, he wrote a huge account of the facts and phenomena he had observed.[85] A Bonapartist officer at the service of the *Grande Armée*, he was exiled after the Restoration, though allowed to return to Paris in 1820, where he launched the decade-long editorial adventure of the *Dictionnaire classique d'histoire naturelle* in seventeen volumes, a combative medium for spreading the Lamarckian theory and Geoffroy Saint-Hilaire's *Philosophie anatomique*.

83 Ibid., pp. 297, 260.

84 B. [Jean-Baptiste Bory de Saint Vincent], "Homme. *Homo*," in *Dictionnaire classique d'histoire naturelle. Tome huitième. H-INV.* (Paris: Rey et Gravier, 1825), pp. 269-346 (in particular p. 270).

85 Jean-Baptiste Bory de Saint-Vincent, *Voyage dans les quatre principales îles des mers d'Afrique, fait par ordre du gouvernement, pendant les années neuf et dix de la République (1801-1802), avec l'Histoire de la Traversée du Capitaine Baudin jusqu'àu Port Louis de l'Île Maurice* (Paris: F. Buisson, 1804). His report occupies three thick volumes plus one of *planches*, introduced by a rhetoric of factual objectivity: "Convaincu qu'un Voyageur ne doit que voir, il n'a jamais émis une opinion positive. Il a rapporté des Faits, écrit ses doutes, et s'en est référé au jugement des Savans qui le liront." (p. xiv). Shortly before, his *Essais sur les Isles Fortunées et l'antique Atlantide, ou Précis de l'Histoire générale de l'Archipel des Canaries* (Paris: Baudouin, 1803) had already come out.

In the aforementioned entry for the *Dictionnaire*, Bory faced a question that had become crucial:

> Nul doute que le cerveau de certains Ethiopiens, tout comparativement plus étroit qu'il puisse être, ne soit aussi capable de concevoir des idées justes, que celui d'un Autrichien, par exemple, le Béotien de l'Europe, et même que celui des quatre cinquièmes de nos compatriots. Dans une seule Antille encore, on voit de ces Hommes, réputés inférieurs par l'intellect, donner plus de preuves de raison qu'il n'existe dans tout la péninsule Ibérique et l'Italie ensemble.[86]

A firm invective against slavery followed: one should wish that the Africans, so perverted on their native continent by contacts with the Europeans, could emancipate themselves in the colonies from that barbarity to which their owners' greed had relegated them. Hopefully, Bory saw that the soil of those islands – sprayed in the past by tears and fattened by blood – had recently fertilized by liberal ideas that were producing a kind of civilization virtually superior to that of the old corrupt Europe.

The entry *Homme* from the *Dictionnaire* was added to the two volumes of an *Essai zoologique sur le genre humain* – paradoxically dedicated to his opponent Cuvier – that not only did vigorously proclaim the existence of distinct *species* within it, but also listed as many as fifteen of them, some of which further articulated in races. The usual nomenclature, of Linnean or Blumenbachian coinage, was enriched with new and sometimes imaginative terms. For the first time the type of hair divided humanity into two large groups, *Léiotriques* and *Ulotriques*: straight and curly, respectively. Eleven species would belong to the first, only four to the second. Tending to highlight a specific singularity of the Ethiopian species, for instance, Bory de Saint-Vincent still had to invoke Soemmerring – narrow brain and big nerves – and then insist on the usual features: protruded face, smaller skull, darker blood and humors, fetid sweat, different size of sexual organs, peculiar diseases not transmissible to the other species. In the state of abjection in which the *Éthiopiens* were reduced by the Europeans, no wonder that they appeared markedly inferior as for intellect, behavior, and sociability. This did not mean, however, that nature had condemned them to the state of beasts of burden, incapable of civilization.[87]

86 B. [Bory de Saint Vincent], "Homme. *Homo*" (cit. note 84), p. 320. See Hervé Ferrière, *Bory de Saint-Vincent: l'évolution d'un voyageur naturaliste* (Paris: Syllepse, 2009).

87 Jean-Baptiste Bory de Saint Vincent, *L'Homme (Homo.) Essai zoologique sur le genre humain. 2e Édition enrichie d'une carte nouvelle pour l'intelligence de la distribution des espèces d'hommes à la surface du globe terrestre* (Paris, Rey et Gravier, 1827). These are the

With a dedication to Étienne Geoffroy Saint-Hilaire – "professeur et administrateur au Muséum d'Histoire naturelle" – in 1832 the anatomist Étienne Serres announced the final results of his researches on organogenesis, undertaken in previous years and already partly published. The *anatomie transcendante* that he practiced – in his mentor's footsteps– would move on, from a comparison of concrete anatomical arrangements to an abstract conception of the laws of the organization. Its French exponents shared a set of assumptions with the German *Naturphilosophen*, first of all the belief that all animals are built upon a single, structural plan, then the idea of a chain of being, and finally the conviction that the development of the human embryo recapitulates, in a transitory manner, the organic configurations, which are permanent in lower forms. The Recapitulation theory had a mixed appeal, it intrigued theologians and politicians as well as scientists, and eventually culminated in Ernst Haeckel's "evolutionary triumph".[88]

As far as the human brain was concerned, Serres claimed to have discovered its more or less rapid embryological transition through stages, which had already been identified in fishes, reptiles, birds, and mammals. A curious consequence came from the *ostéogénie* of the head. During the progressive movement of the bones towards the end that was achieved by the Caucasian race, the skull and face – according to Serres – would transiently assume the characters they permanently presented in the Ethiopian, Malay, American and Mongolian race. As a consequence, one of the effects of imperfection in the development of the Caucasian brain, the *chef-d'oeuvre* of Nature, consisted in its arrested evolution and in its descent to the level of the lower races:

> Les divers degrés de la *microcéphalie* nous présentent cette répétition et la dépassent encore; il n'est pas rare même de la voir se maintenir après la naissance, et de retrouver, dans les traits de ces enfants, les caractères physiques de ces diverses races, empreints sur une figure caucasique.[89]

---

identified fifteen human species: *Japétique, Arabique, Hindou, Scythique, Sinique, Hyperboréenne, Neptunienne, Australasienne, Colombique, Américaine, Patagone, Ethiopienne, Cafre, Mélanienne, Hottentotte.* For the description of the Ethiopian species see tome II, pp. 29-86.

88 See Stephen Jay Gould, *Ontogeny and Phylogeny* (Cambridge, Mass.: The Belknap Press of Harvard University Press, 1977), pp. 47-52.

89 Étienne Serres, *Recherches d'anatomie transcendante et pathologique. Traité des formations et déformations organique appliquées à l'anatomie de Ritta-Christina, et de la duplicité monstrueuse* (Paris: Baillière, 1832), pp. 24-25. Serres had already published the two volumes of his *Anatomie comparée du cerveau, dans les quatre classes des animaux vertébrés, appliquée à la physiologie et à la pathologie du système nerveux* (Paris: Gabon et Compagnie, 1824-1826), and a report of his work still in progress: "Recherches d'anatomie

By the way, the theory of Recapitulation will also be embraced by the anonymous author of the *Vestiges of the Natural History of Creation* – written between 1841 and 1844 while trying to overcome a strong depression – that took British culture by storm. The Scottish publisher and amateur geologist Robert Chambers offered a developmental view of the universe, which also entailed a form of transmutation of the living beings. A chapter was about the early history of mankind, which assumedly sprang "from one stock [...] in a state of simplicity, if not of barbarism". As yet there was no distinct idea of how it had become to be differentiated by peculiar external features, but such processes appeared to Chambers to be explicable in terms of development:

> We have already seen that various leading animal forms represent stages in the embryotic progress of the highest – the human being. Our brain goes through the various stages of a fish's, a reptile's, and a mammifer's brain, and finally becomes human. There is more than this, for, after completing the animal transformations, it passes through the characters in which it appears, in the Negro, Malay, American, and Mongolian nations, and finally is Caucasian. [...] The leading characters, in short, of the various races of mankind, are simply representations of particular stages in the development of the highest or Caucasian type. The Negro exhibits permanently the imperfect brain, projecting lower jaws, and slender bent limbs, of a Caucasian child, some considerable time before the period of its birth. The aboriginal American represents the same child nearer birth. The Mongolian is an arrested infant newly born. And so forth."[90]

A case in point of the widespread concept of those lower races, around the fourth decade of the century, might be what Thomas Hope – the scion of a wealthy and powerful family of merchants and bankers, a patron of the arts and an art collector, the promoter of the new classical aesthetics, a great traveller – wrote in a three-volumes work published a few weeks after his death. Certainly not a significant contribution to science, it was if anything a verbose and eccentric *summa* demolished by Thomas Carlyle as a "monstrous anomaly, where all sciences are heaped and huddled together, and the principles of all

transcendante, sur les Lois de l'Organogénie appliquées à l'anatomie pathologique," *Annales des sciences naturelles*, 1827, 11:47-70.

90 [Robert Chambers], *Vestiges of the Natural History of Creation* (London: John Churchill, 1844), pp. 306-307. The book was a publishing success, and only its twelfth, posthumous edition of 1884 will bear the author's name on the title page. See Secord, "Behind the Veil: Robert Chambers and *Vestiges*" (cit. note 35).

are, with a childlike innocence, plied hither and thither".[91] Despite such denigration, Hope's *Essay on the Origin and Prospects of Man* expressed with perfect clarity the gradual transfer, in the previous half a century, of scientific evaluations into the mind-sets of Europe's educated classes. As to Borneo, Sumatra, and some Polynesian islands, Hope spoke of "tribes of which the resemblance to the baboon is most striking, the superiority over the brute, in mind and body, least perceptible". And more descriptively:

> In them, of all human beings, the organs of vitality most early and fundamental – those named abdominal – still present the greatest expansion; [...] The external seats of those lower senses which still remain most directly connected with the wants of the stomach – the tongue, the jaws, the lips, and the nostrils – still preserve the most disgusting amplitude, openness and spread; the organs of the higher senses, still often exhibit an equally disagreeable and repulsive obliquity and want of room. The face, from a preposterous width above, terminates abruptly underneath in a sharp, chinless point. The organs of intellect still appear in their narrow cramped receptacle wholly underdeveloped. Of the rude ill-formed skull the anterior part – that which, in proportion as it is better arched, gives earnest of organs of thought more expanded, is still so low, so narrow and depressed, that it can hardly be said to form a distinct forehead. The small deep eyes, like those of the baboon, still keep constantly vibrating in their narrow sockets.[92]

Hope's portrait of those distant populations led to emphasize their most unusual and repellent aspects, both physical and moral. The New Zealander's skull, for instance, would present a texture so coarse, a form so contracted, so similar to that of the orangutan, that anatomists had identified the connecting link between animal and man in that inhabitant of the antipodes. Even certain African races of a higher rank would still retain, "as in certain brute races", a much greater development of the olfactory, optic, and fifth pair of nerves: that is, a power of conveying sensations much more acute than that found in higher human beings, and, of course, an affinity with the monkey due to the marked contraction and want of room in their brain, in addition to the thickness of their skull.[93]

---

91 [Thomas Carlyle], "Characteristics," in *The Edinburgh Review*, 1831, *54*:351-383, p. 378.

92 Thomas Hope, *Essay on the Origin and Prospects of Man* (London: John Murray, 1831), II, pp. 391-392.

93 Ibid., pp. 395, 398-399. As regards phrenology, Hope did not believe that Gall had exactly discovered "in which peculiar organ of the brain resides each peculiar faculty of the mind;

4 Weighing Empty, Filled Spaces

Until then, the scientific practice, which aimed at comparatively assessing brains, had not been particularly inventive and remained within the confines of plain, though accurate, anatomical observation. Versatile as he was, the Scottish philosopher William Hamilton, elected to the Chair of Logic and Metaphysics at the University of Edinburgh in 1836, had also studied medicine in 1806-1807 and engaged in bitter disputes with phrenologists. Between 1825 and 1829, he read a few papers *On the Practical Conclusions from Gall's Theory Regarding the Functions of the Brain* before the Royal Society of Edinburgh, intending to collect his arguments under the eloquent title *The Fictions of Phrenology and the Facts of Nature* but never carrying it out. An anonymous pamphlet reporting his harsh criticism was published in 1826 and was soon refuted by the phrenologist Andrew Combe, with the controversy continued in the press. Not only did those *practical conclusions* – Hamilton had argued – consist in spreading materialism, atheism, and fatalism, but also Gall's doctrine was unquestionably false.[94]

In a preface to Alexander Monro's *Anatomy of the Brain* of 1831, Hamilton referred about

> an induction drawn from above sixty human brains, from nearly three hundred human skulls, of determined sex, – the capacity of which, by a method I devised, was taken in sand, and the original weight of the brain thus recovered, – and from more than seven hundred brains of different animals.[95]

---

because he could not have done so without also ascertaining how the different faculties of the mind are connected with each other, and without becoming a good metaphysician; whereas he seems to have been a very bad one [...]." (p. 251).

94 *Sir William Hamilton and Phrenology. An exposition of Phrenology; shewing the complete inefficacy of the objections lately advanced in the Royal Society, and the real ground on which the system ought to be assailed* (Edinburgh-London: Hunter and Duncan, 1826). His views are also exposed in the appendix to William Hamilton, *Lectures on Metaphysics and Logic, edited by the Rev. Henry L. Mansel and John Veitch* (Boston: Gould and Lincoln, 1860), I, pp. 648-658. On the controversy between him and a group of phrenologists see van Wyhe, *Phrenology and the Origins of Victorian Scientific Naturalism* (cit. note 8), pp. 85-92.

95 William Hamilton, *An Account of Experiments on the Weight and Relative Proportions of the Brain, Cerebellum, and Tuber Annulare in Man and Animals, under Various Circumstances of Age, Sex, Country, etc.* in Alexander Monro, *The Anatomy of the Brain: with some Observations of its Functions* (Edinburgh: J. Carfrae, 1831), pp. 4-8; also in Hamilton, *Lectures on Metaphysics and Logic* (cit. note 94), pp. 658-660.

That was one of the first attempts, marking the start of a long story, to weigh the brain indirectly through the emptiness of the cranial capacity. The use of a filling material such as sand allowed him to make a series of inferences, first of all that the "encephalos" of the adult Scotsman would be heavier than the female one, with an average difference of 4 oz., reaching its full size at about seven years of age. The latter was – Hamilton claimed – a new discovery, given the very uncertain estimates available until then; moreover, despite the usual opinion, there was no adequate evidence that the cranial content diminished in old age. Furthermore,

> The common doctrine, that the African brain, and in particular that of the Negro, is greatly smaller than the European, is false. By a comparison of the capacity of two Caffre skulls, male and female, and of thirteen negro crania (six male, six female, and two of doubtful sex) the encephalos of the African was found not inferior to the average size of the European.[96]

Some other of Hamilton's inductions concerned the growth of the cerebellum, and its greater dimension recorded in females than in males, relatively to the brain proper.

Clinical observation, too, was instrumental in providing new insights. At that time, the London St. Marylebone Infirmary, or "Sick House", formed a portion of the parochial establishment for the relief of the poor in a large and populous parish.[97] The number of patients ranged from 150 in the fine season to 330 and upwards in winter. In 1835, one of its staff doctors, John Sims, read a long report before the Royal Society of Medicine to draw the British physicians' attention to cerebral hypertrophy and atrophy. Firstly he described the pathologists' attempts to explain the circumstances of brains that far exceeded the average size. Secondly, fifteen cases of hypertrophy that had been put under his observation were recounted. Thirdly, Sims set himself the goal of recording the average weight of the brain of persons of all ages, from a very large number of specimens. He complained that almost all previous attempts had been ineffectual and liable to error: Haller, Soemmerring, Joseph and Carl Wenzel were mentioned[98], besides William Hamilton, whose efforts at indirect

96 Hamilton, *Lectures on Metaphysics and Logic* (cit. note 94), p. 659.

97 See John Clendinning, "Report on the Experience of the St. Marylebone Infirmary, since 1827, with respect to Admissions, Duration of Treatment, Mortality, and other Statistical Results, according to Age and Sex," *Journal of the Statistical Society of London*, 1844, *7*: 292-310.

98 Six *tabulae* with various measures taken on animal and human brains (*cerebrum* and *cerebellum*), included weights, can be consulted at the beginning of Josephus et Carolus

measurement was thus completely rejected: "I think the weight of the brain cannot be fixed by ascertaining the capacity of the skull, by any means however accurate; it is open to a source of fallacy". Therefore, Sims added a table of results from 253 dissections, showing the real weight of brains of all ages and both sexes, "who died of diseases of almost every description, whether cerebral or otherwise".[99] The second part of his report had to do with the atrophy of the brain and some other pathological conditions it suffers from.

As a physiological axiom, it was generally admitted that a causal relation existed, in the healthy state, between the material development of an organ and the power of its function. Thus, in 1836, Jean-Baptiste-Maximien Parchappe de Viney – psychiatrist *en chef* at the Asile des aliénés in Rouen – opened his *Recherches sur l'encéphale*, the first of which focused on the volume of the head and brain. In view of that alleged principle, the intellectual and moral superiority of man over animals had been explained by the quantitative pre-eminence of his brain. Thereupon, it had to be assumed that the volume of the skull indicated exactly that of the brain: the *idée mère* of cranioscopy and the essential basis of Gall's doctrine. Since the time of Camper, science seemed to have repeatedly dispelled the popular belief expressed by the proverb *Grosse tête, peu de sens*.

Nevertheless, Parchappe did not unconditionally accept that intensity is proportional to the volume of the brain: indeed true for functions representing a mechanical power – like the muscles – but not for the eye, for instance, where rather the form is correlated to the function. He thought that the question of the influence of encephalic volume on the development of intellectual and moral faculties could not be solved *a priori*, by virtue of the same physiological axiom. Only the study of "facts" – which contain the "truth" as long as they are well observed – could offer a solution. Camper, Daubenton, Gall, Cuvier etc. had taken that relationship for granted, never proving it, and judged *sans mesures*, just by sight and touch, senses that could not evaluate the actual volume of the brain. Measuring through meter and scale (*mètre et balance*) would be the only effective way to determine the amount of that matter.

It is well known that, between the end of the 18th and the first decades of the 19th century, the resources of mathematical calculation seemed inexhaustible in analyzing the physical and moral man, in avoiding mistakes due to subjectivity, and in cutting off the effects of prejudice. The quantitative model was successful and led to indulge in looking for numerous samples and averages, as

---

Wenzel, *De penitiori structura cerebri hominis et brutorum* (Tubingae: Apud Cottam, 1812).

99 John Sims, "On Hypertrophy and Atrophy of the Brain," *Medico-Chirurgical Transactions*, 1835, *19*: 315-380, pp. 351-357.

well as to insist rhetorically on the virtue of numbers and diagrams. Paris in the years of the Restoration and the Orleanist monarchy was one of the main centers, if not the main one, of this new epistemic horizon. Adolphe Quetelet, a young Belgian mathematician, had been sent there in 1823 on a mission to learn the techniques to set up an astronomical observatory in Brussels. And there he met the old Pierre-Simon Laplace, who had identified a beneficial relationship between uniformly-repeated observations of phenomena and a knowledge of the constant causes acting in nature. In 1828 Quetelet began to statistically detect the regularity of several kinds of phenomena, crimes included: large numbers seemed to neutralize the contingent and the random – typical of natural and social reality – so that the only useful knowledge had to do with averages, not with the extreme and misleading individual varieties.

The *homme moyen* which became the operational and fetishistic concept of the new *Physique sociale* proposed by Quetelet in 1835 – at that time perpetual secretary of the Académie Royale de Bruxelles – was both descriptive and normative up to the point of shaping the desirable type of the inhabitant of a "nation". In other words, an individual who could keep at a pace with *medietas* would embody the good and the beautiful. He also recalled a question concerning the existence of a type of beauty in the human species that was associated with the development of intelligence. Comparative anatomy had believed a positive answer could be found in the facial angle and in cerebral volume, conducive to the pre-eminence of the Caucasian race. However,

> il paraît qu'on ne possède jusqu'ici que très peu de données sur la loi de développement du cerveau même, ainsi que sur sa grandeur et son poids aux différens âges, soit dans sa valeur moyenne, soit dans ses valeurs limites.[100]

On his part, Parchappe exposed the method he personally used to measure the volume of the head, which he had found – to begin with – to be considerably smaller in women than in men, given a few dozen samples, not only in absolute terms, but in partial dimensions (diameters and curves) as well: clearly, one of the secondary physical characters of sex in the human species. He then attempted to establish the influence of age, height, and mental pathological conditions on the volume of the head, based on his own and on others' data.

---

100 Adolphe Quetelet, *Sur l'homme et le développement de ses facultés, ou Essai de physique sociale* (Paris: Bachelier, 1835), II, pp. 264-265. As for Gall's doctrine, Quetelet regretted that its principles had not yet been subjected to direct and systematic verification (pp. 263-264).

A curious conclusion, contrasting with what had been thought for the most part, was excluding a necessary link between imbecility or idiocy and smallness of the head. As a matter of fact, out of fifty inspected heads of men with normal intelligence, seven of them were smaller than those of the imbecile with the largest head, and thirteen just slightly larger. A normal intelligence could therefore exist to a normal degree in a head whose volume was lower, equal to or just larger than the volume of an idiot head. Moreover, small but well-shaped heads were compatible with a great deal of intelligence. On such grounds, what was left of Gall's law? Just the statement that a certain volume of the head is a prerequisite for a good organization, without any chance to infer anything about intelligence from variation.[101]

Parchappe took note that real "positive" knowledge of the differences in the volume of the head of human races was still scarce. If credit was given to the facial angle, the volume of heads appeared greater in the Caucasian race, and many observers had confirmed that assumption in various ways. He calculated the averages of the measurements taken on twenty skulls of Caucasian individuals and on nineteen of other races, which turned out to be arranged in the following order: Caucasian, Ethiopian, Mongolian, American, Malay. Yet such induction was not to be considered definitive and needed to be tested by further, more extensive research. Not to mention that – according to Parchappe – the division into five races was tentative, and a good natural history, despite Buffon and Blumenbach, still had to be done.[102]

There was no doubt that, in order to understand the influence of the brain volume on the intellectual and moral phenomena, it should have been directly measured, except that facts were still lacking, while opinions abounded. Brain volume, however, had value only as an equivalent of the quantity of matter: better, then, to resort to weight, if only the differences of density in the brain substance did not vary from individual to individual – an important variation on which observers would have to concentrate. Though aware of such limitations, Parchappe did not give up weighing brains, after severing them from the spinal cord above the medulla oblongata and from the cerebellum. He also insisted that the brain was a composite organ with multiple functions, mainly a sensory and a motor one. For a rigorous comparison, only the parts that were actually responsible for intelligence should have been considered: in 1825, Desmoulins and Magendie had already suggested that the number and perfection

101 Maximien Parchappe, *Recherches sur l'encéphale, sa structure, ses fonctions et ses maladies. Premier mémoire. Du volume de la Tête et de l'Encéphale chez l'Homme* (Paris: Librairie des sciences médicales de Just Rouvier et E. Le Bouvier, 1836), pp. 1-46.

102 Ibid., pp. 46-62.

of the intellectual faculties were in relation to the extension of the cortical surface.[103]

Much caution in surmising and judging was then called for. Another alienist – who worked at the Bicêtre hospital and had recently gained fame by diagnosing Socrates' hallucinatory madness – doubted that a greater encephalic weight always corresponded a higher degree of intelligence. The encephalon of ten idiots and fools weighed by Louis Francisque Lélut were lighter than normal, but not by much, just about one thirteenth; less was the difference in the weight of the brain, around one tenth, and even smaller that of the cerebellum, of only one seventeenth. Based on an analysis of the limited number of observed cases, he concluded that the brain was generally heavier in intelligent men, but with numerous exceptions. Lélut – *médecin philosophe* with a penchant for spiritualism – had just tackled the subject with "un peu de statistique, d'une faĉon bien imparfait, bien incomplète"; but even assuming that he could increase the number of weighed brains, his guess was that statistics could only offer formulas a little more accurate than the available information.[104]

Anyway, at that time, this was roughly known or intuited:

> from the superior nervous activities eminently characterizing our times, it necessarily follows, that we may expect to find, in all individuals who are fair representatives of the higher type, a predominance of the corresponding organic conditions, over those of the lower grade, in the nervous system.[105]

---

103 Ibid., pp. 63-97. The third part of the book criticized Gall's system – "qui tend de plus en plus à se vulgariser" – and phrenology in general, "qui n'est pas la physiologie du cerveau, qui n'est pas une science, comme on l'a ridiculement avancé, mais qui est un système" (pp. 103-104).

104 Louis Francisque Lélut, "Du poids du cerveau dans ses rapports avec le développement de l'intelligence," *Gazette médicale de Paris*, 1837, 5:146-149. His *Du démon de Socrate, spécimen d'une application de la science psychologique à celle de l'histoire* was published in 1836 (Paris: Trinquart), one of the first examples of pathography produced in the 19th century.

105 Robert Verity, *Changes Produced in the Nervous System by Civilization, Considered according to the Evidence of Physiology and the Philosophy of History* (London: S. Highley, 1837), pp. 6-7. The book had a second enlarged edition in 1839. An anonymous reviewer expressed some doubt: "As the style of the writer is none of the clearest, we do not understand very distinctly whether he considers a change in the corporeal nature to give rise to civilization, or whether the once-awakened habits of civilization, however slight, create a beneficial change in the nerves; or whether each cause reciprocally contributes to form a third more potent than either. But, from the weight he attaches to the emigration and amalgamation of peoples, we infer that he thinks blood and body has the most to do with the result." (*The Spectator*, 25 November 1837, pp. 1123-1124).

In his curious, small book of 1837, Robert Verity – a member of the Universities of Edinburgh and Göttingen, a physician converted to homeopathy and involved in phrenology – expressed the conviction that mankind had been undergoing progressive modifications in the long run. It was the peculiar task of the physiologist to demonstrate how, throughout centuries, a process had been enlarging the boundaries of the primitive nervous structures into more perfect outlines of proportion and form. Without this organic process, there would be no imaginable means of understanding the successive additions of civilized activity, and of transmitting enhanced natural aptitudes and capabilities to posterity. A historical display of more grandeur and importance than the movement of races and nations towards accomplishing their own destiny could hardly be seen. Verity interpreted the whole course of Western civilization by assuming that "a certain physical evolution in the nervous system is produced correlatively with each successive step of civilization". So, for example, in the first periods of pure feudalism "the higher kinds of nervous function were overborne and swallowed up by the inferior" and all, therefore, was dark and barbarous; while a series of progressive stages, each rising above its predecessor "with a higher character and greater extent of nervous function", would lead up to a wonderful time full of knowledge.[106]

Perhaps the most singular attempt to address a topic considered "of great importance in the natural history, anatomy, and physiology of Man; interesting also in a political and legislative point of view" was made by a German anatomist who, as a Foreign Member, sent a paper to the Royal Society in 1836. Friedrich Tiedemann's father, Dietrich, had been an anti-Kantian philosopher and psychologist, committed among other things to keeping a journal of his son's development – sensory-motor, linguistic, emotional, and cognitive – for thirty months.[107] "Ce fut au milieu de cette atmosphère philosophique – Pierre Flourens wrote in his *Éloge* – que grandit l'intelligence du jeune Tiedemann"[108]. He then studied in Marburg, where he graduated in medicine, in Göttingen, Würzburg, and Paris, attracted by the teachings of Gall, Blumenbach, Cuvier, Schelling, whose *Naturphilosophie* – in his own words – healed him forever from the temptation to abandon the path of empirical inquiry. In 1807, he was recommended by Soemmerring, whom he had met in Frankfurt, to the Landshut medical school for his first academic appointment as professor of anatomy and zoology.

106 Verity, *Changes Produced* (cit. note 105), pp. 25-26. 40-43.

107 See Dietrich Tiedemann, "Beobachtungen über die Entwicklung der Seelenfähigkeiten bei Kindern," *Hessische Beiträge zur Gehlersamkeit und Kunst*, 1787, 2:313-333, 486-502.

108 Pierre Flourens, "Éloge historique de Frédéric Tiedemann, l'un des huit associés étrangers de l'Académie," *Gazette médicale de Paris*, 1861, *31*:815-827, p. 815.

At the end of his Bavarian period, Tiedemann dedicated "unserm Blumenbach" some anatomical researches on the embryological development of the brain, mindful of his father's interests. In this regard, two methods were put into practice by him: first, the developmental one – neglected by Gall and others – with the description of the structures of a growing brain at each stage of gestation, month by month. In the expansion of no other system, as in the brain and nerves, would such a perfect sequence of steps take place, from a simple to a composite construction. Therefore – from the perspective of the Recapitulation theory – the human brain represented a fully developed form, preceded by numerous animal versions that had stopped at specific, incomplete moments. The second method consisted in extending Gall's approach and detailing the differences in nervous patterns, part after part, between man and animals. Tiedemann asserted that the brain structure and functions could certainly be discovered, but by no means the very reason thereof ("keinsewegs aber das Ansich, der letzte Grund dieser selbst").[109]

From 1816 to his retirement in 1849, Tiedemann taught anatomy, physiology, and zoology at the University of Heidelberg, where he also collaborated with the chemist Leopold Gmelin in studying the digestive system. His social sensitivity and political awareness were already apparent in 1829 when he appealed to the humanity of the judicial authorities, calling them to oppose to beheading as a punishment. Advocating for a more humane treatment – certainly not the abolition of the death penalty – implied the reformist feelings in him, although not so openly shown. Given the progress of mechanics, he presumed that it would be easy to invent a machine capable of separating a head from the body, quickly and accurately. However – he added – suggestions of this kind did not befit a physiologist whose endeavor should be directed to the study and preservation of life.[110]

The same liberal tendency became mainly evident in the aforementioned paper that he sent to the Royal Society in 1836. To understand its genesis, it is worth remembering that, after a decades-long abolitionist campaign, rebellions and parliamentary inquiries, on August 28th, 1833 the House of

109 Friedrich Tiedemann, *Anatomie und Bildungsgeschichte des Gehirns im Foetus des Menschen nebst einer vergleichenden Darstellung des Hirnbaues in den Thieren* (Nürnberg: in der Steinischen Buchhandlung, 1816). For the explanations of the two methods, see the *Vorrede* and *Einleitung*, pp. v-vi, 1-6. "Nul anatomist encore – according to Flourens – avait essayé d'étudier jour par jour, et presque heure par heure, la *formation du cerveau humain*." (cit. note 108), p. 818.

110 Friedrich Tiedemann, "Anruf an die Humanität der Höheren Behörden der Gerechtigkeits-Pflege in Deutschland, veranlasst durch eine am 22. October 1827 in Heidelberg vollzogene Enthauptung," *Zeitschrift für Physiologie*, 1829, *3*: 283-288.

Commons discussed an *Act for the Abolition of Slavery throughout the British Colonies*, which came into force the following year and promised compensations to slave-owners. In reality, only slaves below the age of six were freed in the Caribbean colonies, while the older ones were renamed *apprentices*, their emancipation having to go through two successive phases, until 1840. Moreover, the Act did not apply to the territories owned by the East India Company, Ceylon and Saint Helena, and such exceptions were not repealed until later.[111]

Again Flourens' *Éloge* can help introduce "le beau travail de notre anatomiste sur le cerveau du nègre":

> On se rappelle la fortune qu'avait faite la fameuse *ligne facial* de Camper. Camper, examinant le crane du nègre, vu de profil, y avait trouvé quelque rassemblance vague avec celui de l'orang-outang; on exagéra bientôt les choses, selon l'usage, et l'on finit par conclure que le nègre tenait une sorte de milieu entre l'homme blanc et le singe; qu'il avait donc plusieurs espèces humaines, qu'il y en avait de supérieures, qu'il y en avait d'inférieures, et que, par ses derniers anneaux, l'homme tenait aux bêtes.[112]

For his eulogist, one of Tiedemann's best titles consisted precisely in the complete and absolute refutation of the error that from Jacob de Bondt's 'humanization' of the orangutan onwards, some had committed in the last two centuries (included Linnaeus and Buffon, *malgré lui*) when they shortened the distance between the apes and man. Flourens, an uncompromising dualist, could not but stress the unity of the human species, as well as exclusivity in terms of bipedalism and language.

What was the actual core of Tiedemann's paper, first written in English to honor Great Britain's passing of the Slave Trade Act? His polemical targets were the opinions upheld by great anatomists (Camper, Soemmerring, Cuvier) and exaggerated by others "of less authority", who reputed "the Negroes as a race inferior to the European in organization and intellectual powers, having much resemblance with the Monkey". Were they proved to be correct, the Negro would occupy a position in society different from that "so lately been given him by the noble British government":

---

111 Pamphlets published by both abolitionists and slavery proponents are collected in *The Slave Trade Debate. Contemporary Writings For and Against. Introduction by John Pinfold* (Oxford: Bodleian Libray, 2007) in commemoration of the 1807 Slave Trade Act. For the following period, see Keith Hamilton, Patrick Salmon (eds.), *Slavery, Diplomacy and Empire: Britain and the Suppression of the Slave Trade, 1807–1975* (Brighton and Portland: Academic Press, 2009).

112 Flourens, "Éloge historique de Frédéric Tiedemann" (cit. note 108), p. 823.

> I propose in this treatise to examine more minutely the most important part [...] namely, the structure of the brain, the noblest part of the human body, in reference to its functions. A comparison between the brain of the Negro and that of the European and the Orang-Outang, hitherto much neglected, appeared to me most worthy of attention. I shall first of all try to answer the following two questions.
>
> 1st, Is there any important and essential difference between the structure of the brain of the Negro and that of the European? and
>
> 2ndly, Has the brain of the Negro more resemblance to that of the Orang-Outang than the brain of the European?[113]

Had he proven these two statements to be false, they should have been considered "a mere literary fancy". Tiedemann had visited the most celebrated anatomical museums, both on the Continent and in Great Britain, to determine the weight and dimensions of the European brain, then that of the Negro, and to compare them together.

He judged the anatomists' assessments of the size and weight of human brain as generally very uncertain, and the old belief of Aristotle, Pliny, Galen, and others, for many centuries regarded as correct, namely that it would be absolutely and relatively larger than that of any other animal, as erroneous. As for the absolute value, the examples of the elephant and the whale refuted that belief, and also small birds and apes, or rodents had larger brains, in proportion to the size of the body. Moreover, his own and others' researches had corroborated the comparative conclusion, first reached by Soemmerring, about the greater dimension of the human brain in relationship to the size and thickness of the nerves. Yet what the anatomists of the past had said on brain weight was completely unsatisfactory, and only lately had some "valuable investigations" been done by William Hamilton.

Making use of the apothecary or troy weight, Tiedemann weighed 52 brains of Europeans, 35 from male and 17 from female bodies, and listed the results in a table with the height and weight of many of them. From these data he drew a few conclusions, for instance that the female brain is lighter in absolute terms, but not relatively compared with the body, or if anything larger; that, on average, the brain grows to its full size around seven or eight year of age; and,

---

113 Tiedemann, "On the Brain of the Negro" (cit. note 70), p. 498. For the subsequent, expanded German edition see *Das Hirn des Negers mit dem des Europäers und Orang-Outangs verglichen* (Heldelberg: Karl Winter, 1837). Konrad Schmutz has edited a *Faksimileausgabe* of the latter, with an introduction and annotations (Marburg: Basilisken-Presse, 1984).

most importantly, that there is "a very close connexion between the absolute size of the brain and the intellectual powers and functions of the mind".

At the time, all but few observations were available on the weight of the Negro brain, which Tiedemann deemed not to support the assumption about its quantity being less than that the European. He examined and had Honoré's brain drawn: a short, thin African man, twenty-five years of age, killed by small-pox at Liège and there dissected by his son Heinrich and by his son-in-law Vincenz Fohmann, who had been his student in Heidelberg. Once preserved in alcohol and sent to him, it weighed 2 lbs. 3 oz. 2 dr.; moreover, Honoré's spinal cord, medulla oblongata, cerebellum, general form and subdivision were for the most part similar to the European ones. Nothing could be said about the controversial question of the color of the medullary and cortical substance, given the liquid in which the brain was immersed. In the internal structure Tiedemann did not notice any difference, and considered it superfluous, therefore, to give a verbal description of it, opting instead for the "exact representation" of a few plates drawn by his son Heinrich. As to Soemmerring's claim about a greater thickness of the nerves in the Negro compared to the European, he could only say that in Honoré (and in two other African specimens seen in Paris) no such peculiarity was visible.

Even the data he collected on the cranial capacity – by filling the cavity with dry millet-seed through the foramen magnum – revealed that the brains of the five races classified by Blumenbach were roughly equivalent, as evidenced by five tables of data. Why then so many naturalists had sworn that the Negro had a smaller skull and brain? Their mistaken notion – Tiedemann explained – had arisen from the application of Camper's angle to a sample of few skulls of Africans who had lived on the coast, had been degraded by slavery and ill treatment, that could not be received as universal. Different, for example, was the countenance met in other regions of the continent, as reported by "credible travellers" and "accurate observers" who supported what Blumenbach had attested thirty years earlier about the outward appearance of Blacks gradually approaching to that of other races, and acquiring their fine features by degrees.

In disagreement with some of his predecessors – Edward Tyson and Buffon especially – Tiedemann had seen how the brains of the Orangutan (*Simia Satyrus*) and the Chimpanzee (*Simia Troglodytes*) differed in many respects from the human: smaller and lighter, shorter and narrower, with fewer gyri and sulci. "Exact plates" with representations of the brain of two apes at the Hunterian Museum should visually convey evidence of that hiatus, although granting that many ugly and degenerate Negro tribes on the coast had some similarity with the ape in their outward and inward structures. Instead, most of the Africans who lived at the heart of the continent were well made, with handsome

PROFESSOR TIEDEMANN ON THE BRAIN OF THE NEGRO. 501

	Age.	Height of the body.	State of the body.	Weight of the body.	Weight of the brain.	The weight of the brain compared with that of the body.
	I. Male Bodies.					
		′ ″ ‴		lbs. oz. dr. gr.	lbs. oz. dr. gr.	
1.	New-born child	0 18 6	Lean	6 2 6 50	1 2 3 30	as 1 : 5·15
2.	The same	0 20 2	Well nourished	7 3 2 8	1 1 1 10	1 : 6·63
3.	Boy two years old	....	The same	28 5 0 0	1 11 3 0	1 : 14·58
4.	Boy two years and a half old	....	The same	......	2 2 1 0	
5.	Boy three years old	....	The same	41 2 0 0	2 3 2 28	1 : 18·008
6.	Boy six years old	....	......	......	2 11 5 0	
7.	Boy fifteen years old	4 6 0	Well nourished	100 7 0 3	4 6 0 0	1 : 24·75
8.	Young man seventeen years old	....	......	......	4 2 1 0	
9.	Man twenty-two years old	....	......	......	4 2 1 0	
10.	Man twenty-eight years old	....	Thin	108 11 4 44	3 11 2 0	1 : 27·67
11.	Man thirty years old	....	Very thin	......	3 11 7 0	
12.	Man thirty-one years old	....	Lean	100 10 1 22	3 10 5 0	1 : 25·95
13.	The same	....	......	136 7 0 0	3 10 1 0	1 : 35·53
14.	Man thirty-two years old	5 1 6	Muscular	162 9 0 0	4 7 5 0	1 : 35·11
15.	The same	5 2 0	The same	169 8 2 0	4 7 0 0	1 : 37·02
16.	Man thirty-two years old	....	Well nourished	148 0 0 0	3 2 0 20	1 : 46·68
17.	Man thirty-three years old	....	Lean	......	3 9 4 0	
18.	Man thirty-five years old	....	......	......	3 2 7 0	
19.	Man thirty-six years old	....	Muscular	166 2 7 19	4 0 5 30	1 : 8·97
20.	Man thirty-eight years old	....	The same	160 7 6 50	4 4 0 6	1 : 36·54
21.	Man thirty-eight years old	....	The same	162 0 4 57	4 6 6 0	1 : 35·51
22.	Man thirty-nine years old	5 1 6	Very muscular	173 1 4 0	4 4 3 0	1 : 39·66
23.	Man forty years old	5 0 6	Not muscular	133 0 6 0	4 4 2 0	1 : 30·56
24.	Man about forty years old	....	......	......	3 11 2 0	
25.	The same	....	Very muscular	185 9 0 0	4 1 2 40	1 : 45·18
26.	Man forty-six years old	5 1 6	Lean	107 3 4 0	3 10 1 0	1 : 27·91
27.	Man forty-six to fifty years old	....	Well nourished	164 7 6 20	3 7 4 0	1 : 45·42
28.	Man fifty years old	5 0 2	Thin	132 8 4 35	3 10 7 5	1 : 33·96
29.	Man about fifty years old	5 0 2½	Muscular	181 8 2 0	4 1 0 10	1 : 44·47
30.	Man about fifty years old	....	Well nourished	141 1 0 0	3 8 1 40	1 : 37·76
31.	Man fifty-five years old	5 5 6	Very muscular	182 1 7 37	4 5 1 30	1 : 41·09
32.	Man sixty-one years old	....	Lean	......	3 7 4 0	
33.	Man sixty-four years old	5 2 0	Well nourished	157 0 0 0	3 11 4 0	1 : 39·66
34.	Man sixty-four years old	5 4 6	Lean	141 0 5 0	4 0 4 0	1 : 34·89
35.	Man eighty-two years old	....	The same	124 10 7 30	3 2 3 0	1 : 39·06
	II. Female Bodies.					
36.	New-born child	0 17 3	Lean	4 11 0 0	0 9 3 0	as 1 : 6·29
37.	The same	0 18 5	Well nourished	7 2 0 0	1 0 4 40	1 : 6·83
38.	Girl three years old	....	......	......	2 2 3 0	
39.	Girl five years old	....	......	......	2 4 1 50	
40.	Girl eight years and eight months old	3 4 6	Well nourished	49 0 2 51	3 5 5 0	1 : 14·13
41.	Girl thirteen years old	....	The same	63 2 6 23	3 6 2 30	1 : 17·93
42.	Girl sixteen years old	....	The same	......	3 10 2 0	
43.	Girl about twenty years old	....	The same	......	3 8 6 0	
44.	Woman twenty-five years old	....	The same	......	2 11 0 0	
45.	Woman about thirty-four years old	4 10 6	Well nourished	133 7 0 0	3 7 2 0	1 : 37·06
46.	Woman about thirty years old	....	Lean	......	3 11 0 0	
47.	Woman about thirty years old	....	Well nourished	123 4 2 25	3 7 0 0	1 : 34·42
48.	Woman thirty-eight years old	4 11 6	Muscular	153 6 5 35	3 5 0 20	1 : 44·89
49.	Woman forty-eight years old	....	......	......	2 8 5 50	
50.	Woman about fifty years old	....	Well nourished	134 6 2 57	3 4 0 40	1 : 40·27
51.	Woman sixty years old	....	The same	135 11 0 0	3 5 5 0	1 : 39·18
52.	Woman about eighty years old	....	......	......	2 9 1 0	

FIGURE 2.6 *Male and female bodies*, table in Friedrich Tiedemann, "On the Brain of the Negro, compared with that of the European and the Orang-Outang," in *Philosophical Transactions of the Royal Society of London*, 1836, *126*: 497-527, p. 501.

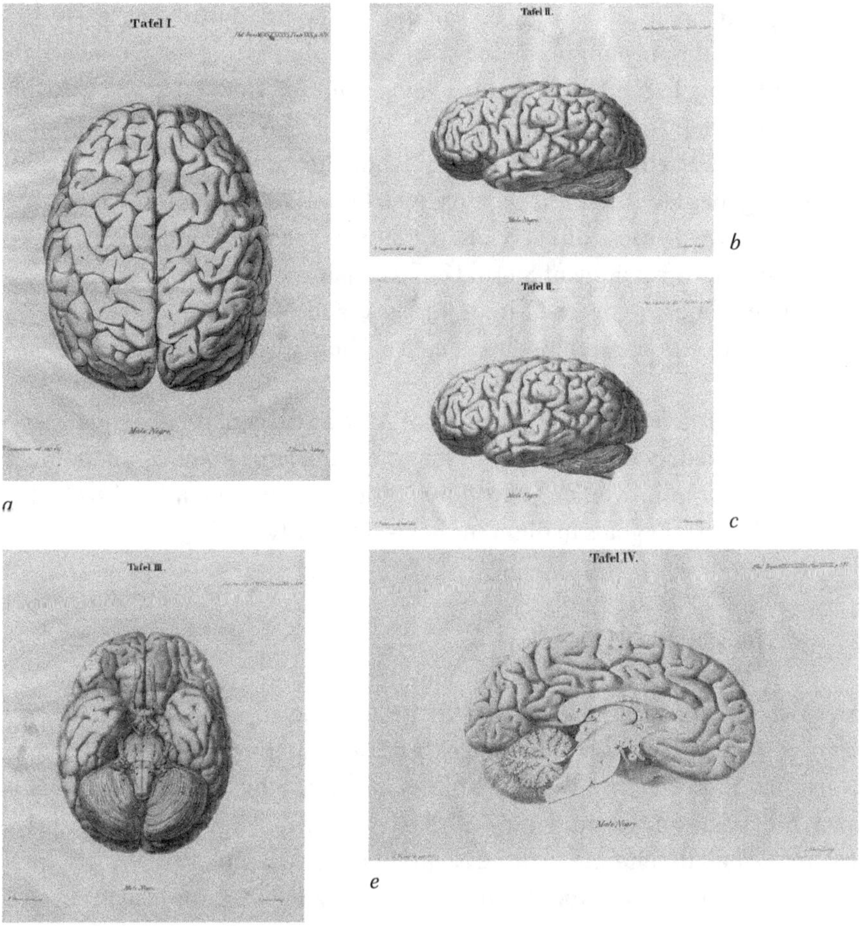

a b c d e

FIGURE 2.7A-E *Views of a Negro's brain*, Tafeln I-IV, in Friedrich Tiedemann, *Das Hirn des Negers mit dem des Europäers und Orang-Outangs verglichen* (Heldelberg 1837).

features. Tiedemann then proposed a further step: having argued about the lack of a well-marked and essential divergence between the brains of the Negro and the European, he inferred that no innate difference could exist in their intellectual and moral faculties. So degrading was the nature of the slave trade and regime that any fortitude of mind got necessarily lost and its free agency restrained. With great confidence, he argued that no natural formation or disposition would substantiate the notion of a predestined subservient state:

> The principal result of my researches on the brain of the Negro is, that neither anatomy nor physiology can justify our placing them beneath the Europeans in a moral and intellectual point of view. How is it possible, then, to deny that the Ethiopian race is capable of civilization? [...] Great Britain has achieved a noble and splendid act of national justice in abolishing the slave trade. The chain which bound Africa to the dust, and prevented the success of every effort that was made to raise her, is broken. Hayti and the colony of Sierra Leone can attest that free Negroes are capable of being governed by mild laws, and require neither whips nor chains to enforce submission to civil authority.[114]

Tiedemann wished to assign to peoples of African descent original qualities of character, ruined by their century-old poor conditions, as well as an undoubted potential for improvement, as attested by a good number of Blacks who had achieved important goals in different fields of activity.

5 The Will to Differentiate

Over time, Tiedemann's egalitarian position attracted much and harsh criticism; as early as 1837, the phrenologist Andrew Combe confessed to no small reluctance in correcting the "amiable error" made by Tiedemann "with so much philanthropic warmth and with such hearty zeal in the cause of the Negro". Yet he had to reiterate that the Negro brain was definitely inferior in intellectual power to the European. The German physiologist was also accused of taking the absolute size of the brain as a measure of intellectual power only, without determining whether such power lies in the extent of intellect, in strength of moral feeling, or in the force of passion or affection:

> A brain of four pounds' weight may be large in the anterior lobe, and smaller in the middle and posterior lobes; or its chief size and weight may be in the posterior lobes, and the anterior portions be actually small. In both cases, Tiedemann would infer equal "intellectual" power; whereas, the phrenologist would perceive at a glance that, in the former, the intellectual ability would far preponderate, while, in the latter, the power of

114 Tiedemann, "On the Brain of the Negro" (cit. note 70), pp. 525-526.

> mind would consist entirely in intensity of feeling, and the intellect, properly so called, be rather weak than strong.[115]

A comparative example served to further clarify the point: in the entire cerebral mass the Charib outweighs the Hindoo: from Tiedemann's viewpoint, the former would be equipped with more intellectual talent, but the reverse was actually true, and an explanation could only be provided by distinguishing the regions of the brain. In the Charib the anterior lobe was very small, in perfect harmony with his poverty of intellect, while the posterior and basilar regions appeared as large as his ferocity and energy of passion. The opposite held true for the Hindoo.

The same distinction occurred every day in social life: a criminal might have a big brain, along with a low intellect and strong destructive propensities; conversely, an amiable member of society might show a smaller brain in absolute size, but being far superior to the former in the size of its anterior lobes and thinking power. To render his conclusions worth anything, besides measuring the equality of the two brains in absolute size, Tiedemann also had to demonstrate an equal development in their anterior portions. This was the real question of interest for Combe, who lamented that a physiologist of great talent and merited reputation had failed in a demanding investigation with such a benevolent purpose. Had Tiedemann availed himself of the aid of Gall's doctrine, he would have realized at once the futility of any study based on considering the whole brain as the organ of intellect, as if its different parts did not perform different functions. Was it intentional or just forgetful? In both cases, he had to be almost equally blamed.

Tiedemann's paper defied conventional views of race and gave voice to a small "egalitarian" tradition that – according to Stephen J. Gould – had emerged at the end of the 18th century and survived for a while, here and there. In doing so, the "great physiologist of Heidelberg" undertook his comparative researches on the brain "with his mind already set (or at least strongly inclined) to a verdict of equality".[116] For Jeannette Eileen Jones, Tiedemann's interest in the

---

115 Andrew Combe, "Remarks on the Fallacy of Professor Tiedemann's Comparison of the Negro Brain and Intellect with those of the European," *The Phrenological Journal and Magazine of Moral Science*, 1837-1838, *11*:13-22, p. 15.

116 Stephen J. Gould, "The Great Physiologist of Heidelberg," *Natural History*, 1999, *108/7-8*:26-29, 62-64, 68-70, where it is also noted that Tiedemann's tables list each skull individually without presenting any summary statistics for groups. However, the figures can easily be calculated from his raw data, and indicate a largest average value for Caucasian brains and a lowest for the Ethiopian. The alternative posed by Gould is as follows: "Did Tiedemann calculate these means and not publish them because he sensed the confusion that

topic was no "fleeting pastime" and his conclusions were based on empirical research that required onerous visits to London, Edinburgh, Groningen, and Kassel. During the rise of transatlantic abolitionism, he became aware of how proslavery agitators were exploiting the works of Camper and of his mentors: "his own sense of liberalism moved him to discredit them."[117] A decade later, Tiedemann's life was to be tragically marked by the revolution of 1848-1849 in Baden: his eldest son Gustav Nicolaus, a soldier who had joined the insurgents, was tried by a Prussian martial court and executed. Instead, his younger son Heinrich, having married into a revolutionary family, was forced to emigrate overseas and settled for a medical practice in America. Almost seventy years old, Tiedemann retired from Heidelberg to live his last years in Frankfurt am Main and Munich.

In 1847, the secretary of the Société d'Ethnologie, Gustave Séligmann d'Eichthal – a Hellenist Greek scholar and a geographer – introduced the question that was to be discussed during the April 23rd session: "Quels sont les caractères distinctifs de la race blanche et de la race noire, et les conditions d'association de ces deux races?". He proclaimed this "la plus importante que l'Ethnologie puisse se proposer", since all groups of the human species could essentially be traced back to those two fundamental races. The white fraction included not only the Caucasians, but rather oddly also the Mongolians, Americans and Polynesians; the black one subdivided instead into Africa and Oceania. There had been weak contacts with each other until the age of geographical discoveries, conquests and the slave trade, which had created relations of power and domination, which were still prevalent, even if the philosophical and religious sentiments tried to mitigate them.

The confrontation set up by d'Eichthal between the two races was merciless: progress and culture on the one side, "impuissance scientifique et civilisatrice" on the other, to such an extent that a widespread opinion denied to Blacks the most essential attributes of humanity, as if they were just a lower species destined to serve. There was nonetheless an opposite opinion, which attributed

would then be generated – a procedure that I would have to label as indefensible, however understandable? Or did he never calculate them because he got what he wanted from the more obvious data on ranges and then never proceeded further – the more usual situation of failure to recognize potential interpretations as a consequence of unconscious bias? I rather suspect the second scenario as more consistent with Tiedemann's personal procedures and the actual norms – as opposed to the stated desirabilities – of scientific study in general. But I cannot disprove the first conjecture." (p. 69).

117 Jeannette Eileen Jones, "«On the Brain of the Negro». Race, Abolitionism and Friedrich Tiedemann's Scientific Discourse on the African Diaspora," in Mischa Honeck, Martin Klimke, and Anne Kuhlmann (eds.), *Germany and the Black Diaspora. Points of Contact, 1250-1914*, edited by (New York-Oxford: Berghahn, 2013), pp. 134-152, p. 142.

their inferiority to external and accidental circumstances: their improvement would be enough to achieve a development similar to that of the white race. A third, conciliatory opinion was based on the fact that the absence of initiative, found in the Negro, had also affected white women for a long time: "l'homme noir est à l'homme blanc ce que la femme est à l'homme". It seemed however that a role was then attributed to the woman, if not one of absolute equality, at least one of equal importance in the organization of the human couple, exactly just like that of the Black in mankind. During the next *séance*, the skull of a Bantu (Caffre) was presented by a traveler, Adolphe Delgorgue, who also ventured to draw a rough correspondence between brain and intelligence, cerebellum and will, face and *sensibilité*. Needless to say, the African skull was distinguished by a particular development of the third element. A very lively discussion among the members of the Société d'Ethnologie continued in the following sessions, fueled by the ongoing *tournant colonial*.[118]

In the meantime, the Swedish anatomist Anders Adolf Retzius, whose main purpose was racial classification, tested a new geometric instrument on the skulls that were stored at the Karolinska Institute of Stockholm. He adopted Blumenbach's *norma verticalis* – looking from above – to record variations of the cranial oval, its temporal or parietal swellings, the relative length of the anteroposterior diameter. But observing was not enough, only measuring the ratio of length to width of the head could guarantee sharpness and scientific accuracy. Thus, each skull was ranked on a scale from dolichocephaly (long-headed) to brachycephaly (short-headed), and the same applied to the average cranial index of the human races. This binary classification model had doubled since Retzius introduced the jaw shape as a second criterion: here the orthognathism (straight and vertical facial skeleton) and prognathism (protruding jaws in front of the braincase) were placed at the opposite ends of the scale. The final result, therefore, included four classes produced by crossing the two perspectives, giving further impetus to craniometry.[119]

118 "Séances du 23 avril au 9 juillet 1847," *Bulletin de la Société Ethnologique de Paris*, 1847, *1*:65-246. During the long discussion a provisional conclusion was summarized as follows: "La civilisation du noir se développe surtout par son contact avec les blancs. Elle se manifeste principalement par le développement des qualités morales, la docilité du caractère et la propension aux jouissances de luxe." (p. 90). See Thomas Bernon, "La science de races: la Société Ethnologique de Paris et le tournant colonial (1839-1848)," *La Révolution française. Cahiers de l'Institut d'histoire de la Révolution française*, 2018, *15*:1-56.

119 A presentation of his method was held at a meeting of the Scandinavian naturalists in 1842 (*Om Formen af Nordboerner Cranier*), published in their *Forhandlingar*, then translated in German and French: see Anders Retzius, "Mémoire sur les formes du crâne des habitants du Nord," *Annales des sciences naturelles*, 1846, *6*:133-172. In 1846 he included other races into his grid: see "Ueber die Form des Knochengerüstes des Kopfes bei den

In that age of 'positive' science – so proud of its adjective – it is hard to imagine anything less akin than the multifaceted activity unfolded by the scientific romanticism of Carl Gustav Carus, professor of obstetrics and gynecology, anatomist and physiologist, explorer of a reality called unconscious (*Unbewusstsein*), a traveler but also a draftsman, a painter, and an art critic. In 1838, he went so far as to fancifully classify two main *Stämme*, the Ethiopian and the Caucasian, that is, the nocturnal and diurnal sides (*Nachtseite*, *Tagseite*) of mankind. Next to these, another two crepuscular sides (*Dämmerungseiten*), Eastern and Western respectively, corresponding to Asia and America. The original lands of those peoples – climate and geography – wrought indelible effects on their anatomy, especially on skull sizes and brain formation, marking them with different intellectual attitudes. In 1849, for the centenary of Goethe's birth, Carus further elaborated on his theory about unequal qualifications of the various human stocks. First of all, it was obvious that the spatial relations of the brain – an organic workshop of thought – were generally found to be quite different in the four main *Stämme* of mankind. Even so, it was not only the physical size of the brain that determined the power of the mind. At the same time the regularity of its structure, the greater or lesser perfection in the formation of its fibers and cells, as well as the relation with the development of the spinal cord and the nerves, would considerably tip the scales at a higher or lower mental capacity.[120]

Among the sources cited by Carus, Samuel George Morton occupied a prominent place, for the outstanding mass of data available in his writings. As a matter of fact, it had taken "some years of toil and anxiety" to complete, in 1839, a task undertaken by a Philadelphia-born physician and naturalist who, coming from a respectable Irish family and raised as a Quaker, had also studied in Edinburgh and Paris. Afterwards, he was one of the founders of the Medical College of Pennsylvania, and served as the secretary of the Academy of Natural Sciences of Philadelphia. Morton started to collect skulls around 1830, when he was preparing some lectures on human anatomy. Nine years later, the great volume of *Crania Americana* offered the results of his untiring research. An

---

verschiedenen Völkern. Vorgetragenen in der vierten Versammlung der scandiv. Naturforscher," *Archiv für Anatomie, Physiologie und wissenschaftliche Medicin*, 1848, 263-284.

120 Carl Gustav Carus, *System der Physiologie umfassend das Allgemeine der Physiologie, die physiologische Geschichte der Menschheit, die des Menschen und die der einzelnen organische Systeme im Menschen, für Naturforscher und Aerzte* (Dresden und Leipzig: Gerhard Fleischer, 1838), I, pp. 123-132; Id., *Denkschrift zum hundertjährigen Geburtsfeste Goethe's: Über ungleiche Befähigung der verschiedenen Mencheitsstämme für höhere geistige Entwicklung* (Leipzig: Brockhaus, 1849).

opening letter to John S. Phillips, a friend and member of the Academy, declared a shared interest:

> You and I have long admitted the fundamental principles of Phrenology, viz: That the brain is the organ of the mind, and that its different parts perform different functions: but we have been slow to acknowledge the details of Cranioscopy as taught by Dr. Gall, and supported and extended by subsequent observers. We have not, however, neglected this branch of inquiry, but have endeavored to examine it in connection with numerous facts, which can only be fully appreciated when they come to be compared with similar measurements derived form the other races of men. Yet I am free to acknowledge that there is a singular harmony between the mental character of the Indian, and his cranial developments as explained by Phrenology.[121]

Morton would have liked to compose a chapter on phrenology, but aware of his own limits he commissioned it to a colleague who could not write it because of an illness. The work was ready for the press when George Combe began his American tour and agreed to provide the desired essay, which was then added as an appendix.[122]

In general, philosophers had investigated the differences between national characters – as remarked by the Scottish spokesman of Gall's doctrine – without knowing the various parts of the brain. Phrenologists had avoided the mistake, but did not publish a separate work specifically devoted to the topic. Regarding the different quarters of the globe, one was struck by the extreme dissimilarities in the attainments of the human varieties that inhabit them. In short, the Europeans had always shown to have a strong tendency for moral and intellectual improvement; the people of Asia had reached a comparatively lower and insuperable point; the African races exhibited one unbroken scene of moral and intellectual desolation, although some of their tribes had advanced beyond the savage condition, with manufactures, agriculture,

---

121 Samuel George Morton, *Crania Americana, or a Comparative View of the Skulls of Various Aboriginal Nations of North and South America: to which is prefixed an Essay on the Varieties of the Human Species Illustrated by Seventy-eight Plates and a Colored Map* (Philadelphia: J. Dobson; London: Simpkin, Marshall & Co., 1839). The lithographic plates, crucial for the success of the *in folio* book, represent North and South American skulls, and the map inserted at the opening of the work shows the geographical distribution of the human species all over the world.

122 George Combe, "Phrenological Remarks in the Relation between the Natural Talents and Dispositions of Nations, and the Developments of their Brains," in Morton, *Crania Americana* (cit. note 121), pp. 269-291.

commerce, government, and laws. Combe saw the status of Native Americans as still more deplorable than that of Africans, enveloped as they were in all their original primitiveness: wandering, houseless and lawless hordes. Soil and climate could exert a certain influence on these inequalities of development, but were not able to fully explain the phenomenon. The phrenologist was convinced that only the dimensions of the brain and the variable proportions of its regions would provide the true key. Without going into the details of the matter, Combe's essay continued by simply repeating the specific organs associated with the cortical surface, with a description of their respective faculties.

Morton's own introduction to *Crania Americana* contains an analysis of the varieties and geographical distribution of the human species, whose different physical characteristics, preserved through numberless generations and dissimilar circumstances, had occasioned speculations on its origin. Beyond all the conjectures made in this regard, he believed that each race could have adapted since the beginnings to its peculiar local destination, and that therefore its distinctive traits were not dependent on external causes. Previous classificatory efforts seemed to form just an "uninviting chaos", so that he adopted Blumenbach's arrangement in five great divisions: though imperfect, it was still appreciated as the most complete. Consequently, a small table listed the internal capacity in the Caucasian, Mongolian, Malay, American, and Ethiopian cranium. What Morton added was a further subdivision of the five varieties into twenty-two families, with an illustration of each of them, relying on the usual, available ethnological reports.[123] Notably, he considered the American race to differ from all others, not excepting the Mongolian, and to be composed of two families, with a small skull,

> wide between the parietal protuberances, prominent at the vertex, and flat at the occiput. In their mental character the Americans are averse to cultivation, and slow in acquiring knowledge; restless, revengeful, and fond of war, and wholly destitute of maritime adventure.[124]

As for his personal material, Morton had spent years measuring the volume of empty skulls by filling them with mercury, white pepper seeds and buckshot, and producing reams of tables with anatomical measurements. The writing of *Crania Americana* had been preceded by lot of collecting, helped by a wide

123 Morton, *Crania Americana* (cit. note 121), pp. 1-95.

124 Ibid., p. 6. For a detailed description of the American race see pp. 62-86.

RACES.	No. of skulls.	Mean internal capacity in cubic inches.	Largest in the series.	Smallest in the series.
Caucasian.	52	87.	109.	75.
Mongolian.	10	83.	93.	69.
Malay.	18	81.	89.	64.
American.	147	80.	100.	60.
Ethiopian.	29	78.	94.	65.

FIGURE 2.8 *On the internal capacity of the cranium of different races of man*, table in Samuel George Morton, *Crania Americana, or a Comparative View of the Skulls of Various Aboriginal Nations of North and South America* (Philadelphia 1839), p. 260.

network of contacts with travelers, diplomats, soldiers and doctors.[125] The re-

125 The literature on Morton is rather vast: see for instance Ann Fabian, *The Skull Collectors. Race, Science, and America's Unburied Dead* (Chicago-London: The University of Chicago Press, 2010), pp. 9-45. Drawing on archival sources from both United States and Britain, James Poskett has tried to chart the "transatlantic context" in which Morton operated: see his "National Types: the Transatlantic Publication and Reception of *Crania Americana*," *History of Science*, 2015, *53*:264-295. Stephen Jay Gould's *The Mismeasure of Man* (New York-London: Norton & Company, 1981, 1996[2]), pp. 30-72, discussed Morton's craniology as an early example of the various attempts at measurement that in the last two centuries had served to build racist visions. According to Gould, Morton had selected and manipulated data – perhaps driven by an unconscious bias – in order to support his hierarchical racist views. A few years later, John S. Michael re-measured a certain number of those *crania* and criticized Gould's interpretation: see "A New Look at Morton's Craniological Research," *Current Anthropology*, 1988, *29*: 349-354. More recently six anthropologists have thoroughly reexamined the whole affair, remeasured the volume of some 300 skulls in Morton's collection, and demonstrated that he cannot be pointed as an example of scientific misconduct: see Jason E. Lewis [et al.], "The Mismeasure of Science: Stephen Jay Gould versus Samuel George Morton on Skulls and Bias," *PLoS Biol*, 2011, *9*(*6*): e1001071. <https://doi.org/10.1371/journal.pbio.1001071>. Again in 2018, newly discovered handwritten documentation has thrown more light on the controversy, leading to the proposal that Gould incorrectly blamed Morton for inaccurate measurements, but was nonetheless correct as for his implicit racial prejudice. In particular, "Tiedemann and Morton independently produced similar data about human brain size in different racial groups but analyzed and interpreted their nearly equivalent results in dramatically different ways: Tiedemann using them to argue for equality and the abolition of slavery, and Morton using them to entrench racial divisions and hierarchy. These differences draw attention to the epistemic limitations of data and the pervasive role of bias within the broader historical, social, and cultural context of science.": see Paul Wolff Mitchell, "The fault in his seeds: Lost notes to the case of bias in Samuel George Morton's cranial race science," *PLoS Biol*, 2018, *16*(*10*): e2007008. <https://doi.org/10.1371/journal.pbio.2007008>; Id. and John S. Michael, "Bias, Brains, and Skulls. Tracing the Legacy of Scientific Racism in the Nineteenth-Century Works of Samuel George Morton and Friedrich Tiedemann,"

sults of all those efforts were witnessed by the Swiss naturalist Louis Agassiz, who, after his arrival in Boston in October 1846, made a tour of the Northeast for the purpose of introducing himself to the American scientific community. He spent some time in Philadelphia, often in the company of Morton, whose collections of skulls and books are praised in a letter to his mother and friends:

> Figurez vous une série de 600 crânes, la plupart d'indiens de toutes les tribus qui habitent et qui ont habité jadis l'Amerique toute entière. Il n'existe nulle part ailleurs rien de pareil. Cette collection à elle seule vaut un voyage en Amerique. M. Morton a publié un grand ouvrage avec planches in folio représentant tous les types de sa collection qui'il a eu l'obligeance de me donner.[126]

In the same letter, Agassiz communicated his "impression pénible" at his contact with the black servants of the hotel where he resided, a feeling that contradicted every idea of brotherhood of the human species: "ils ne sont pas du même sang que nous". What a misfortune for the white race, in certain countries, to have tied its existence so closely to that of the *nègres*! Philanthropists who wanted to make them full citizens forgot their natural limits: they were human – he would soon be arguing – but did not belong to the same species as Whites. No wonder that, during a lecture held at the Literary and Philosophical Society of Charleston in December 1847, Agassiz pleased his Southern audience by equating the Negro's adult brain with the still imperfect one, of a seven-month-old fetus in the womb of a white woman.[127]

Morton's work was widely read in the following decade, and his collection of skulls – bought through a subscription and donated to the Academy of Natural Sciences – continued to grow even after his untimely death in 1851, reaching up to 1225 items. For a while, the Antebellum South welcomed it for having scientifically proved the inferiority of Africans, although some churchmen took issue with his anti-biblical position on the multiple origins. In 1849, Morton submitted the results of his internal measurements of 623 skulls to the

---

in Jamie A. Thomas, Christina Jackson (eds.), *Embodied Difference. Divergent Bodies in Public Discourse* (Lanham MD: Lexington Books, 2019), pp. 77-98.

126 Letter of 2 December 1846, Ms. Am 1419, Houghton Library, Harvard University. See Christoph Hirmscher, *Louis Agassiz. Creator of American Science* (Boston-New York: Houghton Mifflin Harcourt, 2013).

127 Ten years before, a French *émigré* had already read to the Charleston Society portions of his translation of Virey's *Histoire naturelle du genre humain* about the Negroes – "incapable of governing themselves" – together with a selection of Soemmerring's essay on the comparative anatomy of the Negro and European: see J. H. Guenebault, *Natural History of the Negro Race. Translated from the French* (Charleston, S.C.: D. J. Dowling, 1837).

Philadelphia Academy; measurements "made with a view to ascertain the relative size of the brain in the various races and families of Man", stressing the accuracy of "these multitudinous data" and the importance of his contribution to Ethnology, which still was in its "infant state". A table made the data easily comparable and also collected those provided by *Crania Americana*: in particular, the Teutonic or German race possessed the largest brain of any other people, with an average of 92 cubic inches; the nations having the smallest heads belonged to the "American Group", with an impressive 13 cubic inches less; the Malay and Negro groups stood in intermediate positions (respectively 85 and 83 cubic inches).[128]

Nearly twenty years after communicating the results of his own practice of brain volume measurement, Morton's table of data negatively struck William Hamilton, who judged it patently unreliable because it did not distinguish male from female crania, and therefore made it impossible to really compare the races to each other. The conclusions drawn by Morton were contrary to Tiedemann's larger, and to his own, minor, inductions, "which concur in proving, that the Negro encephalos is not less than the European, and greatly larger than the Hindoo, the Ceylonese, and sundry other Asiatic brains". He also claimed to have been the first to invent the method for assessing cranial capacity through pure siliceous sand, so that Morton's was merely "a clumsy and unsatisfactory modification of mine".[129]

Strong opposition to the polygenist views also came from a renowned naturalist, John Bachman, a minister of St. John's Lutheran Church in Charleston, who in 1850 published a series of notes, read before the local Literary Club, on the unity of the human species. For him – albeit a slave-holder believing in the biblical foundation of slavery – the enemies of the monogenist doctrine were not in possession of such materials as would furnish them with the means of disproving it. With the exception of Morton's extensive skull collection, no institution or medical college in America owned any series of skulls of the various races of man, so that one had to depend on the information provided by European anatomists and physiologists. Bachman's heroes were obviously Tiedemann and Blumenbach, abundantly quoted with Hamilton, all of them able to debunk the opinion of naturalists such as Camper, Soemmerring, Cuvier, Lawrence, and Virey, "that the negro has a smaller skull and brain than the

128 Samuel George Morton, "Observations on the Size of the Brain in various Races and Families of Man," *The Edinburgh New Philosophical Journal*, 1849-1850, *48*:262-265.

129 William Hamilton, "Remarks on Morton's Tables on the Size of the Brain," *The Edinburgh New Philosophical Journal*, 1849-1850, *48*:330-333.

Size of the Brain in various Races of Man. 263

Table, shewing the Size of the Brain in Cubic Inches, as obtained from the Internal Measurement of 623 Crania of various Races and Families of Man.

Races and Families.	No. of Skulls.	Largest I. C.	Smallest I. C.	Mean.	Mean.
Modern Caucasian Group.					
Teutonic Family:					
Germans	18	114	70	90	92
English	5	105	91	96	
Anglo-Americans	7	97	82	90	
Pelasgic Family: Persians, Armenians, Circassians	10	94	75	84	
Celtic Family: Native Irish	6	97	78	87	
Indostanic Family: Bengalees, &c.	32	91	67	80	
Semitic Family: Arabs	3	98	84	89	
Nilotic Family: Fellahs	17	96	66	80	
Ancient Caucasian Group.					
From the Catacombs. *Pelasgic Family:* Græco-Egyptians	18	97	74	88	
From the Catacombs. *Nilotic Family:* Egyptians	55	96	68	80	
Mongolian Group.					
Chinese Family	6	91	70	82	
Malay Group.					
Malayan Family	20	97	68	86	85
Polynesian Family	3	84	82	83	
American Group.					
Toltecan Family:					
Peruvians	155	101	58	75	79
Mexicans	22	92	67	79	
Barbarous Tribes: Iroquois, Lenapé, Cherokee, Shoshoné, &c.	161	104	70	84	
Negro Group.					
Native African Family	62	99	65	83	83
American-born Negroes	12	89	73	82	
Hottentot Family	3	83	68	75	
Alforian Family: Australians	8	83	63	75	

FIGURE 2.9
Table in Samuel George Morton, "Observations on the Size of the Brain in various Races and Families of Man," *The Edinburgh New Philosophical Journal*, 1849-1850, *48*:262-265, p. 263.

Europeans." Among many skulls of Negroes and Europeans that Bachman examined,

> We find some where the two races approach each other so nearly that it requires much attention and a practised eye to distinguish between them; – and were we to give the white colour and straight hair of the

> Caucasian to some of the skulls of the negro, the most practised anatomist and physiologist might be easily deceived.[130]

The prevailing American turn of mind, however, was fed by activists like John H. Van Evrie, a New York publisher with a medical degree, proprietor of the *Weekly Day Book*, a newspaper that promoted white supremacy for thirty years. He wrote profusely to justify the slave condition of Blacks, and one of his pamphlets, issued in 1853, became a massive book in 1861. Ignorant of the real nature of the negro, the North – he proclaimed – continued to pity the imaginary sufferings of the slaves; conversely, the South knew from actual experience that their present condition ensured to the black race a greater measure of happiness than ever before in its history. Beside every other physical difference, Van Evrie was certain that, by virtue of a specific nature, the Caucasian brain measured 92 cubic inches, relatively predominating over the cerebellum, the center of the animal instincts; while the Negro brain did not exceed 65 to 70 cubic inches, with a hegemonic cerebellum. Therefore, only the first kind of brain was capable of indefinite progression and could transmit the acquired knowledge from generation to generation. Deluded people who persisted in trying to improve on the works of the Creator by adding 25 or 30 per cent to the Negro brain would have inevitably failed.[131]

---

130 John Bachman, *The Doctrine of the Unity of the Human Race Examined on the Principles of Science* (Charleston: C. Canning, 1850), p. 231. An ally of Bachman was Reverend Thomas Smyth, who – in a book devoted to the same topic – pointed out that "differences as great in reference to the shape of the skull, the size of the brain, and mental endowments, are found to exist among individuals and families of the same nation, as between the Caucasian and other races": *The Unity of the Human Races, proved to be the Doctrine of Scripture, Reason, and Science. With a Review of the Present Position and Theory of Professor Agassiz* (New York: George P. Putnam, 1850), pp. 307-308. On the social and ideological context of these mid-century disputes, see Lester D. Stephens, *Science, Race, and Religion in the American South: John Bachman and the Charleston Circle of Naturalists, 1815-1895* (Chapel Hill: University of North Carolina Press, 2000). More generally, on the so-called American School of Anthropology, see C. Loring Brace, *"Race" is a Four-Letter Word. The Genesis of the Concept* (New York-Oxford: Oxford University Press, 2005), pp. 76-143.

131 John H. Van Evrie, *Negroes and Negro Slavery. The First, an Inferior Race – The Latter, its Normal Condition. Introductory Number: Caused of Popular Delusion on the Subject* (Baltimore: John D. Toy, 1853), pp. 28-29. The book with the same title, but without subtitle (New York: Van Evrie, Horton & Co., 1861), contained a chapter on the brain (pp. 123-131) that praised the scientific merit of Gall and Spurzheim, taking for granted the distinction between cerebrum and cerebellum as organs, respectively, of the intellectual capacity and of the animal instincts. Despite the little that was known about it, for Van Evrie the case of the Negro highlighted "an eternal sameness, a perpetual oneness, the same color, the same feature, same size of the body, and the same volume of brain." (p. 127). The increase in cerebral volume that the abolitionists expected from emancipation would

A portrait of Morton opened, in 1854, a hodgepodge book of over 700 hundred pages, compiled by two of his most fervent followers, the Alabama-born medical practitioner and surgeon Josiah Clark Nott along with the English-born American Egyptologist George Robins Gliddon. A strange couple held together by the mission to spread the word that different acts of creation had produced "many primordial human Types, or Species" within the genus *Homo*. Their *Types of Mankind* became a lucky editorial enterprise, with ten successive editions, and a conspicuous visual apparatus:

> Above three hundred and sixty wood-cuts, besides many lithographic plates, adorn this volume, and upon them, to some extent, depend its value and success. The reader can well imagine the immense labor and heavy expense required to prepare a series of illustrations of this kind, wherein minute accuracy is so indispensable, and where such accuracy can be attained only through long-continued and patient industry combined with high artistic skill. So great indeed, were the difficulties to be overcome, that the authors could never for a moment have entertained the idea of publishing a work like "Types of Mankind," had it not been for the aid generously proffered by Mrs. Gliddon, the accomplished lady of my colleague. To her amateur pencil are we indebted for the drawings of more than three hundred of our wood-cuts [...] I venture to say that no scientific work in our language presents as long a series of illustrations more reliable for faithfulness to originals.[132]

Sharing the workload between them, Nott was in charge of writing the scientific chapters, so Gliddon could deal with the archaeological part and the biblical commentary. Incidentally, even the terms "monogenism" and "polygenism"

---

have been a chimera, "for such a brain could no more be born of a negress than an elephant could be!" (p. 131).

132 Josiah C. Nott, George G. Gliddon, *Types of Mankind: or, Ethnological Researches, based upon the Ancient Monuments, Paintings, Sculptures, and Crania of Races, and upon their Natural, Geographical, Philological, and Biblical History: Illustrated by selections from the Inedited Papers of Samuel George Morton, M.D., and by Additional Contributions from Prof. L. Agassiz, LL.D.; W. Usher, M.D.; and Prof. H. S. Patterson, M.D.* (Philadelphia: Lippincott, Grambo & Co., 1854), pp. xii-xiii. The book contained a hagiographic "Notice of the Life and Scientific Labors of the Late Samuel Geo. Morton, M.D.", by Henry S. Patterson (pp. xvii-lvii), a sketch by Louis Agassiz "On the Natural Provinces of the Animal World and their Relation to the Different Types of Man" (pp. lviii-lxxvi), and a chapter by William Usher, M.D., on "Geology and Palaeontology, in Connection with Human Origins" (pp. 327-372).

seem to have made their first appearance in a chapter written by Gliddon for a coauthored book in 1857.[133]

That the brain could not be expanded or altered in form, seemed to Nott – who astutely quoted Prichard's "herculean work" in this regard – a truth admitted by every anatomist, corroborated by all of past history, and backed by every monument. Morton had proven that the Negro races had about nine cubic inches less of brain than the Teuton, and, unless some facts in history taught how these deficient inches could be artificially added, the Africans would remain substantially "in the same benighted state wherein Nature has placed them, and in which they have stood, according to Egyptian monuments, for at least 5000 years."[134] No doubt, then: the size and shape of the brain had played a major role in the advancement and destiny of races. Once emerged from barbarism, for instance, the large heads of Teutons and Celts could reach the higher pinnacles of civilization, and everywhere outstripped and ruled over the small-headed races of mankind. Moreover, the brain and nervous system, in animal nature, would be so influenced by crossing, as to make instincts and senses partake of intermediate characters:

> The same law applies to human white and black races; for the mulatto, if certainly more intelligent than the Negro, is less so than the white man. His intelligence, as a general rule, augments in proportion to the amount of white-blood in his veins. This is inevitably the case in the United States.[135]

Riding the wave of success met by *Types of Mankind*, Nott commissioned a Swiss immigrant printer, Henry Hotz, to produce an abridged translation of Joseph Arthur de Gobineau's *Essai sur l'inegalité des races humaines*, which was published in four volumes between 1853 and 1855. With a fifty-page appendix added by Nott about the unity or plurality of species, the book appeared in

133 George R. Gliddon, "The Monogenists and the Polygenists; being an exposition of the doctrines of schools professing to sustain dogmatically the unity or the diversity of human races etc.," in Josiah C. Nott and George R. Gliddon, *Indigenous Races of the Earth; or, Few Chapters of Ethnological Enquiry; including Monographs on Special Departments of Philology, Iconography, Cranioscopy, Palaeontology, Pathology, Archaeology, Comparative Geography, and Natural History: contributed by Alfred Maury, Francis Pulszky and J. Aitken Meigs. Presenting fresh Investigations, documents, and Materials* (Philadelphia: J. B. Lippincott & Co., 1857), pp. 402-602. The attribution to Gliddon of the paternity of the two neologisms is in Claude Blanckaert, *De la race à l'evolution. Paul Broca et l'anthropologie française* (Paris: L'Harmattan, 2009), pp. 35-36 n.

134 Nott, Gliddon, *Types of Mankind* (cit. note 132), p. 189.

135 Ibid., p. 402.

1856, but was a disappointment to the author, who complained of the treatment received by Americans readers, whose polygenist and pro-slavery dictates he did not accept.[136]

That anti-slavery and racist positions were compatible with each other is well exemplified by the case of the Scottish anatomist Robert Knox, who had experienced some misadventures as a young army surgeon in South Africa, followed by his involvement in the notorious Edinburgh West Port murders in the late 1820s.[137] The bizarre and famous man of science was almost sixty years old when he published a "Fragment" that had cost him "much thought and anxiety", claiming that its views were so wholly at variance with long received doctrines and stereotyped prejudices. All in all, in 1850 he took a kind of common sense from the previous decades to its extreme, by announcing that "with me, race, or hereditary descent, is everything; it stamps the man." The book opened with a glorification of the Saxon or Scandinavian race – "the only absolutely fair race on the face of the globe" – which was expected to become the dominant one. There was no way to alter its fine structure by crossing it with other breeds: a natural "antipathy" would have prevented any hybrids from being fertile.[138]

Knox credited to Hippocrates with the hypothesis that alterations in body and mind as a result of external influences – particularly in the brain – could become permanent over time and transmissible by hereditary descent. From this point of view, he had no doubt that the dark races were generally much inferior to the Saxon and Celt ones, both in physical strength and brain size. Truth to tell, he also confessed to his own "extremely limited experience", so

136 *The Moral and Intellectual Diversity of Races, with Particular Reference to their Respective Influence in the Civil and Political History of Mankind from the French of Count A. de Gobineau: with an Analytical Introduction and Copious Historical Notes by H. Hotz. To which is added an Appendix Containing a Summary of the Latest Scientific Facts upon the Question of Unity or Plurality of Species. By J. C. Nott* (Philadelphia: J. B. Lippincott & Co., 1856). On Gobineau's reaction see Reginald Horsman, *Josiah Nott of Mobile: Southerner, Physician, and Racial Theorist* (Baton Rouge: Louisiana State University Press, 1987), pp. 205-206.

137 For a report of Knox's early anthropological observations see his "Inquiry into the Origin and Characteristic Differences of the Native Races inhabiting the Extra-Tropical Part of Southern Africa," *Memoirs of the Wernerian Natural History Society*, 1823-24, 5:206-219, where he derived all existing races from the Caucasian, and arranged the Bosjeman with the Mongolic, distinguishing them from the Ethiopian Kaffre nations. About the corpse sales affair see Lisa Rosner, *The Anatomy Murders Being the True and Spectacular History of Edinburgh's Notorious Burke and Hare and of the Man of Science Who Abetted Them in the Commission of Their Most Heinous Crimes* (Philadelphia: University of Pennsylvania Press, 2010).

138 Robert Knox, *The Races of Man. A Fragment* (Philadelphia: Lea & Blanchard, 1850), pp. 13, 36-58.

that he merely simply relayed old or recent beliefs on the darker color of the brain, a greater fibrousness of its white part, and a remarkable symmetry of the hemispheres that would distinguish black races. On the other hand, Knox did not deny the Negro "qualities of a high order" that might have made him also capable of reaching a certain point of civilization.[139]

6 Early Doubts

Almost at the end of his career, in 1852 the psychiatrist Gottlob Heinrich Bergmann, director of the Hildesheim asylum and author of anatomical texts on the nervous system, made a quick review of the studies thitherto carried out on the weight of the brain, a field of practice that had already taken shape, even more so in the following years. He himself presented in three tables of personally obtained data, some of which supported the assumption that size and weight of the brain could have not have absolute value.[140] This assumption – as a certainty or a doubt – constantly reappeared even later.

Soon after, Jules Baillarger, who worked at the hospice of the Salpêtrière, explained his procedure of stretching (*déplisser*) the cortical surface in order to measure its extension, by replacing the inaccurate and destructive action of the fingers with a long and meticulous dissection that avoided any alteration. The second part of his article criticized Desmoulins for having asserted, three decades earlier, that the number and perfection of the intellectual faculties were in relation to that extension, without having ever roughly measured any brain surface. To find out how things stood, Baillarger carefully weighed a human brain and those of some animals (mutton, pig, dog, cat, rabbit), then moved on to measure the area of the respective cortical surfaces, and compared the results. It seemed clear that smaller brains had the greatest cortical extension, and that the relative surface of the hemispheres was inversely proportional to their weight. Something opposite to Desmoulins' old proposition, and simply obedient to the mathematical law about body volumes and surfaces corresponding to the cube and the square of their diameters, respectively. Therefore, in proportion to its volume, the surface of the human brain would be much smaller than that of the brain of the lower mammals. In other words, it could not be used as an indicator of intelligence; this conclusion, however,

139 Ibid., pp. 63, 151-152, 163.
140 Gottlob Heinrich Bergmann, "Einige Bemerkungen über das Gewicht des Gehirns nebst drei Tabellen," *Allgemeine Zeitschrift für Psychiatrie und psychisch-gerichtliche Medicin*, 1852, 9:361-376.

did not rule out the possibility that the latter might be directly related to the number and the development of the convolutions: an issue that Baillarger thought was still to be examined, not without considerable difficulties.[141]

If the spiritual activities had to be localized somewhere, the convolutions were their appropriate place: the more tortuous their gyri, the deeper the sulci between them, the more impressions and branches, which meant a greater perfection of the nervous functions. Of this Emil Huschke – professor of anatomy at the University of Jena and still sensitive to the sirens of *Naturphilosophie* – was sure when in 1854 he introduced a book that had cost him nine years of research, using new methods to lay the foundations of physiological psychology. Skull, brain, and soul (*Seele*) were at stake in its high-sounding title page, which also mentioned the addition of six tables with lithographs, made for the first time from photographic images: only this way of representing brains in their natural size could attain the most natural truthfulness, as prescribed by the rhetoric of objectivity that followed the recent diffusion of photography.[142]

All the observations made by Huschke, or those he borrowed from other sources, led him to try to decipher the cortex as a system of signs. He maintained that the frontal lobe was greater in man than in woman, whose parietal lobe instead prevailed. On the other hand, the *Neger* seemed to have less complete convolutions, and a Sylvian fissure slightly more oblique than that of the chimpanzee, while in the European it was horizontal. Many other details indicated that the Africans' brain resembled that of the infant or female of a European, and morphologically approached that of the higher monkeys. If it were not for the new anatomical expertise with which Huschke drew his conclusions, they only echoed a comparative differentiation that had been successful for over half a century.[143]

---

141 Jules Baillarger, "De l'étendue de la surface du cerveau et de ses rapports avec le développement de l'intelligence," *Annales medico-psychologiques*, 1853, 5:1-9. The young Baillarger – a former student of Esquirol – had already presented some researches showing that six alternately gray and white layers made up the substance of the convolutions. In his opinion, the analogy with the disposition of a "pile galvanique" reinforced the thesis that the nervous action was proportional not to the masses but to the surfaces (see his "Recherches sur la structure de la couche corticale des circonvolutions du cerveau," *Mémoires de l'Académie Royale de Médecine*, 1840, 8:149-183).

142 Emil Huschke, *Schädel, Hirn und Seele des Menschen und der Thiere nach Alter, Geschlecht und Raçe. Nebst sechs Steintafeln mit photographischen Abbildungen* (Jena: Friedrich Mauke, 1854), pp. v-vi.

143 Ibid., pp. 152-158. It is worth noting that a chapter of the book was dedicated to the brain as an *elektrischer Apparat* (pp. 168-174). Especially after Du Bois Reymond's *Untersuchungen über thierische Elektricität* (1849), Huschke assumed that the nervous system was the source of an electric current.

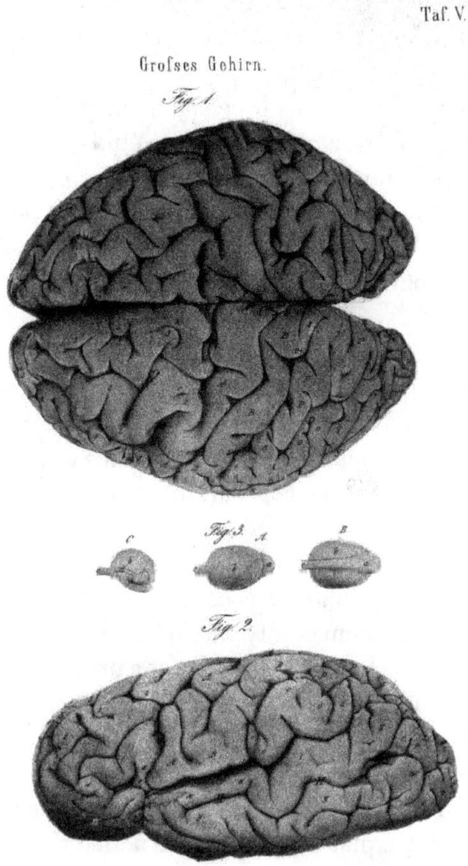

FIGURE 2.10
Fig 1-2: *Grosses Gehirn eines 29 Jahres alten Deutschen, von oben und von links* (The brain of a 29 year old German, from above and from the left). Tafel v in Emil Huschke, *Schädel, Hirn und Seele des Menschen und der Thiere nach Alter, Geschlecht und Raçe* (Jena 1854).

At that time, great effort was made to explore the brain folds – in humans and primates – by the head of anatomical works at the Muséum d'Histoire naturelle, Pierre Gratiolet, a former *protégé* and assistant of the zoologist Henri de Blainville, who had died in 1850. He started his report with a methodological warning: it is dangerous, in scientific research, to reach a conclusion too quickly. When one is reasoning on the basis of an insufficient number of observations, it becomes easy to build some general system, and since the observers differ from each other, it is not rare to see a hundred different hypotheses rise on the same facts. Such games of finesse can seduce the imagination and produce ghosts, but a healthy method must reject them. Only by concentrating gingerly on a particular order of facts and forces, useful outcomes can be attained. Gratiolet claimed to have done this in his arrangement of the cerebral folds, from man to the *Hapalidae* monkey family, thanks to the rich collection of brains owned by the Muséum. He was deeply impressed by the singular analogies existing among them, but, as a Catholic, he also retained the separation between human and non-human primates. Every natural family of animals had their own cerebral type, which he intended to study in the future:

> Je serais heureux si, en appelant l'attention sur ce sujet, j'inspirais aux voyageurs et aux psychiâtres la pensée de reprendre cette étude, et de la completer, en ce qui touche l'espèce humaine, par une étude approfondie des variétés que les plis cérébraux présentent, aux différents âges de la vie, dans les différentes races de la famille humaine, et dans ces cas de monstruosité si frequents, hélas! parmi nous.[144]

Gratiolet proved to be aware of the long scientific neglect of the convolutions, as well as the series of investigations carried out since the late eighteenth century up to the present; he appreciated the work of Vicq d'Azyr, Luigi Rolando, Baillarger, and others, and he complained that Gall, too busy with his vain system, had not deepened the comprehension of the cortical structure. The study of the human brain appeared to him as a fruitful and entirely new treasure-trove. In his opinion, Tiedemann's *On the Brain of the Negro*, though so highly celebrated, had not helped much. Actually, anatomists could not believe that filling the skull with dried seeds, weighing them empty and full, and thus judging the equivalence of brain development in different races was a process worthy of the science of the time. Under its apparent anatomical form, Tiedemann's text concealed a questionable political intent.[145]

After showing the main characters that the gyri of humans and monkeys had in common, Gratiolet focused on what distinguished the brains of the different "races ou espèces humaines", and explained that he had added the second term, because the unity of the species was less certain than that of the *genus*, without contradicting the biblical text. He could supposedly be a fervent believer and a polygenist at the same time. To his disappointment, that kind of work was extremely deficient due to the scarcity of observational materials. Unfortunately, travelers had collected a large number of skulls, leaving out the crucial issue of the brains. Gratiolet just compared the brain of a white porter – probably of ordinary intelligence – with that of the *Vénus hottentotte* made known by Cuvier in 1817.[146] Her brain was "une de pièces les plus

144 Pierre Gratiolet, *Mémoire sur les plis cérébraux de l'homme et des primatès* (Paris: Arthus Bertrand, 1854), p. iii.

145 Ibid., p. 10n.

146 Georges Cuvier, "Extrait d'observations faites sur le cadavre d'une femme connue à Paris et à Londres sous le nom de *Vénus hottentotte*," *Mémoires du Muséum d'Histoire naturelle*, 1817, *3*: 259-274. On Sarah Baartman, the young slave brought to London in 1810 from the Cape of Good Hope, and her sad destiny see *La Vénus hottentotte entre Barnum et Muséum coordonné par Claude Blanckaert* (Paris: Publications Scientifiques du Muséum d'Histoire naturelle, 2013).

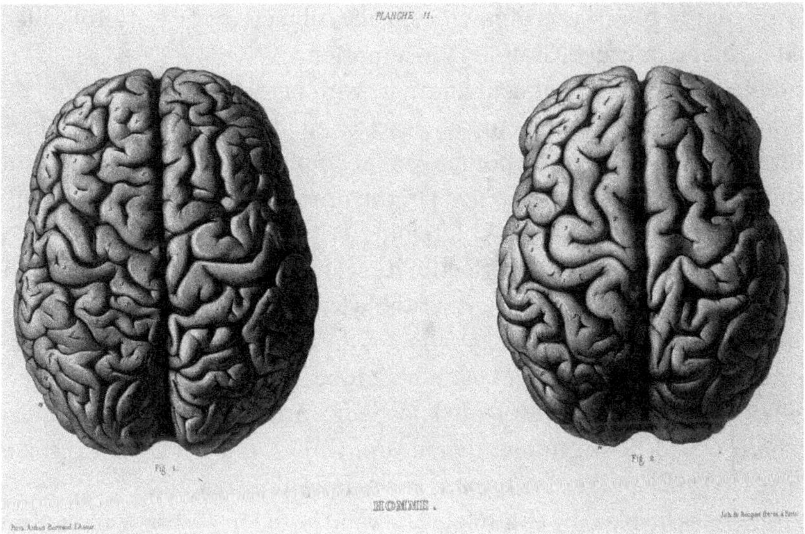

FIGURE 2.11 *Cerveaux d'un homme blanc (portefaix) et de Vénus hottentote.* Planche II in Pierre Gratiolet, *Mémoire sur les plis cérébraux de l'homme et des primatès* (Paris 1854).

précieuses de notre cabinet": despite having belonged to a subject who was not idiot at all, its folds were relatively uncomplicated:

> Mais, ce qui frappe surtout, c'est la simplicité, l'arrangement régulier des deux plis qui composent l'étage supérieur du lobe frontal. Ces plis, comparées de droite à gauche sur les deux hémisphères, présentent [...] une symétrie presque parfait, et que n'offrent jamais les cerveaux normaux de la race caucasique. Ils présentent, à la région frontale, un espace que l'oeil distingue aussitôt. On le dirait, en effet, découpé sur le reste du cerveau. Cette regularité, cette symétrie, rappellent involontairement la régularité, la symétrie des plis cérébraux dans les espèces inférieures.[147]

Gratiolet tried to highlight the difference in a *planche* reproducing a lateral view of both brains – not drawn on the same scale – and regretted that he could not compare the brain of the African woman with those of some white idiots. And this is because he assumed that the peculiarities of her central nervous organ might depend on a limited development, possibly also observable in the *arrêts de développement* of the superior races. In April 1853, during one of his lectures, Baillarger had shown the brain of an idiot whose anterior lobes resembled, although with a smaller volume, the forms already seen in the

147 Gratiolet, *Mémoire sur les plis cérébraux* (cit. note 144), p. 65.

Vénus. Apparently, brain casts of microcephalic subjects in the Esquirol collection had further confirmed Gratiolet's assumption.

The *Vénus* was no idiot: being in harmony with her whole organization, her small brain resulted in a sufficient, albeit weak intelligence. A brain like hers, in a white subject, would correspond to an incomplete form (*avortement*) and to a condition of idiocy or imbecility. On the other hand, the sufficiency of a brain such as hers implied the absolute inferiority of her race, or *sous-espèce du genre humain*. Since infinite perfectibility seemed to him a dangerous sophism, Gratiolet wanted to escape the charitable fantasies of the philanthropists, who pursued the redemption or exaltation of the wild races. Going back to the origins of history, as Joseph de Maistre had done, one could see the different degrees with which the races aspire to knowledge, art, and freedom. In some of them, there would be an innate dynamism, while others tended to isolate themselves in small groups, having war as the prevailing instinct.

The principles invoked by Gratiolet were valid both for the brain and for the other organic systems. In all of them, when a lower form suffices to survive, the inferiority of the zoological type is clear: "les espèces inférieures sont comme des embryons des plus élevés". Nevertheless, it was not fair to say that Blacks, Bushmen, and Botocudes were imperfect or unfinished creatures: they just reached their natural and innate level, as can happen to distinct species. From such a hierarchy within the *genre humain*, however, nothing odious or offensive to morality could derive. All of them were men with intelligence and speech, and, as such, respectable. The law of humanity that maternally protects the most monstrous Caucasian idiots should be extended to include every race, the same way as the charity was to be provided to the weak.[148]

Two years passed, and at a session of the Académie des sciences Gratiolet submitted a *Mémoire* aimed at carefully determining the modifications undergone by the shape of the skull, from birth to adulthood. That kind of research had also allowed him to track the subsequent obliteration of the sutures, and to record morphological differences between the wild and civilized races. The white one, in particular, showed a late ossification of the sutures, while this was precocious "dans les races ethiopienne et alforienne"; moreover, at first the former closed posteriorly, while the latter closed anteriorly. Others – Baillarger included – had already seen a premature ossification in microcephalic idiots, as well as the absence of *fontanelles* at their birth. In this regard Gratiolet hazarded a guess:

> La longue persistence des sutures dans la race blanche aurait-elle quelque rapport avec la perfectibilité presque indéfinie de l'intelligence dans les

148 Ibid., pp. 68-71.

> hommes de cette race? Cette durée d'une des conditions organiques de l'enfance ne semble-t-elle pas indiquer que le cerveau doit, chez les hommes perfectibles, demeurer capable d'un accroissement lent, mais continu? [...] Mais, chez les idiots et dans les races abruties, le crâne se ferme sur le cerveau comme une prison. Ce n'est plus un temple divin, pour me servir de l'expression de Malpighi, c'est une sorte de casque capable de résister a des coups de massue.[149]

Afterwards, Gratiolet picked up the thread of the discourse when he completed and published a great work which had been started by the anatomist and alienist François Leuret, one of the founders and editors of the *Annales d'hygiène publique et de médicine légale*, who died in 1851. A first volume of the *Anatomie comparée du système nerveux considéré dans ses rapports avec l'intelligence* had been published in 1839, authored by Leuret, with an atlas of 33 *planches in-folio*, after ten years of observations and studies.[150] It was actually a sort of long-lasting work, which culminated in the two volumes with the atlas, bearing the names of Leuret and Gratiolet side by side, as its authors, in 1857.

In his investigation Leuret had included all the animal series, invertebrates and vertebrates, preferring to proceed as if nothing had yet been established on the relationship between intelligence and the nervous system. Special attention was paid by him to dissections and microscopic experiences: brain, cerebellum, medulla oblongata, spinal cord had been measured, weighed, and analyzed, both morphologically and chemically, by means of reagents. The description of the cerebro-spinal system begins with mollusks and rises through the articulated animals, fishes, reptiles, birds, up to mammals. Leuret dealt with the 'psychic' state of each type and drew a table of their faculties. Over one hundred kinds of mammals were ranked by complexity of their convolutions, which would correspond to a parallel increase in intelligence. The last three pages tried to challenge some theses that had been put forward by Gall and were still upheld by phrenologists. Here, *inter alia*, Leuret plainly stated that "le volume absolu du cerveau n'est pas dans un rapport nécessaire avec le développement de l'intelligence." It's no coincidence that the book's *incipit* had asserted that the nature and conditions of that crucial relation remained entirely unknown.[151]

---

149 Pierre Gratiolet, "Mémoire sur le développement de la forme du crâne de l'homme, et sur quelques variations qu'on observe dans la marche de l'ossification de ses sutures," *Comptes rendus hebdomadaires des séances de l'Académie des sciences*, 1856, *47*:428-431.

150 François Leuret, *Anatomie comparée du système nerveux considéré dans ses rapports avec l'intelligence* (Paris: J.-B. Baillière, 1839).

151 Leuret, *Anatomie comparée du système nerveux* (cite note 143) p. 588, vii.

The second volume of the *Anatomie comparée* was entirely written by Gratiolet, who considered two opposite ways of philosophizing equally exaggerated, "deux écueils à éviter": on the one hand, making animals and humans too similar; on the other, separating them too radically. A third way was taking in consideration both their common faculties and those peculiar to man alone, which placed him in a separate and superior kingdom. At a certain point of this big volume of seven hundred pages, the figure of the Hottentot Venus reappeared, with her simpler brain; along with an invitation to look, first, for the type of beauty that belonged to each individual race or species: "cette étude convenablement faite, renverserait bien des préjugés".[152] Along the same line, here Gratiolet's most original proposal consists in the distinction between parietal (Mongolian), occipital (African), and frontal (Caucasian) races, the shape of their skull taken as a differential criterion, to explain its relationship with the development of the brain:

> La capacité des fosses cérébrales de l'occipital étant assez faible, un grand accroissement relatif de cette vertèbre ajoute peu de chose à l'ampliation du crâne en general, et de sa région cérébrale en particulier. Au contraire, l'ampliation du frontal est tout entière au profit du cerveau; et l'ampleur de cette vertèbre est telle, que ses moindres dilatations apparentes ajoutent notablement à la capacité du crâne. Ainsi, et surtout à l'égard du cerveau, une grande saillie de l'occipital donne beaucoup moins qu'une égale saillie du frontal.[153]

With Caucasian skulls taken at random in his hands, Gratiolet could not fail to notice the great dilatation of their frontal vertebra, a fundamental and allegedly advantageous character of the white race, which would be invariably present both in the most abject heads, and in that of Schiller.

In 1855, a doctor and hygienist from Geneva, who had studied in Paris and lived as a youth in Greece, having joined the struggle for independence from the Turkish domination, dealt with the artificial deformations of the skull in newborns, a curious and rather mysterious practice that had begun to attract much scientific attention. Without accepting all the details of the phrenological doctrine, Louis-André Gosse believed that the anterior and superior cerebral lobes were especially related to intellectual faculties, and took up the idea – recently expressed by Carl Gustav Carus – of a functional specialization of

152 François Leuret, Pierre Gratiolet, *Anatomie comparée du système nerveux considéré dans ses rapports avec l'intelligence* (Paris: J.-B. Baillière, 1839-1857), II, pp. 113-114, 262, 328.

153 Ibid., p. 298.

the three sections of the brain: an anterior organ for the "erkennenden, vergleichenden und urtheilenden Geistesleben"; an intermediate ganglion charged with impressions and feelings (*das Gemüth*); a posterior part, the seat of instincts, passion, and will (*Region des Triebes*).[154] From this premise, Gosse deduced that the development of intellectual faculties would suffer in the event of an anterior cranial compression; if on the contrary the occipital region underwent a similar operation, then the activity of the instinct would be attenuated. Such deduction could be easily brought to a comparative level:

> Les individus, les nations, les races, dont le front est normalement large et relevé, sont plus en harmonie avec le progrès de la civilisation intellectuelle et morale que ceux qui presentent des conditions opposés. De même les individus ou les nations dont la partie postérieure de la tête est proportionellement prédominante, sont plus disposés aux penchants instinctifs et aux passions irréflechies.[155]

Furthermore, it was widely believed, and Gosse agreed, that the deformations caused to the body at birth could be transmitted to a certain extent by inheritance, and then become more or less stable if the two procreative agents were subjected to the same deformations. And besides, the experiments made to create new permanent varieties of domestic animals were based on that principle.

After classifying and reviewing, thanks to a large number of sources, sixteen types of deformed heads – in various times and peoples –, Gosse speculated that a practical application could be contrived for the purpose of perfecting mankind. As a desirable consequence of the observed facts and of the resulting theory, he suggested the possibility of restoring balance, "à l'aide de la dépression occipital artificielle", between intellectual faculties and impulsive passions, in those races to which nature had given a compressed front and an overdeveloped occipital region. It seemed clear to him that, for instance, Africans and the inhabitants of the so-called Nouvelle-Hollande had a hereditary

154 Carl Gustav Carus, *Symbolik der menschlichen Gestalt. Ein Handbuch zur Menschenkenntnis* (Leipzig: Brockhaus, 1853), pp. 117-120.

155 Louis-André Gosse, *Essai sur les deformations artificielles du crâne* (Paris: J.-B. Baillière, 1855), p. 6. A few years earlier a discussion on the topic had been held during a session of the Société ethnologique de Paris, inspired by one of Morton's articles on some artificially modified American skulls: see *Bulletin de la Société ethnologique de Paris*, 1847, 1: 262-273. Among those who intervened, Achille Foville had already published a work on *Déformation du crâne résultant de la méthode la plus générale de couvrir la tête des enfans: Influence des vêtemens sur nos organes* (Paris: Madame Prevost-Crucius, 1834).

cachet of intellectual and moral inferiority, due to the structure of their brains, as well as to a lack of education. To remedy the latter, however, would have been possible only by adapting the organ of intelligence:

> Il peut être matériellement très efficace pour faire saillir en avant la voûte du crâne, et pour contrebalancer, jusqu'à un certain point, par le développement plus considérable des lobes antérieures du cerveau, le moindre volume total de cet organe, ou la prédominance, naturelle chez eux, de la partie postérieure. [...] Le redressement de la forme du nez ne présenterait pas advantage de difficultés, et pourrait amener peut-être une modification dans la face prognate. On appliquerait cette pratique immédiatement à la génération naissante, et exercée invariablement sur les deux sexes pendant plusieurs générations successives.[156]

While Gosse was thinking about how to best perform the regenerative change, another believer in the inheritance of acquired characters, Herbert Spencer – still at the beginning of his long polymath career – tried to explain why "the large-brained European differed from the small-brained savage", having from twenty to thirty cubic inches more, say, than the Papuan. In 1855, he wrote that

> The brain represents an infinitude of experiences received during the evolution of life in general: the most uniform and frequent of which, have been successively bequeathed; and have thus slowly amounted to that high intelligence which lies latent in the brain of the infant — which the infant in the course of its after life exercises usually strengthens or further complicates — and which, with minute additions, it again bequeaths to future generations.

Thus it happened that, out of savages who could not even count up to the number of their fingers, and spoke a rudimentary language, came at length "our Newtons and Shakespeares". That was all about evolution and progress, the law and cause of which Spencer was to enunciate in a famous review essay in 1857.[157]

For Ludwig Büchner, the nature of the physical and corresponding mental differences between the various human races was so firmly taken for granted that only a brief mention was required in a chapter about brain and mind of

156 Gosse, *Essai* (cit. note 155), p. 142.

157 Herbert Spencer, *The Principles of Psychology* (London: Longman, Brown, Green, and Longmans, 1855), pp. 465, 583; Id., "Progress: Its Law and Causes", *The Westminster Review*, 1857, 67: 445-485.

his *Kraft und Stoff*, one of the pugnacious *manifestos* of scientific materialism, addressed to a large public (*In allgemein-verständlicher Darstellung*). The book caused such an outcry that he was forced to resign from his position as a lecturer at the University of Tübingen. A rhetorical question introduced the topic: who had not seen, either in reality or in a picture, the narrow sloping skull of Africans, similar to that of the ape, and compared it with the noble and broad one typical of the Caucasian? No doubt that, accordingly, the brain of the Negro was smaller, more brutal, less convoluted; not to mention the inhabitants of New Holland, in whose brain the upper parts seemed almost absent. Büchner just reported the stories told by some travelers, who had compared Blacks to children, or who had blamed the worst behaviors on those living in Cuba. The chapter on *Gehirn und Seele* was opened by a quote from Justus Liebig's *Chemische Briefe*, the thirtieth one: the effects of the brain must be in relation to its mass (*Die Wirkungen des Gehirns müſſen im Verhältniß ſtehen zu der Maſſe des Gehirns*).[158] Liebig's verb (*müſſen*) is suggestive of a sort of wishful thinking, as it were, and it was just around this *vexata quaestio* – in its various and changing forms – that much search and controversy grew for a few decades.

158 Ludwig Büchner, *Kraft und Stoff. Empirisch-naturphilosophischen Studien. In allgemein-verständlicher Darstellung* (Frankfurt a. M.: Meidinger Sohn & Cie., 1855), pp. 129-130, 118.

CHAPTER 3

Climax

1 Uncertain Certainty: Paris on Stage

Recently appointed Superintendent of the Natural History Departments at the British Museum, in 1857 Richard Owen proposed to his "fellow-labourers" of the Linnean Society a new classification criterion for mammals he had been working on for a long time. The brain is – he argued – that part of the organization, which by its superior development distinguishes the *Mammalia* from all the inferior classes of *Vertebrata*, and, as such, the organ "that by its modifications marks the best and most natural primary divisions of the class". From that perspective, he outlined a list of four subclasses and fifteen orders, among which *Homo* represented a separate subclass (*Archencephala*) with a single order (*Bimana*). Man has such a peculiar brain, characterized by a series of anatomical traits – for instance the exclusive presence of the *Hippocampus minor*, or the number and depth of convolutions – to be necessarily configured as the solitary summit of a large zoological class. Peculiar mental powers would be associated with this highest cerebral type of Mammalia, oddly portrayed by Owen as two figures of the brain of a Negro.[1] As is known, the primacy of *Archencephala* soon stirred the controversial verve of Thomas Huxley, who had started to defend Darwin's theory in some speeches addressed to widely different audiences and then collected in his successful *Evidence as to Man's Place in Nature* (1863). In 1860, he had managed to refute three of Owen's statements about the human exclusivity of the third cerebral lobe, lateral ventricle, and *Hippocampus minor*, with a slew of further confrontations between the two.[2]

---

1 Richard Owen, "On the Characters, Principles of Division, and Primary Groups of the Class Mammalia," *Journal of the Proceedings of the Linnean Society*, 1858, 2:1-37, pp. 19-20. See also Id., *On the Classification and Geographical Distribution of the Mammalia, being the Lecture on Sir Robert Reade's Foundation, Delivered before the University of Cambridge, in the Senate-House, May 10, 1859* (London: John Parker, 1859). Owen's Hunterian lectures, held at the Royal College of Surgeons, "On the Comparative Anatomy and Physiology of the Nervous System," were reported in the *Medical Times* (1842-43), and already illustrated the leading modifications of the mammalian brain and their value in classification.

2 Thomas H. Huxley, "On the Zoological Relations of Man with the Lower Animals," *Natural History Review*, 1861: 67-84. As for the Owen-Huxley dispute see Charles G. Gross, "Hippocampus Minor and Man's Place in Nature: a Case Study in the Social Construction of Neuroanatomy," *Hippocampus*, 1993, 3:403-416; Leonard G. Wilson, "The Gorilla and the Question of Human Origins: The Brain Controversy," *The Journal of the History of Medicine and Allied Sciences*, 1996,

© KONINKLIJKE BRILL NV, LEIDEN, 2020 | DOI:10.1163/9789004431881_005

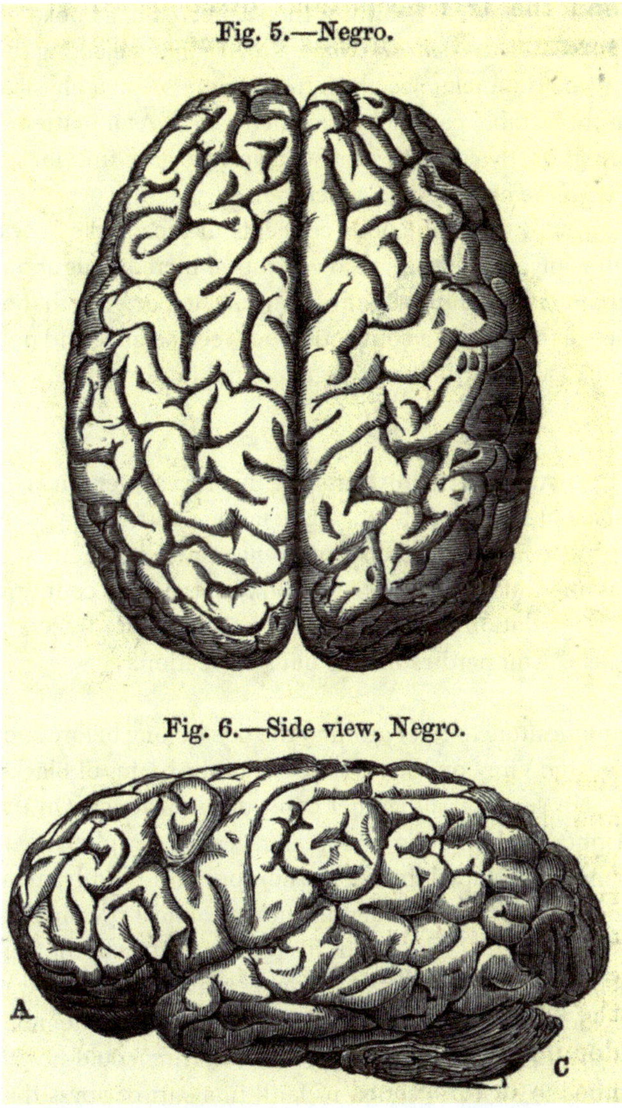

FIGURE 3.1
Two figures of a Negro brain, in Richard Owen, "On the Characters, Principles of Division, and Primary Groups of the Class Mammalia," *Journal of the Proceedings of the Linnean Society*, 1858, 2:1-37, p. 20.

While in Britain the discussion about Darwin's "one long argument" was growing, on the other side of the English Channel a relatively young *professeur agregé* of anatomy, and a surgeon in the Paris's hospitals, set up a line of research that led him to systematize a new discipline. Since 1858, Paul Broca, who at that time was skeptical of any form of evolutionism, had published

51:184-207; more in general Nicolaas A. Rupke, *Richard Owen. Biology without Darwin, a Revised Edition* (Chicago-London, The University of Chicago Press, 2009).

three *Mémoires* on the phenomenon of hybridity, then collected in a book, the second part of which focused on the *genre humain*, and began by rejecting the belief – shared by Arthur de Gobineau, Josiah Nott, and others – in a physical and moral degradation ineluctably caused by racial crossings. As a matter of fact, some of these were perfectly *eugénesiques*, while others gave inferior results, up to the infertility of the first mestizo generation.[3]

Everything revolved around the ambiguous concepts of race and species, closely scrutinized by the polygenist Broca, who was rather incredulous about the possibility of fixing the original number of human species, or even that of the current ones. Above all, the idea of a connection between science and politics had to be debunked: to affirm a different origin did not imply a racial subordination, rather

> on peut dire que la doctrine polygéniste assigne aux races inférieures de l'humanité une place plus honorable que ne le fait la doctrine opposée. Être inférieur à un autre homme soit en intelligence, soit en vigueur, soit en beauté, n'est pas une condition humiliante. On peut rougir au contraire d'avoir subi une dégradation physique ou morale, d'avoir descendu l'échelle des êtres, et d'avoir perdu son rang dans la création.[4]

In his work, Broca did not hesitate to refer again to the thesis, long before supported by Meckel, Le Cat, and Virey, on the darker color of the brains of Blacks, commenting that it did not seem an altogether constant occurrence. On the other hand, it was indisputable that in the Ethiopian races the peripheral nervous system was much more developed than in the European ones, with a relatively smaller central nervous system. A magnificent anatomical preparation at the Muséum d'Histoire naturelle highlighted this predominance of the nervous *cordons*. It was as if – according to Virey's image – the Negro's brain was poured into his nerves, increasing the animal life at the expense of the intellectual one. Among all the authorities he quoted, Soemmerring could not be missing: after insisting on his own profession of faith in a sort of abyss that separates men from apes, Broca still thought that the physical conformation of the Negro was something in between them, in several physical features. These varied greatly, between Africans and Europeans, not only on the body surface, but also in the deepest parts of their organization. However, he did not dare

3 Paul Broca, *Recherches sur l'hybridité animale en général et sur l'hybridité humaine en particulier, considérées dans leurs rapports avec la question de la pluralité des espèces humaines* (Paris: Imprimerie de J. Claye, 1860). The *Mémoires* were originally published in the first three volumes (1858-1860) of the *Journal de la Physiologie de l'homme et des animaux*.

4 Broca, *Recherches sur l'hybridité animale* (cit. note 3), p. 664.

push the argument up to the psychological phenomena, about which the partisans of slavery were spreading so many exaggerations, but it remained incomprehensible to him that those two human types, so different from each other, had been generated by a single species. On this subject, Broca, with various arguments, engaged in a dispute with his monogenist opponents.[5]

On a real fact he could not contradict them:

> C'est que les peuples les plus intelligents, les plus civilisés et les plus perfectibles, sont ceux qui possèdent en moyenne le cerveau le plus volumineux et le crâne le plus caucasique. Pour nous restreindre au parallèle des Ethiopiens et des Européens, nous dirons d'abord que les auteurs sont loin de s'accorder sur la capacité relative du crâne dans le deux types.[6]

Tiedemann and Hamilton had pretended that the difference was almost nothing, in spite of the data produced by Soemmerring a few decades before. Virey and Palisot de Beauvois had weighed nine ounces of liquid more in the Caucasian skull than in the Negro, and the results of measurements taken on specimens from the Morton collection approximately confirmed that result. For Broca this seemed revealing, since the difference in volume presumably concerned the cerebral lobes that were responsible for thought, as the animal life of the Negro, with its central organs, was similarly developed.

Broca's memories about hybridization – a fierce polygenist *plaidoyer* – aimed at invalidating the criterion that defined the zoological species by the constant success of crossings between its varieties, and were coldly received by his colleagues at the Société de Biologie. This was one of the reasons that accelerated the establishment of the Société d'Anthropologie in 1859, joined by members of the Paris medical school, the Muséum d'Histoire naturelle, and the Collège de France, as well as other professionals. Soon, a laboratory, a school, and a periodical helped to draw an institutional perimeter, within which the identity of anthropology might take a clearer shape. It was a matter of rigorously continuing the construction of the *histoire naturelle de l'homme*, a Buffonian discipline that had been rather florid in France during the first part of the century, and an essential part of which was devoted to the study of races. With 19 founding members, the Société had grown to 676 in 1880, when Broca died. His apotheosis had occurred two years earlier, at the Congrès international des sciences anthropologiques at the Palais du Trocadéro, as part of the Exposition Universelle Internationale. Broca's opening speech began by

---

5 Ibid., pp. 504-510.

6 Ibid., p. 510.

extolling that weak being who in distant times had prevailed over other species thanks to two wonderful instruments: "le cerveau qui commande et la main qui execute."[7]

One of the founders of the Société d'Anthropologie was Pierre Gratiolet, who at a session in 1861 showed the internal cast of the skull of a Mexican Totonac, illustrating the general configuration of the brain, next to that of a Caucasian. A curious and important peculiarity was evident: the second cast had only some slight undulation, without protrusions, an almost smooth surface, without any trace of convolutions. On the contrary, protuberances separated by tortuous fissures were visible on the cast of the Totonac skull. Gratiolet could not say if he had discovered a normal character of the two races, but he wondered what the meaning of that imprint was. It might seem a sign of superiority, since the brain was supposed to be all the more perfect when the cortex was more convoluted.

On second thought, the conclusion was the opposite: more developed convolutions rise to the same level, divided by fissures so narrow as to leave no marks on the bones of the skull. On its part, embryogenesis had identified a gradual complication of the cortical surface in human fetuses; when an arrest or a defect of development determined idiocy or microcephaly, the convolutions retain the typical simplicity of those of the fetus two months before birth all their life, such as to produce a depression on the inner face of the skull. An arrangement that was quite similar to that observed in the Totonac cast, such as to suggest inferiority, though that Amerindian race was known for its propensity to welcome European civilization and education, but also for its inclination to suddenly commit acts of unbelievable cruelty in the typical cold-blooded manner of savages. Gratiolet also pointed to the volume of the Totonac brain being equal to the Caucasian one, at the same time declaring himself skeptic of all measurements of cranial capacity.

The Muséum owned a skull that had presumably belonged to Descartes, religiously preserved in Sweden by a Cartesian family, later sold at auction and bought by Jöns Jacob Berzelius who had donated it to France. Gratiolet informed the members of the Société about the smallness of the philosopher's brain, admirably formed but not as big as one might have expected: "C'est un de plus beaux types de crâne caucasique qu'il soit possible de voir et, pour me résumer, je dirai que c'est la forme et non le volume qui fait la dignité du

---

7 Paul Broca, [Discours d'ouverture], in *Congrès international des sciences anthropologiques tenu à Paris du 16 au 21 août 1878* (Paris: Imprimerie Nationale, 1880), pp. 17-23, p. 17. See Jean-Claude Wartelle, "La Société d'Anthropologie de Paris de 1859 à 1920," *Revue d'histoire des sciences humaines*, 2004, 1/10: 125-171; Claude Blanckaert, *De la race à l'évolution. Paul Broca et l'anthropologie française (1850-1900)* (Paris: L'Harmattan, 2009).

cerveau."[8] Gratiolet's *communication* was followed by a discussion involving three parties. Hastily and erroneously, according to Ernest Aubertin – who played a role in the brain localization of language – the speaker had solved the question of the relationship between the development of the brain mass and that of intelligence. The latter largely depended on the former, though not entirely: had it not been found that many men of genius – except Descartes – had a huge brain? On the other hand, the functional complexity of the organ meant that its volume, alone, could not give an exact measure of the degree of intelligence. Gratiolet replied that there was certainly a limit of weight and volume below which the human brain ceases to be intelligent, but when its weight in eminent men is seen to vary from 1200 to 1400 grams, then one felt entitled to say that that variable is of little importance. Especially in the comparative study of races, one should not attach too much importance to cranial capacity and to the cerebral weight. In fact, the attitude to civilization depended on a peculiar instinct that was not related to the extent of intelligence, an instinct of which even the most mediocre of Caucasians was in possession. Thinking that intelligence could be measured by measuring brains was totally delusional.

Gratiolet's criticisms were not entirely groundless – as Broca said in his speech – but nevertheless it seemed that settling for cranial capacity was still better than having no idea about the cerebral volume. While waiting for anatomists to find a way to analyze the brains of the different races "à l'état frais", those who had studied skulls with praiseworthy perseverance were to be thanked. Both Gratiolet and Broca, although with different purposes, brought into play a recent book by Rudolph Wagner, the combative anti-materialist anatomist and physiologist, heir to Blumenbach's chair in Göttingen, who had opposed to the revolution of 1848.[9] It was the first part of a preliminary work (*Vorstudien*) on the morphology and physiology of the human brain as an organ of the soul (*Seelenorgan*). Gratiolet used it to reinforce his own thesis about the lack of constant relations between intelligence and encephalic mass. Instead, Broca drew attention to the fact that the 964 brains of which Wagner had supplied a table of weights belonged mostly to pathological cases, and

8 Pierre Gratiolet, "Sur la forme et la cavité crânienne d'un Totonaque, avec réflexions sur la signification du volume de l'encéphale," *Bulletins de la Société d'Anthropologie de Paris*, 1861, 2/1:66-81, p. 71.

9 "Adversaire des matérialistes quarante-huitards qui pensaient, avec Carl Vogt, que le cerveau sécrète la pensée comme les reins produisent l'urine, Wagner proclamait en 1854 que si les sciences naturelles déniaient l'âme immortelle, elle mineraient, du même coup, les fondements moraux de l'ordre social.": see Claude Blanckaert, *De la race à l'évolution*, (cit. note 7), pp. 148-149.

therefore not so relevant; but in that list, ranking from the heaviest to the lightest, Wagner had also included two healthy brains in third and fourth place, those of Cuvier (1861 g.) and Byron (1807 g.): "deux grands hommes, deux hommes de genie, qui marchent de front avec les premiers de notre siècle, sont ceux qui on eu le plus de masse cérébrale."[10]

Wagner's *Vorstudien* of 1860 asserted above all that the convolutions of the hemispheres (simpler and more complex, normal and pathological) represented fairly narrow variations of a constant basic type, which shared some general features with monkey brains, though different in other respects. A parallel could be found between the constant meanderings of the latter and the earlier stages of formation of the embryonic human brain, although here, too, in the developmental course of the convolutions, radical differences were apparent. After treating the phenomenon of microcephaly, Wagner argued that the measurement of the surface of the individual cerebral lobes showed no clue to a definite relationship between them and some particular psychic activities, nor to the general development of intelligence. By contrast, stronger or more numerous furrows occurred in more intelligent brains, but such a statement could only be made with great caution. At any rate, the architecture of the cortical surface, in the human and quadrumanous brain, could justify a creationist hypothesis (*Schöpfungshypothese*), against Darwin's theory, disclosed just a few months earlier. The only certainty, Wagner claimed, was that a certain development of the brain mass was necessary for the formation of that kind of intelligence that separates man from animals.[11]

The most original part of Wagner's *Vorstudien*, however, consists in their attempt to verify two physiological theorems. As for the first: mentally gifted men are supposedly characterized by a greater size of the skull cavity and a correspondingly more developed brain, especially in the hemispheres and the frontal lobes. The absolute weight of the brain, as well as the relative weight of the hemispheres to the other parts, would be more marked in them than in 'normal' people. The second theorem refers to the fact that in very intelligent men the surface had exhibited more windings and deeper furrows. An

10 Gratiolet, "Sur la forme et la cavité crânienne" (cit. note 8), pp. 76-77.

11 See the *allgemeine Ergebnisse* in Rudolph Wagner, *Vorstudien zu einer wissenschaftlichen Morphologie und Physiologie des menschlichen Gehirns als Seelenorgan. Erste Abhandlung. Über die typischen Verschiedenheiten der Windungen der Hemisphären und über die Lehre vom Hirngewicht mit besondrer Rücksicht auf die Hirnbildung intelligenter Männer* (Göttingen: Verlag der Dieterischen Buchhandlung, 1860), pp. 84-86. A second volume had as a subtitle *Über den Hirnbau der Mikrocephalen mit vergleichender Rücksicht auf den Bau des Gehirns der normalen Menschen und der Quadrumanen* (Göttingen: Verlag der Dieterischen Buchhandlung, 1862).

opportunity for a rigorous study of the question had been offered to Wagner a few years after the death of the mathematician – and his colleague at the University of Göttingen – Carl Friedrich Gauss, whose brain he could examine. He complained that no exact descriptions or drawings of brains belonging to excellent men were yet available, and concentrated on the issue, inferring that outstanding individuals did not seem to have heavier brains than ordinary ones.[12]

At a meeting held in Göttingen, September 1861, Wagner gave a general demonstration of the brain from a comparative anthropological perspective by means of some anatomical preparations and casts, partly to establish a more rigorous terminology of convolutions, on which Foville, Leuret, Hutschke and Gratiolet had already fruitfully worked. He was convinced that they were constant in the individual, and that the inter-individual differences were due to a greater or smaller folding of the cortex, which resulted in two main and extreme cerebral forms, characterized by wealth or poverty of furrows. Moreover, the less wrinkled brains should be brought back to a previous, more embryonic stage of development, like in the seventh or eighth month of life; in this respect, the simplest and scarcest of folds belonged to females and to the so-called lower races, "Neger und Hottentotten", yet the material available in the various collections was still insufficient for a reliable estimate.[13]

The 1860s were a crucial decade, when systematic studies multiplied and for a while Paris aspired to take center stage. Towards the end of 1859, Paul Broca examined the brain of a black man died at the Hôpital de la Pitié in the August heatwave that caused the first signs of putrefaction in his corpse. The cerebral pulp being too soft to allow a study of the convolutions, Broca could only observe its color, compared to that of the brain of a White who arrived at the

12 See Michael Hagner, *Geniale Gehirne. Zur Geschichte der Elitegehirnforschung* (Göttingen: Wallstein Verlag, 2004), pp. 140-154. For his published contributions see Rudolph Wagner, "Kritische und experimentelle Untersuchungen über die Functionen des Gehirns. Siebente Reihe. Über die angeblichen Verhältnisse des Gewichts und des Windungsreichthums des menschlichen Gehirns zur Intelligenz," *Nachrichten von der G. A. Universität und der Königl. Gesellschaft der Wissenschaften zu Göttingen*, 1860, *n.7*:65-80; "Notiz über das Hirngewicht von Lord Byron mit einigen Berichtigungen und Bemerkungen über das Gehirn Cromwell's, Cuvier's und Dupuytrens, als Nachtrag zur siebenten Reihe der kritischen und experimentellen Untersuchungen über die Funktionen des Gehirns," ibid, *n.12*:125-130; "Über das relative Hirngewicht der Hemisphären des grossen Gehirn's zum kleinen Gehirn und Hirnstamm mit besonderer Rücksicht auf geistige Begabung und Arbeit," ibid., *n.16*:176-180.

13 *Bericht über die Zusammenkunft einiger Anthropologen im September 1861 in Göttingen zum zwecke gemeinsamer besprechungen, erstattet von Karl Ernst von Baer und Rudolph Wagner* (Leipzig: Leopold Voss, 1861), pp. 31-32.

anatomical amphitheater on the same day. Certain points of the pia mater and the grey matter of the first brain presented a brownish color that was totally absent in the second. Once immersed in alcohol, after three days the difference in pigmentation between the two brains had become even greater. The two organs, deprived of the membranes and drained, differed by 8.3% (1003 in front of 925.5 grams): a fact that was almost insignificant *per se*, in Broca's eyes, if it had not confirmed the data already in possession of science. On that occasion, news was given of a lecture by Adolphe Gubler, to be published in full in the *Mémoires de la Société d'Anthropologie*. The eclectic physician – associate professor at the Faculté de médecine and working at the Hôpital Beaujon – dealt with the blackish color that he had found in the brains of some brown individuals of white race: "il serait peut-être avantageux de les designer sous le nom de *Négrinos* par opposition à celui d'*Albinos*, aujourd'hui généralement adopté". So, an alleged peculiarity of the black brain was somehow losing the distinctive meaning that had been associated with it.[14]

At the beginning of 1861, a great traveler of Bavarian origins, Franz Ignaz Pruner, read a *Mémoire sur les Nègres* at a session of the Société d'Anthropologie, of which he would become president in 1865. His wandering life took place between Europe, Middle East, East and Africa; in Munich he had studied medicine, in Pavia specialized in ophthalmology, and in Paris he encountered anthropology. In 1849, he was appointed personal physician of the pasha Abbas of Egypt, with the title of Bey proudly added to his surname. Pruner-Bey's *Mémoire* emphasized in detail the admirable adaptation of the black organism to its original environmental context. Several physical and moral traits of the *race nigritique* resembled those present in the fetus or child of the *race aryenne* at different stages of development; other traits, however, recalled certain characteristics that were typical of old Whites. The curve of human development and its decline were wide enough to assess the differences identified in the black race, which – as potentially close to the animal – was neither the most ferocious, nor the most dangerous, nor the most evil, but almost always the most patient and useful: "en tout cas une honorable médiocrité lui est échue en partage".[15]

In a gallery of the Muséum d'Histoire naturelle, Pruner-Bey had seen an anatomical preparation of a brain with the peculiarity observed by Soemmerring

14 Paul Broca, "Études sur le cerveau d'un nègre," *Bulletins de la Société d'Anthropologie de Paris*, 1860, *1*:53-55; Adolphe Gubler, "Sur la coloration noirâtre des centres nerveux chez le individus de race blanche, remarquables par l'abondance du pigment extérieur," *Mémoires de la Société d'Anthropologie de Paris*, 1860-63, *1*:57-60, p. 60.

15 Franz Ignaz Pruner-Bey,"Mémoire sur les Négres," *Mémoires de la Société d'Anthropologie de Paris*, 1860-63, *1*: 293-336, p. 336.

in the physical structure of the *Négre*, that is, nerves proportionally larger than the cerebral volume. He went on to offer a meticulous description of other detected differences: narrow and elongated brain, brownish tint in various of its parts, simplicity and greater hardness, shallow and flattened convolutions with respect to the "véritable labyrinthe" of the Aryan brain, similarity with that of the fetus for the rounded tip and the slightly developed posterior lobe, or to the female one due to the prominence of its parietal lobe. Regarding the weight, unfortunately too few observations had been worked out, and the extremes generally seemed too far apart to be even minimally trustworthy.[16]

The questions of weight, volume, and shape of the brain engaged the members of the Société d'Anthropologie for a while: in 1861, Broca stated that no other issue discussed by them until then was more important, so much that it deserved a thorough treatment.[17] First of all, he repeated that the direct study of the brain would be infinitely preferable to the measurement of cranial capacity, but things should be taken as they were. While waiting – as Gratiolet had called for – for zealous travelers from every part of the world to help, the investigation of skulls could still provide precious, albeit approximate, notions on the brain.[18] Broca had words of praise for the outstanding contributions given by his colleague about the degree of complication of the cranial sutures and on the period of their closure, which shed light on the psychology of human races. Nevertheless, the eulogy was followed by the criticism of two beliefs expressed by Gratiolet, especially in his recent speech on the Totonac cast: first, that the volume of the brain, in individuals and races, has almost no meaning, since the form and not the mass is in relation to intelligence; second, that the brain is unitary just like the mind of which it is an organ, and that its various parts have no differentiated attributes. As soon as such beliefs were accepted, the racial study of brains and skulls would have lost much of its interest and utility.

Broca pointed to a contradiction between the two propositions: if the development of intellectual faculties depended only on the shape of the brain, the logical consequence would be that not all cerebral regions had the same

16 Ibid., pp. 312-314.

17 Paul Broca, *Sur le volume et la forme du cerveau suivant les individus et suivant les races* (Paris: Typographie Hennuyer, 1861), also in *Bulletins et mémoires de la Société d'Anthropologie de Paris*, 1861, 2/1:139-207.

18 The skull/brain relationship will not cease to torment Broca: see, for instance, "Sur la topographie cranio-cérébrale et sur les rapports anatomiques du crâne et du cerveau," *Revue d'Anthropologie*, 1876, 5:193-248, and his "Revue critique. Topographie cranio-cérébrale. Position relative des diverses parties du crâne et des diverses parties du cerveau," ibid., pp. 278-279.

functions. Indeed, Gratiolet, in his *Anatomie comparée du système nerveux*, had divided the human races into frontal, parietal and occipital. Before clarifying if intelligence has definite relations with weight and volume, it was convenient to remind that the encephalon is not merely responsible for its higher functions; some of its structures do not participate in the elaboration of thought, so it would have been essential to separate them from the cortex and its convolutions, finally to be weighed as *cerveau pensant*. Desmoulins had been the first to discover the existence of a relationship between the extent of convolutions and the development of intelligence: for Broca too, the folds of the cortex were, *ceteris paribus*, proportional to the degree of intelligence. But why did this happen? The answer was simple: if the brain were unfolded, the mass of the gray matter that covers the convolutions would be equal to the surface, multiplied by the thickness of the cortex, which varies in each species, races and even in each individual, though with minor differences in mankind.

As to cortical complexity, Broca resorted to comparing the brain of the Hottentot Venus with the Causasian one, or the brain of a child with that of an adult, and making reference to microcephaly and idiotism was convenient too. The anatomists who attended Cuvier's autopsy had easily noticed a brain covered with complicated and deep convolutions. Too many issues were at stake: thickness, flexuosity, secondary divisions, and uneven depth of the folds; therefore using scales as a testing method could not deliver reliable results. In other words, "il ne peut donc venir à la pensée d'un homme éclairé de mesurer l'intelligence en mesurant l'encéphale". On the other hand, however, "il serait difficile de ne pas admettre l'existence d'une relation bien déterminée entre la masse du cerveau et la puissance de l'intelligence."[19] Additional evidence of this fact was the reduction in the brain, and especially in the anterior lobes, which occurs in old age. An uncertain certainty, one would be tempted to say. Howsoever, it had been sufficiently shown – by Parchappe and Huschke – that the female brain, at any age, weighed less than the male one, even considering the size of the woman's body. Broca thought it perfectly reasonable to postulate a correspondence between such relative smallness with a physical and intellectual inferiority: "il ne faut pas perdre de vue que la femme est *en moyenne* un peu moins intelligente que l'homme".[20] After all, involved in the struggles of social life, the human male is constantly urged to make efforts of intelligence, while the woman, enclosed within the boundaries of domestic life, has fewer opportunities to use her faculties.

19 Broca, *Sur le volume et la forme du cerveau* (cit. note 17), pp. 10, 16.
20 Ibid., p. 15.

Broca did not attach much importance to the *immense tableau* that had recently been compiled by Wagner, with its list of 964 cases: disparate data on heterogeneous brains, normal and pathological. In the past, brains of remarkable men had often been inspected, but rarely weighed. The German anatomist had presented a few specimens of that kind, characterized by an equally remarkable weight: even though Gratiolet was aware of this, he still thought that no relationship existed between the encephalic weight and the development of intelligence, even adducing the example of Descartes' quite small skull. A respectable relic – Broca commented – but of dubious authenticity and, above all, difficult to evaluate. Again, for him that relationship was undeniable, even if not strictly definable:

> Certes, jes suis bien loin d'en conclure qu'il y ait un rapport rigoureux entre le développement de l'intelligence et le poids de l'encéphale. Les conditions qui font varier ce poids en plus ou en moins sont complexes comme l'encéphale lui-même, et on ne peut s'attendre à trover une solution simple pour un problème où des éléments aussi divers sont étroitement combinés. Mais ce qu'on peut dire, c'est que le développement de l'intelligence est une des conditions qui exercent le plus d'influence sur le point de l'encéphale, et réciproquement.[21]

The relationship between those two elements in the human races presented no less difficulty. As in the case of the female brain, Broca decided to start from an assumed truth, namely the superiority of Europeans over Africans, American Indians, and the Blacks of Oceania. Prognathism and early closure of the sutures were well known phenomena in the inferior races. Unfortunately, anatomists had mostly Caucasian brains available by then, and the gap concerning other races would be hardly filled, not to mention that the different structure of their brains – thickness of the membranes, volume of the sinuses, amount of the cephalic-rachidian liquid – caused the ratio between cranial capacity and encephalic volume to vary significantly. Quite misleading Broca deemed the procedure followed by Tiedemann, who had filled a few dozen skulls with millet-seeds and then weighed the amount contained in each of them, pleased to announce that the brain volume was equal in all races.

He thought a better technique had been devised by Morton, who had used small lead shots to measure cranial capacity and collected data on 663 skulls of different races: if 100 was the average volume of the brain of Blacks, the figure was 111.8 for Germans, a result similar to that reached by Virey and Palisot de

21 Ibid., p. 31.

Beauvois more than forty years earlier. If one descended to the lowest degree of the human scale, and compared the unhappy Australian race to the European, the respective skulls would be in a proportion of 100 to 124.8. So, Africans came to occupy a position halfway between Australians and Europeans: their figures were a very accurate match for the intellectual hierarchy of the three races, though this did not mean that Broca agreed with the assumption of an absolute relationship existing between intelligence and cranial capacity. *Toutes choses égales d'ailleurs* – his often-repeated cautionary clause –, an indubitable relationship had to exist between the two elements, the material and the mental.

In this regard, Broca gave some credit to the scientific reform triggered by Gall, who had raised awareness of the great principle of localizations, albeit wrongly applied: an ingenious system built on too fragile foundations to resist the shock of criticism. Moreover, Gall and Spurzheim, though first-rate anatomists, had overlooked the comparative study of convolutions. In his own days – Broca stated – a few followers of phrenology were still around, but no longer among men of science. On the other hand, the principle of localizations was thriving and inspired research into the particular functions performed by the wide range of cerebral organs, for example the five lobes of each hemisphere, which in the human species could be further split in numerous groups of folds. In the apparent disorder of convolutions, a surprising regularity had been discovered, which reduced the *hasard*, inspiring a typically positivistic metaphor: "c'est un dieu de passage, qui recule comme tant d'autres à mesure que nous avançons, et là où nous découvrons l'ordre, nous pouvons affirmer la loi".[22]

Broca's speech to the Société d'Anthropologie could not but open a lively discussion, which lasted for a few sessions – from February to June 1861 – and engaged several parties.[23] Gratiolet defended himself from the accusation of having fallen into a contradiction, and then tried to explain his position. Again, he recalled Wagner's thesis in support of there being an uncertain relationship between encephalic weight and intelligence: a thesis rejected by Broca and others in the ongoing debate. Above all, Gratiolet insisted on the scarcity of facts, which made it premature and imprudent to found any theory, but even more than that, he ridiculed the worthlessness of weight as an evaluation criterion:

22 Ibid., pp. 53-62.

23 "Discussion," *Bulletins de la Société d'Anthropologie de Paris*, 1861, 2/1:204-233; "Reprise de la discussion sur le volume et la forme du cerveau," pp. 238-279, 283-322, 421-449.

> Nous aurions des intelligences de 1000 grammes, de 1500 grammes, de 1900 grammes; malheureusement, cela n'est pas tout à fait aussi simple. Pensez-vous, par example, que, pour apprécier la valeur réelle d'un mineral d'or ou d'argent, on pût se contenter de le peser? Non certainement: l'analyse chimique interviendrait d'abord; elle séparerait les éléments plus ou moins hétérogènes du mineral, l'emploi de la balance ne serait invoqué qu'au debut et à la fin de l'expérience [...] Mais, quand'il s'agit du cerveau, quelle analyse a précédé nos pesées? [...] Je m'élève donc, messieurs, et de toute la force de mes convictions, contre les conclusions qu'on voudrait tirer des pesées encéphaliques en masse.[24]

As for absolute weight, the biggest brains would not necessarily be the smartest: think of the elephant that, despite its hemispheres being three-times heavier than the human ones, with considerably complicated convolutions, does not speak and has no self-consciousness. Precisely because it was a different "type" of brain, weight was not everything, and "type" was not weighable.

Speaking instead of relative weight, its ratio to the weight of the body, as proposed by Cuvier, was not enough: after all, "c'est l'enfance de l'art". It also involved embarrassing consequences, such as an elephant being inferior to a shrew, or a man to a squirrel monkey. Soemmerring had rightly observed more voluminous nerves in Blacks than in Europeans, and had warned that the "king of brains" should not be separated from its empire; therefore, for the sake of consequency, the marrow and the nerves should also be weighed. Moving a step further, Gratiolet urged to shift attention to the cerebral form and its natural kind of perfection, with a view to establishing a sort of aesthetics and physiognomy. A form is perfect when it expresses the fulfillment and natural completion of a harmonious development; or, in other words, the beauty of a visible thing is nothing but perfection made intelligible. From this point of view, the more developed the brain, the more the parts above the Sylvian fissure prevail over the lower ones.

This principle also applied to the embryological dimension since, in the transition from infancy to adulthood, the path of cerebral growth proceeds from the occipital to the front, and it was no less applicable to human races. The most abject of them – "les Australiens, les Alfouroux, les nègres inférieurs" – suffered from occipital dolicocephaly, while in Asia, America, in the Oceanic archipelagos, Gratiolet saw parietal races dominating in activity and intelligence. Finally the white races were conspicuous for their frontal eminence, although the measure of this relative development required the most meticulous

24 Ibid., pp. 250-251.

precautions. However, none of this allowed anyone to propose manipulative practices – as Gosse had done, supported by Broca. In fact, compressing the occipital would only have moved the parietal forward and, subsequently, the frontal to the point of atrophying it, the dignity of which was undoubted: "la fleur du cerveau". But if form was above weight, Gratiolet placed "l'énergie vitale, la puissance intrinsèque du cerveau" even higher up.[25]

Causes hard to define, irreducible to measure and regulated by hidden laws, affected mental performances, to such an extent that "la météorologie du microcosme cérébral" might remain in profound darkness forever. It was, so to speak, a magnetic organ, its substance guided by unknown currents and communications, capable of contributing to a common goal. Final causes often peeked out of the anatomist's discourse. Gratiolet conceded that, on average, the finer a race, the higher it rises away from its fetal forms, the more complicated and magnified its brain. In a randomly-selected group of one thousand white people, there was certainly more brain than in a thousand Bushmen. Nevertheless, if two individuals belonging to these two different races had the same cranial capacity, nothing could be inferred about their similarity: even with the same brain weight, the White would remain White and the Bushman a Bushman. It was known that, in the latter, the degree of complication of the gyrus was lower, and yet sufficient for his own degree of perfection; instead, for a White, such a simplified form would result in a degrading imperfection.

Stephen J. Gould has commented that Gratiolet was "a confirmed monarchist, no egalitarian", who merely sought most effective ways to affirm the superiority of white European males.[26] Even more than that, he professed to believe in the existence of the soul, without pretending to touch it by hand: not a phenomenon, but a primary condition of existence, a mystery that escapes the senses and the imagination. In vain the body and the brain were studied, the scalpel would never explain the soul. Then, as a declared spiritualist, Gratiolet had no interest in fighting the brain localizations, since the more distinct the organs appeared, the more striking the indivisibility of the immaterial self. Though sharing Pierre Flourens' position on the unity of the brain, he was nonetheless inclined to favor certain groups of convolutions, whose special development gave man his excellence, i.e. "le lobe frontal, dans le quel réside, en quelque sort, la majesté du cerveau humain".[27]

---

25 Ibid., p. 258.

26 Stephen J. Gould, *The Panda's Thumb. More Reflections in Natural History* (New York-London: W. W. Morton & Company, 1980), p. 145. See also the whole chapter on "La mesure de l'intelligence" in Blanckaert, *De la race à l'évolution* (cit. note 7), pp. 143-174.

27 "Discussion," (cit. note 23), p. 274. At the end of the century, in an encyclopedic article on the brain, Léonce Manouvrier thus interpreted the dispute between Broca and Gratiolet:

Shortly thereafter, Constant Sappey – an anatomist who made the lymphatic system the main object of his research – talked at a session of the Société de Biologie on cranial and cerebral differences between man and woman. On that subject, he listed three possible opinions: in agreement with Aristotle, Parchappe believed in a larger volume of the male brain, as opposed to Soemmerring. Most of the anatomists and physiologists, however, abstained from taking up a position and remained doubtful. For his part, Sappey had undertaken a study of 32 individuals, equally distributed between men and women and randomly chosen among those who had been brought to the École pratique. However small, the sample allowed him to affirm – and to show in tables of figures – that both the cranial capacity and the corresponding encephalic weight were greater in men than in women. Moreover, the average weight difference (102 grams) was for the most part due to the brain (94 grams), while the cerebellum, the protuberance, and the bulb did not vary much between the two sexes. On the other hand, Sappey was also keen to point out that inter-individual differences were so wide that they overruled sexual ones:

> bien que l'encéphale soit plus volumineux dans le sex masculin, il faut donc admettre comme un fait également démontré qu'un très grand nombre de femmes peuvent avoir et ont en effet une masse encéphalique d'un volume supérieur à celui de beaucoups d'hommes.[28]

2 An Intense Decade

In 1862, the Harvard anatomist Jeffries Wyman had a chance of dissecting the corpse of a nearly adult Hottentot, who had killed himself in Boston. Unusually tall for his race – 5 feet and 5 inches –, his cranium was somewhat flattened, with a narrow forehead sloping outwards and backwards from a somewhat prominent ridge corresponding to the obliterated frontal suture. Among all the other information on the skeletal system, a short paragraph

---

"le premier croyant que son adversaire refusait toute signification physiologique au poids du cerveau, le second croyant que Broca sacrifiait au poids la *forme* du cerveau. En réalité Broca était loin de méconnaître l'importance physiologique de la forme du cerveau, mais Gratiolet avait peut-être le tort d'introduire dans la question un troisième élément, tout metaphysique, dont nous pouvons accomplir, aujourd'hui, la materialization." See *Dictionnaire de Physiologie par Charles Richet* (Paris: Félix Alcan, 1897), II, p. 671.

28 Constant Sappey, "Recherches sur le volume et la capacité du crâne, sur le volume et le poids de l'encéphale, comparés chez l'homme et chez la femme, lues à la Societé de Biologie, dans sa séance du 18 mai 1861," *Comptes rendus des séances et Mémoires de la Société de Biologie*, 1862, *3*: 109-120, p. 117.

dealt with the interior of the cranium, where the most striking feature was the narrowness and diminutive size of the fossae that accomodated the anterior frontal lobes. Although the brain weighed 3 lbs. 2 oz. – about the average of a European one – the olfactory fossa seemed to Wyman particularly deep, at the expense of other parts. The foramen magnum was rather undersized, differing from Cuvier's estimate of the same part in the Hottentot Venus, the magnitude of which would have indicated, in his opinion, an inferior nature.[29]

Therefore, in any way examined the general subject, one often had to confess to one's doubts and failure to reach a definite conclusion. Only a militant propensity could lead one to proclaim truths, which did not seem to be provable from a strictly scientific point of view. That was the case of the Anthropological Society of London, founded in 1863 by the explorer, orientalist, and literary polymath Richard Francis Burton in collaboration with the young James Hunt, who became its first president. Until then, the latter had continued his father's specialty as a speech therapist and served as a secretary of the Ethnological Society, which had its roots in Quaker and Evangelical philanthropy. Its humanitarian orientation began to weaken after the admission of new members – including Hunt – who shared radical convictions. Toward the end of the 1850s, the times were rapidly changing, and, after the outbreak of the American Civil War, Europe's public opinion took sides with either factions.

In his introductory and presidential address to the newborn Anthropological Society of London, the "sister" of the Parisian one, Hunt tried to distinguish anthropology, as "the science of the whole nature of Man", still in its infancy, from ethnology, which would be "the history or science of nations or races": the former one covering, therefore, a wider field with greater ambitions. It is no accident, though, that the first example he made was a racial one:

> Whatever may be the conclusion to which our scientific inquiries may lead us, we should always remember, that by whatever means the Negro, for instance, acquired his present physical, mental, and moral character, whether he has risen from an ape or descended from a perfect man, we still know that the races of Europe have now much in their mental and moral nature which the races of Africa have not got. We have hitherto devoted our attention almost exclusively to physical Anthropology, which

29 Jeffries Wyman, [no title], *Proceedings of the Boston Society of Natural History*, 1862-1863, *9*:352-357. His report was also republished with the title "Observations on the Skeleton of a Hottentot," *The Anthropological Review*, 1865, *3*:330-335.

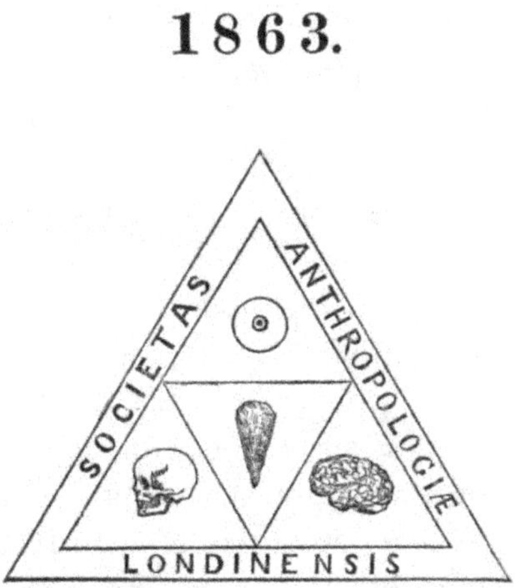

FIGURE 3.2
Logo of the Anthropological Society, in *The Anthropological Review*, 1, 1863, frontispiece.

> Blumenbach first founded. We now require to investigate the mental and moral characteristics of mankind generally.[30]

In addition to the evident physical differences between the European and the African – skin, hair, different color of the white matter of the brain and of the pia mater, etc. – the psychological ones had yet to be thoroughly studied.

Hunt hoped that the new Society would never prostitute itself to the point of supporting the slave trade, with all its abuses; but, at the same time, "we must not shrink from the candid avowal of what we believe to be the real place in nature or in society, of the African or any other race." In particular, the metaphysician and others had attempted to prove the logical necessity of the unity of mankind; but was the question of origins to be solved metaphysically? Hunt saw the fact that Prichard's old works were still the textbooks of the day as no little disgrace to British anthropology. That unity represented such an article of faith with so many respectable people that he did not intend to discuss it, although he said out loud that the theologian had no right whatsoever to

30 James Hunt, "Introductory Address on the Study of Anthropology, Delivered before the Anthropological Society of London, February 24th, 1863," *The Anthropological Review*, 1863, *1*:1-20, p. 3. The triangular logo of the Society, which appeared only on the first year of the periodical, included a skull and a brain among its symbols (see Fig. 3.2). On the reasons for the split that gave birth to the Anthropological Society and subsequent events see George W. Stocking, Jr., *Victorian Anthropology* (London: The Free Press, 1987, pp. 238-273. Efram Sera-Shriar has proposed the idea of "two competing intellectual camps" that vied for control of the nascent discipline relating to human diversity: see his "Observing Human Difference. James Hunt, Thomas Huxley and Competing Disciplinary Strategies in the 1860's," *Annals of Science*, 2013, *70*:461-491, and *The Making of British Anthropology, 1813-1871* (London: Pickering & Chatto, 2013), pp. 109-145.

interfere in scientific achievements. To get facts straight, on such matter Broca had maintained something important – the choice of the accomplice is quite significant – but it was still "in a most unsettled and unsatisfactory state". While refusing phrenology "as a system", Hunt also said in his address that nearly everyone was willing to admit a relationship between form and quality of the brain on the one hand, and the intellectual and moral functions on the other. Many obscurities and contradictions still surrounded the subject: in the minute histological anatomy, or in the chemical constituents of the brain a solution had to be sought.[31]

Not many months passed from that introductory speech, and Hunt already felt the urgency of dealing with the Negro's place in nature. At a meeting of the British Association for the Advancement of Science at Newcastle-upon-Tyne, before a "miscellaneous audience", his statement of a few simple facts was received "with such loud hisses that you would have thought the room had nearly been filled with a quantity of Eve's tempters instead of her amiable descendants".[32] He again presented his further elaborated propositions to the "scientific brethren" of the Anthropological Society, receiving their "cordial and earnest support".[33] What had somehow shocked the non-specialized audience of Newcastle-upon-Tyne was the confinement of the typical woolly-headed Africans into a separate species of the genus *Homo*, due to their physical, mental, and moral characters. The list of distinctions drawn up by Hunt did not overlook any aspect, starting from their height, "less than that of the European", and their heavier skeleton, the thorax and the limbs, through to the flat and low forehead, with a facial angle sometimes reduced up to 65 degrees. Gratiolet was quoted for having shown Negroes' comparatively smaller brain volume and premature joining of the cranial bones, which might entail an arrested development of their mind. Needless to say, Tiedemann's investigations of 1836 were dismissed by Hunt as unsatisfactory and not warranted by the facts that had come to light in the meantime. Numerous other authorities – Lawrence, Morton, Huschke, Pruner-Bey etc. – helped corroborate Hunt's belief that "the Negro brain bears a great resemblance to a European female or child's brain, and thus approaches the ape far more than the European, while the Negress approaches the ape still nearer."[34] Little was yet known as to its

31 Hunt, "Introductory Address" (cit. note 30), pp. 14-17.

32 The funny simile is in the dedication to Richard F. Burton, cofounder of the Anthropological Society: see James Hunt, *On the Negro's Place in Nature* (*Read before the Anthropological Society of London, November 17th, 1863*) (London: Trübner & Co., 1863), p. vi.

33 For a synthesis of Hunt's paper and the following, long discussion see *Journal of the Anthropological Society of London*, 1864, 2:xv-lvi.

34 Hunt, *On the Negro's Place in Nature* (cit. note 32), pp. 16-17.

chemical constituents, even though Broca had recently confirmed the "truth" of the old statement about a "smoky tint" of the whole Negro's brain, and a pia mater containing brown spots. Finally, Hunt referred to the shared idea about the lesser complexity of its convolutions.

Almost nothing original was contained in the first part of Hunt's text, which focused on the anatomical and physiological peculiarities. In particular, Broca and Nott had already belittled the assumed unity of the human species, based on the fertility of the racial hybrids, by arguing that mixed offspring would not be indefinitely prolific. Not to mention the deep immorality often attributed – as Hunt did – to mulattos and others of mixed blood. From a psychological point of view, the statement that the Negro would only require an opportunity for becoming civilized seemed to him patently disproved by history. Black children were indeed precocious, but no advance in education could be made as soon as they arrived at the age of maturity, and any individual exceptions to the rule would be due to some European blood in their veins. The second part was also a jumble of anecdotes, impressions of travelers, and opinions aimed at demonstrating the thesis of an irremediable inferiority. Hunt thought it possible, by observation and experiment, to determine the exact place that the Negro race should hold, while it was absurd and chimerical to try to give it a better one. Despite the initial declaration of neutrality on the question of slavery, no wonder the American edition of his book was introduced by John H. van Evrie, the doctor turned into a white supremacist, a journalist and publisher in New York, who warmly recommended to read it.[35]

Also from the first volume of *Memoirs*, a paper followed Hunt's inaugural speech, where the cardiologist Thomas B. Peacock illustrated the results of his weighing four brains of blacks, with the proviso that their small number prevented any significant statistical evaluation. Furthermore, in forming an estimate of the relative weight in the Negro and in the European, the variation of brain size in the two sexes, at different times of their life, made it mandatory to compare its parts in persons of the same age. At any rate, the few observations offered by Peacock generally tended to support what Hamilton and Tiedemann had thought about the lack of a very marked difference between the ordinary size of the brain in the African and the European; nevertheless, they also indicated that the former was a little smaller than the latter. And here too, the need for an extended series of observations was emphasized.[36]

35 James Hunt, *The Negro's Place in Nature: A Paper Read before the London Anthropological Society* (New York: Van Evrie, Horton & Co., 1864). Van Evrie had written in 1853 a pamphlet with the eloquent title *Negroes and Negro Slavery. The First, an Inferior Race – The Latter, its Normal Condition* (Baltimore: John D. Toy, 1853).

36 Thomas B. Peacock, "On the Weight of the Brain in the Negro," *Memoirs read before the Anthropological Society of London*, 1963-64, *1*: 65-71. Peacock had previously worked out

The year in which Hunt inaugurated his presidency of the Anthropological Society by professing a staunch polygenist creed, saw also the release of one of the major contributions to the early Darwinian debate, over five hundred pages that required three editions in a few months. The name of its author, Charles Lyell, and the subject, the *Antiquity of Man* – just when Darwin was still silent about it – sufficiently explain the uproar aroused by the work. Together with Huxley's *Evidence as to Man's Place in Nature,* it set knowledge about human fossils into a framework and, in the author's career, it marked two memorable "conversions", as he stated himself. In the ninth edition of the *Principles of Geology* (1853), Lyell had expressed, once again, his skepticism at the occurrence of human remains or stone tools in alluvial deposits, due to the high chances of error. In 1859, he went to visit Jacques Boucher de Perthes' sites and collections at Abbeville, and came to regard the Pyramids of Egypt as "things of yesterday" in comparison with the North-French relics. The second "conversion" had to do with the controversial ideas of the "transmutation" of species, which – "so far from having a materialistic tendency" – Lyell finally accepted after long hesitation, though he did not give up on the teleological picture of an "ever-increasing dominion of mind over matter".[37]

In presenting and discussing the fossil skulls of the Neanderthal and Engis Caves, Lyell noted how it was now admitted that

> the differences between the brain of the highest races of man and that of the lowest, though less in degree, are of the same order as those which separate the human from the simian brain.[...] The brain is somewhat less voluminous on the average in the lower races of mankind, its convolutions rather less complicated, and those of the two hemispheres more symmetrical, in all which points an approach is made to the simian type. It will also be seen, by reference to the late Dr. Morton's works, and by the foregoing statements of Professor Huxley, that the range of capacity between the highest and lowest human brain is far greater than that between the highest simian and the lowest human brain.[38]

---

quantitative data, also concerning the encephalon of white people: see his "Tables of the Weights of the Brain and of some of the Organs of the Human Body," *The Monthly Journal of Medical Science*, 1846-47, 7: 101-110, 166-178.

37 So Lyell wrote to the Hispanist George Ticknor on January 9th, 1860: Katharine Lyell, *The Life, Letters and Journals of Sir Charles Lyell* (London: John Murray, 1881), II, p. 330. On his attitude towards Lamarck and Darwin see Chapters XX-XXIV of *The Geological Evidences of the Antiquity of Man with Remarks on Theories of the Origin of Species by Variation* (London: John Murray, 1863), pp. 385-506.

38 Ibid., pp. 90-91.

Lyell, who was not an anatomist, merely echoed a fairly widespread conviction, which someone completely unlike him, the German Carl Vogt, was striving to divulge at that time. A former student of Liebig, he had earned his medical degree in Bern and began to teach zoology at the University of Giessen. His active participation in the revolutionary events of 1848 forced him to move to Geneva, where he taught geology and became a naturalized citizen. Throughout his life, he never tired of preaching an atheistic and materialistic doctrine, which also permeates the large volume of his popular *Vorlesungen über den Menschen*, issued in the Canton of Neuchâtel, then edited by James Hunt in 1864 for the Anthropological Society of London. A German edition had been published the year before, and a French one followed in 1865, widening its potential public readership.[39]

Hunt dedicated the English translation to his friend Broca, whom he thanked for having rendered "incalculable services" to anthropology, "the noblest of all sciences", which included ethnology as a sub-discipline. That was one of the reasons – according to him – that led to the foundation of the new Society, soon overcrowded by nearly five hundred members, and its taking distance from the previous, Ethnological Society, which had the "strange idea" of admitting females to its discussions: a "fatal mistake" capable of hindering the promotion of research. Hunt knew that Vogt's tone might occasionally be offensive, so that, in agreement with him, some passages had been omitted or moved to the appendix, but after all "the Fellows of the Anthropological Society of London are happily neither women nor children", and the editor decided not to encumber the work with notes, though not sparing some criticism in his preface. The two contexts were different: in Germany men of science and theologians looked upon one another with contempt, while in Britain mutual respect prevailed.[40]

In his third Lecture, Vogt complained that most recent racial depictions, whether of living men or of skulls, had little or no scientific value, being very often just caricatures unconsciously performed by the artist. Only two aspects could be relied upon, to get a good image of the human head: the profile and the full front view. In keeping with a widespread rhetoric of objectivity, the

39 Carl Vogt, *Lectures on Man: his Place in Creation, and in the History of the Earth, edited by James Hunt* (London: Longman, Green, Longman, and Roberts, 1864); *Vorlesungen über den Menschen, seine Stellung in der Schöpfung und in der Geschichte der Erde* (Giessen: J. Ricker'sche Buchhandlung, 1863), 2 vols.; *Leçons sur l'homme, sa place dans la creation et dans l'histoire de la Terre* (Paris: C. Reinwald, 1865). For an overview of his activity see *Carl Vogt. Science, philosophie et politique (1817-1895). Actes du colloque de mai 1995 édités par Jean-Claude Pont (et al.)*, (Genève: Georg, 1998).

40 Vogt, *Lectures on Man* (cit. note 39), pp. v-xvii.

photographic medium, in which the ray of light ("and not the erring hand of man") takes the picture, seemed to be the only way to get round one of the greatest obstacles to the study of anthropology and to finally compare objects that are separated in time and space.[41] Geometrical drawing was a necessary technique as well, preferably using the method developed by the anatomist Johann Christian Gustav Lucae, professor at the Senckenbergische Institut in Frankfurt am Main, where his collection of skulls was housed. In the form of a letter to Carl Ernst von Baer, he had tried to fulfill at least part of the need for a comparative *ethnographische Kraniologie*, as pointed out by the famous biologist. To accomplish something dependable, Lucae believed that two circumstances should be taken into account, namely: firstly, more importance had to be attached to getting correct images, so that scattered materials could be made available to everyone; secondly, cranial studies should be extended and carried out more thoroughly than in the past. What the perspective image is to art – as stated by Lucae – the geometric one is to science: the former delivers the object as it appears, the latter makes it as it is. Twelve plates with drawings of different populations and individuals illustrated his method.[42]

Vogt believed that the various cranial measurements taken by compass, tape, and rule, could give but an imperfect idea, and only some of the main dimensions of the head. Other means were to be contrived to measure the internal capacity of the skull, with a view to inferring something about the development of the brain and its components, which was what really mattered. Procedures for evaluating its volume, using different substances, had been devised by Tiedemann, Morton, Huschke; a dispute occurred as to the most suitable substance for making casts of the brain, by filling the cranial cavity to preserve its shape. Vogt himself made wax models of a Carib, a Cossack, etc., Wagner had used gypsum, Lucae glue, but the best substance would be a metallic alloy that melts at the boiling point of water and delivers not just the outer shape, but the weight as well, since its specific gravity is known.[43]

Many attempts had been made to connect the development of the several cerebral lobes with the mental qualification of individuals and races, but – Vogt commented – "with very scarce success". If, by taking advantage of their anatomical peculiarity Gratiolet classified three racial types (frontal, parietal, occipital), not long before Carus had already gone too far with his four *Menschheitstämme*, speculatively connected to different brain structures. In the

41 Ibid., pp. 74-75.

42 Johann Christian Gustav Lucae, *Zur Morphologie der Rassen-Schädel. Einleitende Bemerkungen und Beiträge. Ein Sendschreiben an Carl Ernst v. Baer. Mit zwölf Tafeln* (Frankfurt a.M.: Heinrich Ludwig Brünner, 1861).

43 Vogt, *Lectures on Man* (cit. note 39), pp. 78-80.

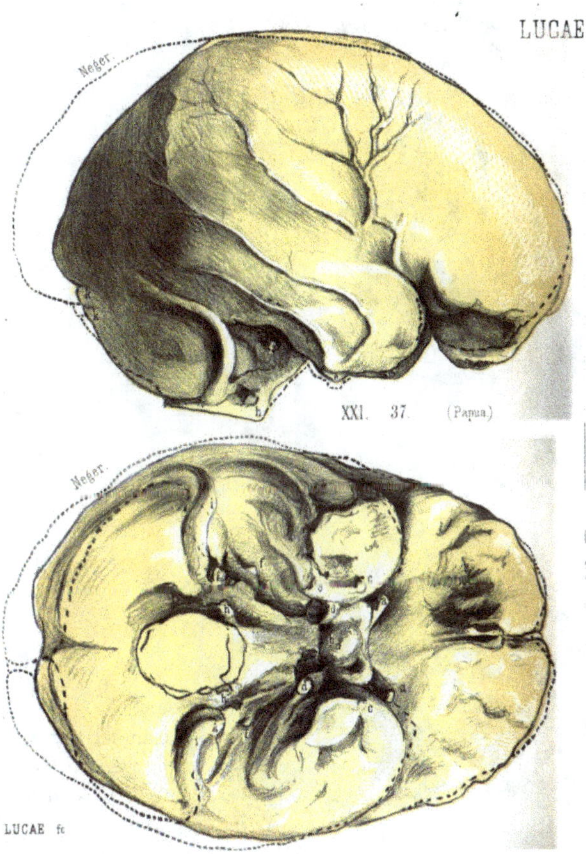

FIGURE 3.3
Skull of a Papua compared to the profile of a Negro skull, Tafel XXI. 37, in Johann Christian Gustav Lucae, *Zur Morphologie der Rassen-Schädel. Einleitende Bemerkungen und Beiträge* (Frankfurt am Main 1861).

attempt to estimate of intellectual capacity, the actual facts seemed only to point to an observation of the frontal lobes and of the convolutions. As, among the superior apes the orangutan, gorilla, and chimpanzee represented three distinct types, so it happened for mankind, and Vogt selected two extreme human groups – German and Negro – to study and compare their characteristics. Because of a whole series of morphological differences, the "simiousness" (*Affenähnlichkeit* in the original text) of the Negro's skull was evident, no less than many other bodily characters.

Once again, after many decades, both Soemmerring's discovery of the quantitative nerve-brain ratio and Meckel's insistence on a darker color in the cerebral substance were referred to as still valid. Not personally owning any Negro's brain, Vogt reported what Huschke and Gratiolet had added to acquired knowledge, re-proposing also the idea of an early closing of the sutures – an alleged obstacle to full cerebral development – and the visual comparison between the profile of two brains, that of the Hottentot Venus and that of Gauss.

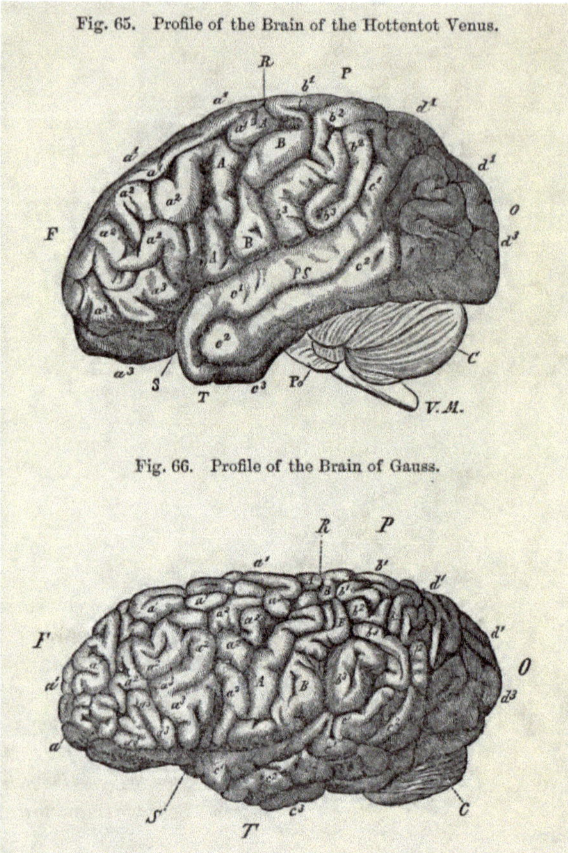

FIGURE 3.4
Fig. 65-66, in Carl Vogt, *Lectures on Man: his Place in Creation, and in the History of the Earth, edited by James Hunt* (London 1864), p. 184.

Moreover, a couple of years earlier, the anatomist Hermann Welcker had stressed the undeniable disparity between the male and female brains, arguing that it would increase with the hierarchical position of the race, so that the male European excelled the female much more than the Negro the Negress. For Vogt, if the statement were proved to be correct, it would mean that the inequality between the sexes grew with the progress of civilisation.[44]

The scantiness of information on the brain of the various races induced John Marshall, surgeon at the University College London, to ask several medical friends living in the British colonies for specimens. In 1863, an entire head

44 Ibid., pp. 171-186, 81-82, 191-194. See Hermann Welcker, *Untersuchungen über Wachsthum und Bau des menschlichen Schädels. Erster Theil. Allgemeine Verhältnisse des Schädelwachsthums und Schädelbaues, normaler Schädel deutschen Stammes* (Leipzig: Wilhelm Engelmann, 1862).

of a Bushwoman, duly prepared according to certain instructions, was sent to him by a former pupil practicing in Cape Town. Shipped in a tin case filled with spirit, the long journey had not spoiled it, and no decomposition was apparent in any part. Having secured a plaster cast and four photographic views of it, Marshall proceeded to expose the left side of the brain, to dissect off its membranes and to take other photographs, "the size of nature", of the successive stages in the removal of the structures. The right half, instead, was weighed and left intact, to serve as a museum specimen or for further investigation. Its weight was added to the weighing results of the already separated parts of the left hemisphere to obtain the total weight. The two halves differed only by 20 grains, the right side being the heavier one, with part of the choroid plexus attached to it. Fissures, lobes, convolutions, commissures, ventricles and ganglionic masses were then studied on the preserved brain and compared with that of the Hottentot Venus, and with those of higher Apes.

After measuring and describing the head and skull in detail, Marshall stated that the entire encephalon of the Bushwoman, hardened in spirit and deprived of its membranes, weighed 21.77 oz. However, weight loss due to the conservation in spirit and the removal of membranes had to be considered, so that it presumably reached 31.5 oz., a decidedly low weight compared to those usually registered in European women. Cerebrum and cerebellum were about equally defective, in a proportion of about 0.78 to 1, "without claiming for these numbers a perfect accuracy". As for qualitative data, the cerebrum presented a long and narrow ovoid form when viewed from above; laterally, the parietal region appeared salient, while the vertex was low and flattened, its highest point being placed far back; shallow the frontal region, narrow the temporal lobe. After recounting many other meticulous observations, Marshall felt he could conclude that the Bushwoman's cerebrum was small but long, defective in width, and especially so in weight; that its outlines and surfaces were angular and flat, instead of rounded and full as in the European. Her cerebellum showed to be most prominent at the sides, and proportionally wider and longer: "quite human and not ape-like in shape", though.[45]

Fissures and convolutions displayed other important differences. The primary fissures far simpler than in the ordinary European brain, although more complex than those shown in Gratiolet's images of the Hottentot Venus: in this greater simplicity and in the more vertical direction of the Sylvian fissure, the Bushwoman's brain approached somewhat the quadrumanous characters.

45 John Marshall, "On the Brain of a Bushwoman; and on the Brains of two Idiots of European Descent," *Philosophical Transactions of the Royal Society of London*, 1864, *154*: 501-558, p. 508-510.

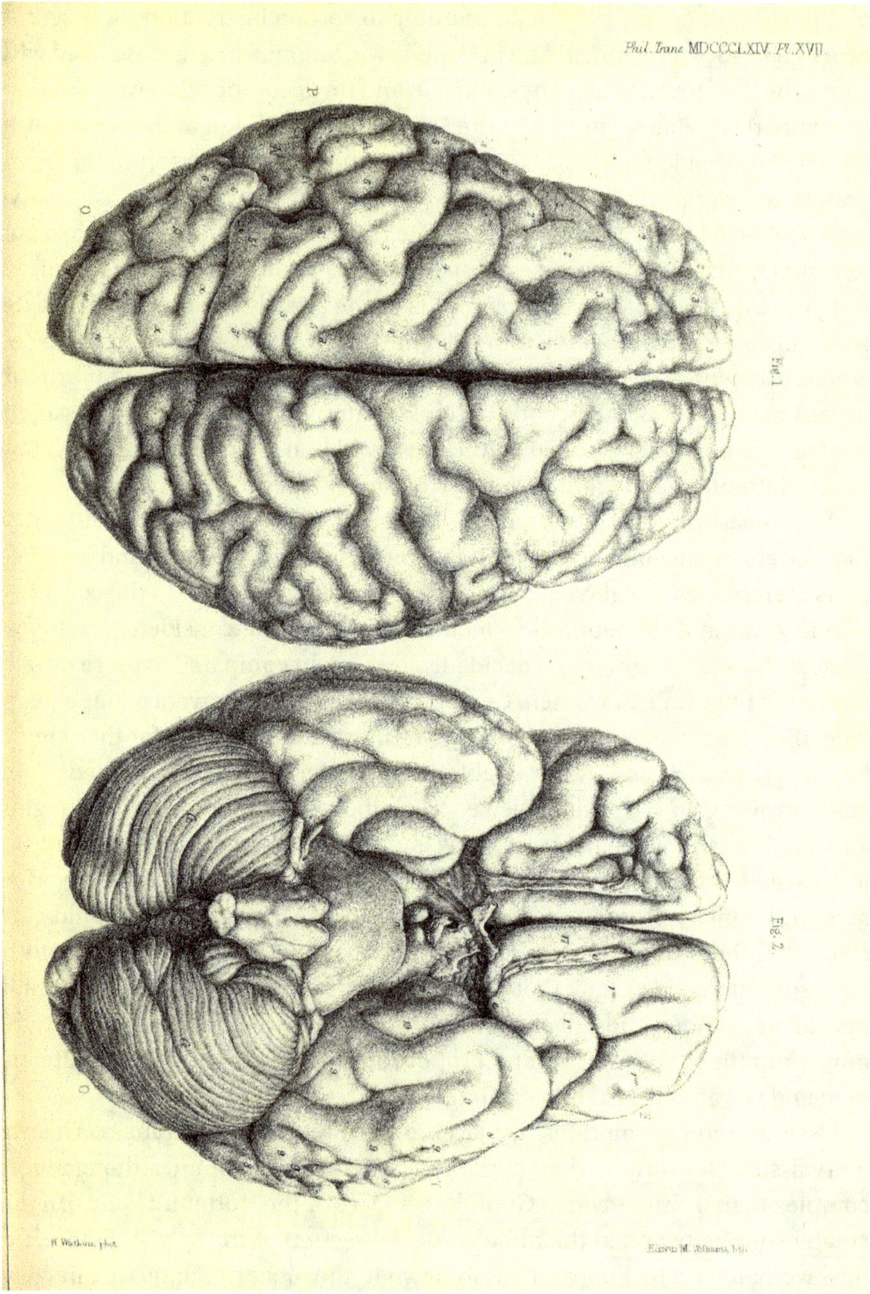

FIGURE 3.5 Plate XVII: Fig. 1. Upper surface of the preserved brain of the Bushwoman; Fig. 2. Base of the same brain, in John Marshall, "On the Brain of a Bushwoman; and on the Brains of two Idiots of European Descent," *Philosophical Transactions of the Royal Society of London*, 1864, *154*: 501-558.

From an accurate analysis of her primary and connecting convolutional patterns, Marshall got the conviction that nothing of the European brain was missing, but also that, in comparison, all of them were smaller and much less complicated, a manifest indication of structural inferiority. Nonetheless, a greater difference was undeniable between the Bushwoman's brain and that of the highest ape, than between it and the European one. On the other hand, her cerebellum was very well developed, far more completely evolved than her cerebrum.[46]

Another brain of a man of the so-called Ethiopian race, coming from Guinea, was examined in 1865 by Luigi Calori – professor of anatomy in Bologna and director of the local Museo Anatomico – and presented at a session of the Accademia delle Scienze. After all the work already done by many of his predecessors, he conceded that his own efforts might have felt redundant: however, some aspects were still controversial and obscure, and anthropologists needed more satisfactory images than those so far created. Both brains visually shown by Tiedemann and Gratiolet were of a woman, moreover belonging to one of the most degraded populations; to make matters worse, their representations differed: which of the two was the most faithful? As for the male brain presented by Tiedemann in 1836, even though Owen and Huschke had reproduced it in their texts, the knowledge gained in the meantime called for a a revision of its characters. For these reasons, Calori did not hesitate to delve into the structure of the brain at his disposal, and especially to portray it with the greatest accuracy in as many views as possible. He also invited other colleagues to follow his example, because solely from multiplied and controllable observations could the true concept of the Negro's brain descend. Only the great skill of Cesare Bettini – painter, lithographer and wax modeler – allowed him to provide eight fine plates drawn from life and life size, with an almost tactile quality, inserted at the end of the text.[47]

46 Ibid., pp. 510-525. Shortly thereafter, in a famous book the psychiatrist Henry Maudsley gave credit to the results of Marshall's research, and summarize them: see his *The Physiology and Pathology of the Mind* (New York: Appleton and Company, 1867), p. 50. In 1874 Marshall returned to the subject and specified that all estimates of the causes modifying the brain weight in Man were liable to error unless the influence of stature be first eliminated: see "On the Influence of Stature on the Weight of the Encephalon and its Parts in Man," *Proceedings of the Royal Society of London*, 1874, 23:564-565.

47 Luigi Calori, "Cervello di un negro della Guinea illustrato con otto tavole litografiche, Mem. letta nella sessione 9 Novembre 1865," *Memorie della Accademia delle Scienze dell'Istituto di Bologna*, 1865, 5:177-212. The collaboration between the two began in 1850, when Bettini had been named *Modellatore Anatomico* for human and comparative anatomy, pathology and obstetrics. See *Cesare Bettini 1814-1885. Disegnatore e modellatore*

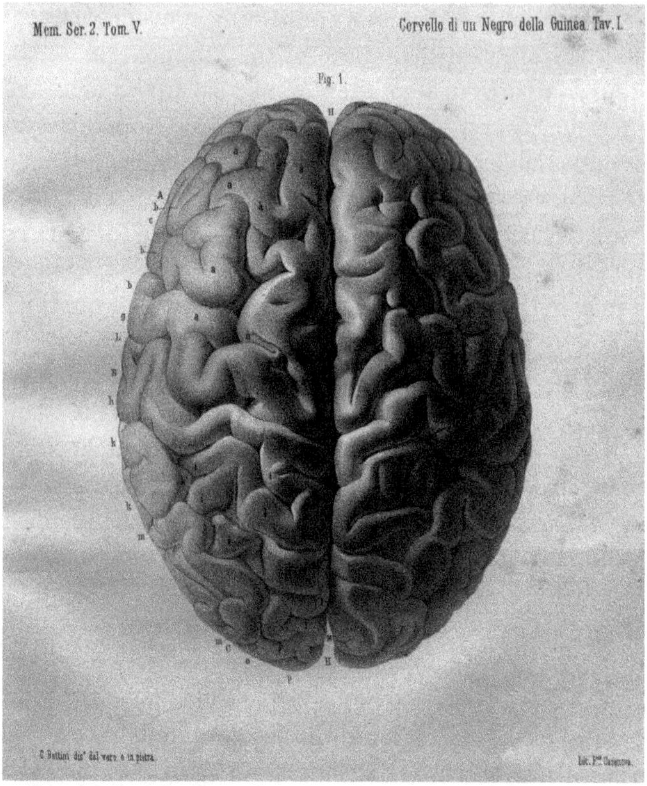

a

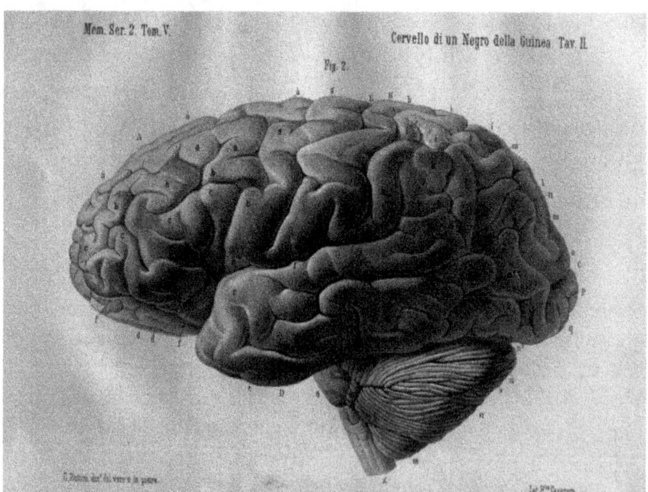

b

FIGURE 3.6A-D Upper face of the Negro's brain; its profile with cerebellum; frontal view; occipital view. First four plates in Luigi Calori, "Cervello di un negro della Guinea illustrato con otto tavole litografiche, Mem. letta nella sessione 9 Novembre 1865," *Memorie della Accademia delle Scienze dell'Istituto di Bologna*, 1865, 5:177-212.

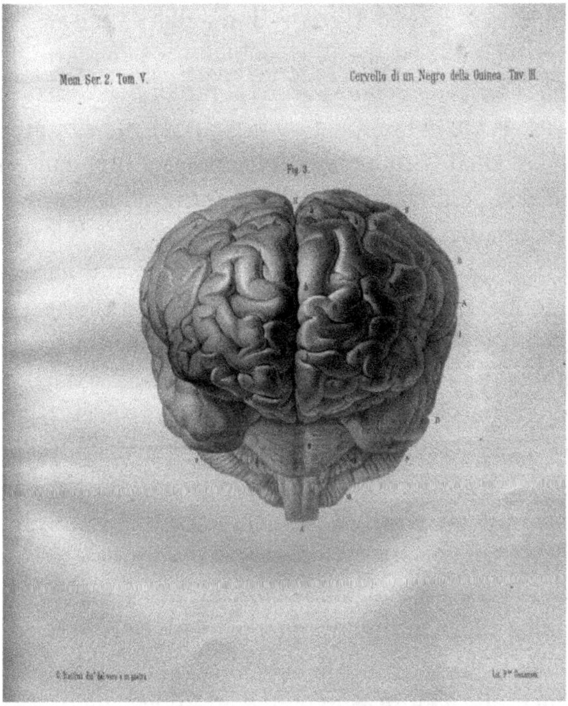
Mem. Ser. 2. Tom. V.
Cervello di un Negro della Guinea. Tav. III.
Fig. 3.

c

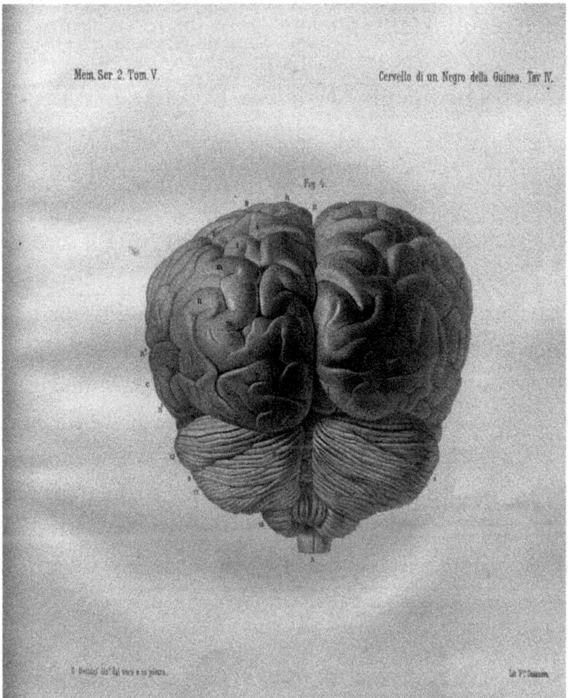
Mem. Ser. 2. Tom. V.
Cervello di un Negro della Guinea. Tav. IV.

d

The dissection of the meninges did not detect any line or dark spots, while the one to be blackened (*nereggiare*) was the venous blood, a peculiarity in which to identify a racial character had become almost commonplace. Except that Calori recalled how it was visible in Europeans too, especially in summertime and in individuals who had died of some breathing disease. Tuberculosis had mostly affected Blacks in Europe, and that was also the case of his Guinean man, so the dark blood might also depend on a morbid effect or the intemperate climate in which it had been observed. A brain of elongated and narrow form was usually attributed to Blacks – two peculiarities clearly visible also in this specimen, 17 cm. long with a maximum width of 11.2. In particular, although a remarkable length was not exclusive of Negroes, it nevertheless prevailed in their occipital area. The weight of the brain under observation, deprived of its meninges, was 1260.488 grams, equal to that of a young Italian who had been beheaded shortly before in Bologna and close to the proportional average of the local brains. Quite rightly, Calori deplored that some chaos was reigning with respect to brain weight in the existing literature.

Meckel first had asserted that the gray substance of the Negro's brain was darker, and the white one gray-yellow. In the intervening century, others had changed that finding in various ways, or denied it. In his Ethiopian brain Calori saw a regular white substance, and a more ash-colored grey one, which again might be a pathological effect, significantly present in other local brains he dissected. Having noted the strong asymmetry of the hemispheres, he then focused on the cortical furrows and fissures. Whereas Huschke had distinguished the Negro from the European for a different inclination in the lateral and central fissures, Calori not only reversed that verdict to his own advantage, but above all he noted how the extreme individual variability of inclination that he found in the degree of obliquity in the European brains might also affect the Ethiopian ones. So, no racial meaning was to be assigned to the morphology of the two main cortical fissures, and the same statement was also applicable to the detected asymmetry of the hemispheres. The thickness of the layer of gray matter in the observed Negro's brain was similar to that measured in local specimens by Calori, who noticed somewhat simpler, yet still very complicated convolutions on its surface. Their total extension was comparatively smaller in the African, but compensated by the greater surface area of those contained in the frontal lobes. Contrary to what had been claimed by Soemmerring and repeated by others, the cranial nerves were not bigger in relation to the weight of the brain. Indeed, a table on two columns compared

---

anatomico, pittore e litografo Bolognese, a cura di Marinella Pigozzi, Alessandro Ruggeri (Bologna: Edizioni TIP.LE.CO., 2010).

		Negro	Uomo nostrale
1.°	Olfattorio nel suo tronco . .	mill. 2¼	2¼
2.°	Ottico	5	5
3.°	Oculomotorio comune	3	3
4.°	Patetico	1	1½
5.°	Porzione maggiore del quinto . .	4	4
6.°	Porzione minore del medesimo . .	2	2
7.°	Oculomotorio esterno	2	2
8.°	Facciale colle fibre intermedie . .	2	2
9.°	Acustico	3	3
10.°	Glosso-faringeo	1	1½
11.°	Vago	3	4
12.°	Accessorio del Willis	2	1¼
13.°	Grande ipoglosso	2	2

FIGURE 3.7 *Comparative table of cranial nerves sizes*, in Luigi Calori, "Cervello di un negro della Guinea illustrato con otto tavole litografiche. Mem. letta nella sessione 9 Novembre 1865," *Memorie della Accademia delle Scienze dell'Istituto di Bologna*, 1865, p. 202.

the sizes of the nerves at the base of the brain, getting mostly identical figures, in three cases higher for the local man and only in one for the Negro. He could not help discarding Huschke's reduction of the brain of Blacks to that of the European woman. All in all, Calori felt compelled to place it on a higher level, that is close to that of the European man.[48]

The chemical analysis commissioned to a colleague in Bologna, Gaetano Sgarzi, to measure the amount of phosphorus contained in the substance of the Guinean man's brain – a metalloid hailed as the "new flame of Prometheus" for cerebral activity – had also given the same result (1.534%) as that of white men. Some research of the same type had already been carried out, abroad and in Italy, where, for example, Antonio Borsarelli, professor of pharmaceutical chemistry in Turin, had found 1.388% of phosphorus in the brain of a

48 Calori, "Cervello di un negro della Guinea," (cit. note 47), pp. 199-202. It should also be remembered that in the following years Calori worked to identify a particular brachycephalic type in the inhabitants of some areas of Northern Italy: "belle testine tonde" whose spheroidal shape should contain a greater brain volume. He conjectured, however, that the composition of the cerebral substances – of which nothing was yet known – and the nourishment of a good education would greatly contribute to the functional result. See his "Del tipo brachicefalo negli Italiani odierni. Memoria," *Memorie della Accademia delle Scienze dell'Istituto di Bologna*, serie II, 1868, *8*:205-234, and "Del cervello nei due tipi brachicefalico e dolicocefalico italiani. Memoria," ibid., 1870, *10*:35-152.

ten-year-old black girl, slightly more than that of her Italian peers. On the other hand – Calori pointed out – the adult pig had on average the same amount of phosphorus as humans. So, such a factor was probably unimportant.[49]

If his results were confirmed in the future, Calori believed that the bestial picture (*pittura bestiale*) of black people painted decades earlier by Virey, and reproduced in recent anthropological treatises with even more acrimonious tones, should be amended. Ironically he called "nice habit" (*bel vezzo*) the recent tendency to take apes as a touchstone for Blacks, inasmuch as all human brains shared a more or less great complication of convolutions, a distinctive character absent in anthropoids. Finally Calori firmly called to educate the Negroes, the only way to let them compete with that "great maker" (*gran fabbro*) who was the white man. If they were given a suitable government, a non-fatalistic religion and freedom, the numerous examples already known, among them, of individuals capable of excelling in various fields would multiply. The anatomist's conclusion was of a political nature: over time providential laws had already abolished the slave trade, and a "holy" war had just ended, for the purpose of cancelling a centuries-old infamy.

A rather critical review of Calori's "Memoria" was published in 1868 by the polygenist *Anthropological Review* and signed J.B.D., the initials of Joseph Barnard Davis, a compulsive collector of skulls and skeletons of various races, which he thought were immutable, in contrast with any evolutionary claim. He began by praising the "careful description" of the "very eminent Italian anatomist" – illustrated by eight "beautiful" lithographic plates – and his "fealty to truth". After offering a summary of its content, however, the reviewer uncovered his true intent and pointed the finger at the "tone of mind" manifested in the political conclusions, not so far from those of Tiedemann and a minority of others, who had investigated or conceived the Negro's brain "under the influence of *a priori* notions". However amiable and pardonable their motives, such efforts were said to be unworthy of confidence. Anthropological knowledge ought to form the real, although remote, basis of all legislation and government, but until then a mere bundle of prejudices had inspired legislators and rulers:

> In spite of all the teachings of all ages, [...] they have assumed the position that mankind are all one and the same, one in organisation, one in faculties, one in capabilities; in fact, that by *education* and *development*

49 Antonio Borsarelli, "Della quantità di fosforo che si trova nella materia del cervello e del ventricolo dell'uomo e di alcuni altri aninali, ed in differenti età," *Annali di chimica applicata alla medicina*, 1861, *32*: 354-357.

> all the various races of man may become equal. The history of every human race in every age proves the utter untruthfulness of this position by giving relief to the peculiarities of each.[50]

Contrary to what Calori said in 1865, "the late sanguinary events across the Atlantic" had already consigned more than a quarter of the four millions of Negroes to destruction: events totally unheeded by the believers in equality and amelioration, "the most cruel and delusive doctrines". Davis also addressed his polemic against John Stuart Mill's enlightened position on the Irish question: the philosopher should go back to school and learn some anthropological lessons, "could his prepossessions be overcome". Irish people *en masse* were unchangeably inferior, and it seemed absurd to hope for their improvement, "except by that objectionable mode which got the name of miscegenation, in the United States of late – i.e. of real degeneration of the higher race".[51]

The speech given by Thomas Huxley in 1865, immediately after the end of the Civil War, is too well known to dwell on his proclamation of an unequivocal racial hierarchy, even if he supported the abolition of slavery on ethical grounds, along with a moderate emancipation of woman:

> It may be quite true that some negroes are better than some white men; but no rational man, cognisant of the facts, believes that the average negro is the equal, still less the superior, of the average white man. And, if this be true, it is simply incredible that, when all his disabilities are removed, and our prognathous relative has a fair field and no favour, as well as no oppressor, he will be able to compete successfully with his bigger-brained and smaller-jawed rival, in a contest which is to be carried on by thoughts and not by bites. The highest places in the hierarchy of civilisation will assuredly not be within the reach of our dusky cousins, though it is by no means necessary that they should be restricted to the lowest.[52]

---

50 J.B.D. [Joseph Barnard Davis], "The Brain of a Negro of Guinea," *The Anthropological Review*, 1868, *6*:279-285, p. 283.

51 Davis referred to a previous anonymous article, in the same periodical, against the schools of "Political Economists and Legislative Reformers", which had as their masters Jeremy Bentham and John Stuart Mill but were also "a far-off reverberation of Democritus and Epicurus": see "Race in Legislation and Political Economy," *Anthropological Review*, 1866, *4*:113-135, p. 115.

52 Thomas Huxley, "Emancipation – Black and White," *Collected Essays. 3. Science & Education* (London: Macmillan and Co., 1893), pp. 66-67. Despite the rivalry between Huxley and Hunt, both were "scientific reformers" with not too distant ideas about the place of women in science and the racial question: see Sera-Shriar, "Observing Human Difference" (cit. note 30). See also Evelleen Richards, "Huxley and woman's place in science: The

Similarly to Huxley' appeal to "nature" to reduce post-war expectations in the United States, so he dismissed fanatical "philogynists" – "who bid the man look upon the woman as the higher type of humanity; who ask us to regard the female intellect as the clearer and the quicker, if not the stronger; who desire us to look up to the feminine moral sense as the purer and the nobler."

In 1869, two papers on the same topic were read at a session of the Anthropological Society of London. As believed by George Harris, a barrister and President of the Anthropological Society of Manchester, nature itself predisposed that, given two different sexes in certain living creatures, "those of one sex must be stronger, those of the other weaker", a disparity maintained "with a jealous care" in all races of man. It was easy to point to the consequences of such specialization: philosophy, literature, art, music, and science offered lots of examples in this respect. While Harris did not mention brain, the second speaker, an opponent of women's voting like James McGrigor Allan – fearful of the dissolution of the family that universal suffrage would have caused – asked some old questions about the differences in the male and female minds, and answered with a repertoire of sources and arguments in which brain played a primary role.

Man had been expressly made for undergoing mental and physical labor, whereas nature had produced woman as a being whose main roles were supposed to be love, leading to gestation, parturition and nutrition – the "grand purpose of her existence". Savage life seemed to be much closer to sexual equality; in Europe and America, almost every woman was steered through life by the "reflecting brain", the strong will, and the protective arm of a husband, a father, a brother, and a son. The female brain, instead, resembled the animal one, in which the organs of sense, as compared with the brain proper, were much larger than in the human species.[53] In the following discussion, however, some criticisms surfaced about the "natural and eternal subordination" of women as argued by the two speeches, which were held – it should be remembered – the same year that saw John Stuart Mill take a clear stance on the subject. Around this he had been jointly working with his wife Harriet Taylor Mill, before her death in 1858, and continued afterwards. He seriously doubted that there was any anatomical evidence of the inferior mental capacity of women: it was by no means established – he wrote in 1869 – that they had a smaller

---

'woman question' and the control of Victorian anthropology," in Moore, James R. (ed.), *History, Humanity and Evolution. Essays for John C. Greene* (Cambridge: Cambridge University Press, 1989), pp. 253-284.

53 George Harris, "On the Distinctions, Mental and Moral, occasioned by the Differences of Sex," *Anthropological Review*, 1869, 7:clxxxix-cxcv; James McGrigor Allan, "On the Real Differences in the Minds of Men and Women," ibid., pp. cxcv-ccxix.

brain. Moreover, the precise relations existing between the brain and intellectual powers were still unclear and a matter of great dispute.[54]

In the debate opened by Harris and McGrigor Allan, John Langdon Down, Superintendent at the Earlswood Asylum for Idiots in Surrey, sided with anti-egalitarianism, aided by both biblical authority and science:

> Langdon Down thought that it might be fairly deduced from the two papers man and woman were distinct creatures, as they were stated to be in the Mosaic account of the creation. They were not alike, and never would be; and the arguments in the papers proved that nature did not intend them to be alike, but differentiated beings, the one the complement of the other.[55]

In the attempt to reform the treatment of developmental disabilities, Down had recently proposed a new classification of the feeble-minded, a "natural system" that should have replaced the preceding ones, "so vague and artificial": among the large number of patients under his observation, a considerable portion could be fairly referred to as "one of the great divisions of the human race other than the class from which they have sprung":

> Of course, there are numerous representatives of the great Caucasian Family. Several well-marked examples of the Ethiopian variety have come under my notice, presenting the characteristics malar bones, the prominent eyes, the puffy lips, and retreating chin. The wholly hair has also been present, although not always black, nor has the skin acquired pigmentary deposit. They have been specimens of white negroes, although of European descent.
>
> Some arrange themselves around the Malay variety, and present in their soft, black, curly hair, their prominent upper jaws and capacious mouths, types of the family which people the South Sea Islands.
>
> Nor has there been wanting the analogues of the people who with shortened foreheads, prominent cheeks, deep-set eyes, and slightly apish nose, originally inhabited the American Continent.
>
> The great Mongolian family has numerous representatives, and it is to this division, I wish, in this paper, to call special attention. A very large

54 John Stuart Mill, *The Subjection of Women* (London: Longmans, Green, Reader, and Dyer, 1869), pp. 119-120.

55 See the synthesis of John Langdon Down' intervention in McGrigor Allan, "On the Real Differences" (cit. note 53), p. ccxvi.

> number of congenital idiots are typical Mongols. So marked is this, that when placed side by side, it is difficult to believe that the specimens compared are not children of the same parents. The number of idiots who arrange themselves around the Mongolian type is so great, and they present such a close resemblance to one another in mental power, that I shall describe an idiot member of this racial division, selected from the large number that have fallen under my observation.[56]

The Mongolian type of idiocy occurred in more than ten per cent of the cases studied by Down, mainly congenital idiots in a degenerate state born to parents suffering from tuberculosis. They did not lack imitative qualities and a weak ability to speak, which could be improved with exercise. The ethnic classification would also have a "considerable philosophical interest" and allowed Down to swim against the current. While the prevailing tendency assumedly denied that human races could be mere varieties of a single original species, in his idiots he had seen examples of a regression or withdrawal from a type to assume the traits of another: a phenomenon that undoubtedly reinforced the monogenist hypothesis.[57]

3 An Urgent Desideratum for Science

John Thurnam, born in a Quaker family, spent his whole professional life in contact with deviance and madness as a superintendent of mental hospitals, first at the York Retreat, established by the Society of Friends, then at the Wiltshire County Asylum. While here, he focused his attention on craniology, and between 1856 and 1862 helped Joseph Barnard Davis – the passionate hunter of racial skulls and skeletons – prepare texts and images of the *Decades* of a laborious treatise *Crania Britannica*, the title-cover of which was adorned with the overlapping profiles of Blumenbach and Morton. Thurnam also contributed to the work with a long chapter on the historical ethnography of Britain that exposed his views as to the existence of successive races in the Celtic period: an earliest race of short stature, short-faced and long-headed; the second one tall,

56 John Langdon Down, "Observations on an Ethnic Classifications of Idiots," *London Hospital Reports*, 1866, 3:259-262.

57 "By the standards of his time, Down was something of a racial «liberal», but his theory lost its alleged rationale when physicians detected it [the syndrome] both in Orientals themselves, and in even lower races, by his classification": see Gould, *The Panda's Thumb* (cit. note 26), pp. 166-167.

large featured, and short-headed.[58] Little time passed, and Thurnam reported having weighed, for many years, the encephalon (cerebrum plus cerebellum) of nearly every patient who had died while under his care at the Wiltshire Asylum. Data concerning 470 cases were thus collected, analyzed and compared with those recorded by previous observers, like Peacock and Wagner, or even Robert Boyd – physician at the Somerset County Lunatic Asylum – who had assembled in 1861 some tables with the weights of bodies and internal organs of a significant number of healthy and unhealthy subjects.[59]

While Tiedemann had supposed that the lesser size of the female brain was only due to her smaller stature, Thurnam agreed with Broca that the relative smallness depended at the same time on a physical and intellectual inferiority.[60] As for the racial factor, he had no doubt that its influence was considerable, but unfortunately very partially investigated until then by actual weighings of the brain. He roughly calculated its average weight indirectly through the assessment of cranial capacity in English and Scotch people, and with less accuracy in French and German ones, all belonging to the Indo-European or Caucasian family. Some internal difference was documented by the available data, albeit much less than that found in more distant races. Using other researchers' data, Thurnam supposed that the male Negro's skull had an average capacity "scarcely if at all exceeding that of the European female"; the diminution in the other lower races approximated such ratio. Needless to say, Tiedemann's figures showed how much he had erred in putting forward his egalitarian thesis.[61] Similarly, the average weight of the brain of the educated, and of those in a higher social class, should be greater, even if no equipment was available back then to provide mathematical evidence thereof. Thurnam could only rely on his own experience at the York Retreat, where the brains of the more cultivated middle class of the insane weighed decidedly above those of paupers in the county asylums of Somerset and Wiltshire. Broca himself had studied the head size of medical students, as compared with those of servants in the Bicêtre Hospital, and had concluded that, "other things being

58 Joseph Barnard Davis and John Thurnam, *Crania Britannica. Delineations and Descriptions of the Skulls of the Aboriginal and Early Inhabitants of the British Islands: with Notices of their Other Remains* (London: Printed for the Subscribers, 1865). The sixty figures of skulls are of life-size, with numerous woodcuts in the text.

59 Robert Boyd, "Tables of the Weights of the Human Body and Internal Organs in the Sane and Insane of both Sexes at various Ages, arranged from 2614 post-mortem examinations," *Philosophical Transactions of the Royal Society of London for the Year* MDCCCLXI, 1862, *151*:241-262.

60 John Thurnam, *The Weight of the Brain, and on the Circumstances Affecting it* (London: J. E. Adlard, 1866), pp. 10-11.

61 Ibid., pp. 16-19.

equal", their volume was greater in the superior than in the inferior classes, whether the result of education or whether hereditary. *A fortiori* this applied to the "distinguished men", fifteen of whom – first and foremost, the inevitable Cuvier, with his exceptional 1830 grams – were included by Thurnam in a table of brain weights.[62]

He also predicted that, when Davis's catalogue of the large collection of skulls would be published, "we shall be able to speak with greater confidence of the cranial capacity of different races". The following year, in fact, his friend's *Thesaurum craniorum* came out, dedicated to "those of different nations and in various parts of the world" who had generously seconded his craving for skulls. As many as 1474 specimens were listed, geographically subdivided, and described.[63] It was January 1868 when Davis sent to the Royal Society a message that invariably insisted on the relatively little work yet done to ascertain brain size in the different races of mankind. Opportunities to inspect exotic brains were very rare, and only by gauging the capacity of trustworthy skulls – and then deducing the volume of the encephalon – could reliable data be obtained. It might be objected, as Peacock had done in 1847, that the indirect craniometrical method was not free from fallacy, but Davis defended it by pointing out, if anything, to the alterations caused in the inner organ by many diseases and by death itself. Therefore, better to handle skulls than brains:

> Practically, this method is more sure to yield an accurate average size of the organ, because we have it in our power to use an unchangeable substance with which to gauge the capacity of the skull. And we thus arrive at conclusions the same in result as if we had the brain in all skulls at a uniform density, which, in fact, is the true basis of comparison.[64]

Clean and dry Calais sand, of a definite specific weight, was the filling material used and recommended by Davis. The great difficulty consisted here in making allowance for the other contents of the cranium besides the encephalon: dura and pia mater, fluids of the membranes and ventricles, blood in the large

62 Ibid., p. 34.

63 Joseph Barnard Davis, *Thesaurum Craniorum. Catalogue of the Skulls of the Various Races of Man* (London: Printed for the Subscribers, 1867). A few years later, a *Supplement to Thesaurum Craniorum* was added, with 300 new specimens that in the meantime had enriched the collection (London: Printed for the Subscribers, 1875).

64 Joseph Barnard Davis, "Contributions towards Determining the Weight of the Brain in Different Races of Man," *Philosophical Transactions of the Royal Society of London*, 1868, 158:505-527, p. 506.

Races.	1.	2.	3.	4.	5.	6.	7.	8.	9.	10.	11.	12.	13.	14.	15.	16.	17.	18.	19.
		Brain-weights of the Skulls of MEN.							Brain-weights of the Skulls of WOMEN.						Mean of Sexes.		Mean of Series.		
	Number.	Heaviest.		Lightest.		Average.		Number.	Heaviest.		Lightest.		Average.						Mean internal capacity.
SUMMARY OF THE SIX TABLES.																			
I. European Races	299	52·68	1493	43·61	1236	48·25	1367	94	46·02	1304	39·56	1121	42·49	1204	45·73	1296	47·12	1335	92·3
II. Asiatic Races	124	50·27	1425	41·57	1178	46·00	1304	86	45·95	1302	37·48	1062	42·13	1194	43·94	1245	44·44	1259	87·1
III. African Races	53	47·37	1342	41·93	1188	45·63	1293	60	43·91	1244	39·59	1122	42·74	1211	43·66	1237	43·89	1244	86
IV. American Races	52	48·16	1365	43·50	1233	46·17	1308	31	45·44	1288	39·13	1109	41·89	1187	44·92	1273	44·64	1265	87·5
V. Australian Races	24	50·86	1441	36·96	1047	42·83	1214	11	42·98	1218	34·77	985	39·22	1111	41·02	1162	41·81	1185	81·9
VI. Oceanic Races	210	49·25	1396	42·90	1216	46·54	1319	95	44·61	1264	41·01	1162	43·00	1219	44·52	1272	45·63	1293	89·4
Numbers and Averages	762	49·76	1410	41·74	1183	45·90	1301	377	44·81	1270	38·59	1094	41·91	1188	43·96	1246	44·58	1263	87·3

FIGURE 3.8 *Summary of the six tables of the computed weights of the brains of different races*, in Joseph Barnard Davis, "Contributions towards Determining the Weight of the Brain in Different Races of Man," *Philosophical Transactions of the Royal Society of London*, 1868, *158*:505-527, p. 514.

vessels. To overcome it, a percent deduction had been fixed at 15% as a tare. Of course, the weight of the sand had to be converted into its equivalent of cerebral matter with a specific gravity of 1040, which was the nearest average datum that had been calculated by several competent observers. Neither Tiedemann – though divesting the brain to be weighed of its membranes – nor Morton had taken such allowance into account, and consequently their data were of little value, while Davis claimed to have been the first to put rigor in the procedure.

According to him, Tiedemann had been wrong in attributing the same cerebral mass to each race, and in insisting on the physical unity of man. It was easy, for Davis, to oppose to egalitarianism by showing six tables of brain-weights measured on 762 skulls of his collection, and divided by race and sex (European, Asiatic, African, American, Australian, Oceanic), with a summary. Laboriously calculated, those averages were the numerical expression of an irrefutable hierarchical order. In a *Postscript*, dated July 1868, Davis expressed satisfaction in seeing that his own computing method had received "an important illustration" from an experiment made (and privately communicated to him) by the young Austrian anatomist Augustin Weisbach, who had taken the encephala out of 115 skulls and carefully weighed them. Subsequently, when the cranium was prepared, he had measured its capacity in each case. His results confirmed in essence the value of Davis's way of calculation, only suggesting that it would probably be desirable to slightly increase the 15% tare.[65]

In 1869, Davis sent a letter to the Editor of the *Anthropological Review* to criticize an article published in the previous issue. The American Sanford B. Hunt – born in the State of New York, serving as an army surgeon in several locations and for a while professor of anatomy at the Buffalo Medical College – had reported the results of weighing the brains of 405 soldiers died in the last war, whose autopsies had been carried out "under the direction of the Surgeon Ira Russell". An investigation of great interest, Davis commented, were it not for some disabling circumstances. First of all, the vague identity of the actual performer; secondly the doubt as to whether those brains had been divested of

65 Ibid., pp. 526-527. In the hospital of the Viennese garrison and in the Bürger-Versorgunghäuser, the military doctor Weisbach had already measured, group by group, 429 skulls and brains belonging to individuals of various nationalities of the Habsburg Empire: see "Die Gewichtsverhältnisse der Gehirne österreichischer Völker mit Rücksicht auf Körpergrösse, Alter, Geschlecht und Krankheiten," *Archiv für Anthropologie*, 1866, *1*:191-218, 285-319. A systematic effort aimed at identifying racial differences within the imperial mosaic continued throughout his life: see a first extended report in his *Beiträge zur Kenntniss der Schädelformen österreichischer Völker* (Wien: Selbstverlag der k. k. Gesellschaft der Aerzte, 1867).

their membranes and drained of fluids, and whether the *medulla oblongata* had been reckoned as a portion to be weighed. The great majority belonged to soldiers of Negro blood, whose brain was assigned an average weight of one to two ounces heavier than those calculated by other observers. Davis was especially worried by Hunt's speculation whether the slightly higher value might be due to living in America. However, the influences to which the Negro of the United States had been historically exposed were not so favorable as to justify a supposed enlargement of the brain. And, moreover, that conjecture made no sense, in Davis's opinion, just because "the weight of the brain, like all the other peculiarities of human races, is a *race character*, appertaining to each race specifically."[66]

Hunt's article proposed an overall assessment of the Negro as a soldier. When, in July 1862, an Act of Congress authorized the recruitment of "persons of African descent", their notorious "passivity" raised doubts as to their potential performance at war: the enlistment of 180.000 negroes was watched, "by friend and foe", with lively interest, and scientists looked upon it as a grand experiment. The Negro's natural imitative faculty, fondness for rhythmical movements, and habit of obedience furthered his aptitude to military drill. His capacity for marching was not inferior to that of the majority of troops, while there were discordant opinions about his endurance of fatigue and hunger. Other aspects considered by Hunt were the powers of digestion and assimilation, and the immunity from, or liability to, certain diseases. As a fervent abolitionist, he believed that what impaired the Negro's performance as a soldier was not intrinsic to his race, but to a great extent educational and therefore expected to disappear "under the energizing influences of freedom and the teacher".[67] A comparison between the free Negro recruits from the North and the grossly ignorant slaves enlisted from the Southern plantations proved that expectation to be correct.

Finally, Hunt addressed the topic of the intellectual capacity, which had prompted Davis's critical reaction. For him, centuries of ignorance and servitude made it impossible to define the Negro's position in a natural scale of races, and it would have been unfair to rate him on his present manifestations of mental acuteness. Till then, three modes of judging the relative value of races had been devised, with reference only to physical facts and depending solely on the observer's honesty and accuracy. The first of these, based on external measurements of the cranium, was essentially faulty, in that it made no

66 Joseph Barnard Davis, "On the Weight of the Brain of the Negro," *The Anthropological Review*, 1869, 7:190-192, p. 191.

67 Sanford B. Hunt, "The Negro as a Soldier," *The Anthropological Review*, 1869, 7:40-54, p. 46.

Ethnographical Table,

Derived from 405 Autopsies of White and Negro Brains. Made under the direction of Surgeon Ira Russell, 11th Massachusetts Volunteers.

	Number of Autopsies.	Grade of Colour.	Average weight of brain.	Maximum weight of Brain.	Minimum weight of Brain.	Brains, 60 ounces and over.	Brains, 55 and under 60 ounces.	Brains, 50 and under 55 ounces.	Brains, 45 and under 50 ounces.	Brains, 40 and under 45 ounces.	Brains, 35 and under 40 ounces.	Brains less than 35 ounces.
			oz.	oz.	oz.							
	24	White.	52-06	64	44¼	1	4	11	7	1	...	...
	25	¾ "	49.05	61	40	1	...	10	12	2	...	...
	47	½ "	47.07	57	37¾	...	2	13	19	12	1	...
	51	¼ ,	46·54	59	38½	...	2	10	22	11	6	...
	95	⅛ "	46·16	57	34½	...	1	15	50	21	7	1
	22	1/16 "	45·18	50½	40	...	...	3	10	9	...	...
	141	Black.	46·96	56	35¾	...	5	42	51	38	3	...
	405	...	...	...	...	2	14	104	171	94	17	1
Autopsies of Clendenning, Sims, Reid, and Tiedemann,	278	Whites, collated from various sources,	49½	65	34	7	28	99	97	39	7	1

FIGURE 3.9 Table, in Sanford B. Hunt, "The Negro as a Soldier," *The Anthropological Review*, 1869, 7:40-54, p. 51.

allowance for its thickness. A second method rested on the supposition of a direct ratio of the mental to the cubic capacity of the cerebral mass. The third way, "the more reliable", consisted in directly weighing the brain at post-mortem examinations. And there Hunt put together an "ethnographical table" from the 405 autopsies of White and Negro brains, carried out "under the direction" of Surgeon Ira Russell. Its figures demonstrated that the standard weight of the Negro brain was four ounces less than that of the White, and that one-half infusion of white blood (mulatto) determined an increase in weight. However, assuming the matter of weight to be a prerequisite for intellectuality, Hunt's optimistic reformism – relying on brain plasticity – was not so widespread as he thought it was:

> The crucial experiment of the effect of freedom and education has only just begun. We cannot judge the ultimate capacity of the Negro from that which he has thus far manifested. [...] In the present state of science, we can only refer to general opinion, which leans to the belief that it is

> within the power of educational causes to modify the form and size of the human brain to a considerable extent, and that the competitive success of the freedman of this country rests upon the effort that may be devoted to their mental and moral elevation.[68]

To get an idea of the common scientific feeling on such issue at the end of the decade, read the classic textbook by Henry Maudsley, a young physician to the West London Hospital who had worked in mental hospitals before. Rising through the mammalian ranks, he found that not only the hemispheres gradually increased in size, but the grey surface was also further extended by folding. The lower mammals are entirely destitute of such convolutions, which are most developed in apes and men. An exact relationship between the development of the convolutions and the degree of intelligence in different animals was not available yet, but an animal was certainly more intelligent than another one of the same size, if its brain had more convolutions and deeper sulci. As stated by Maudsley, that proposition was true of man too. He found proof of this in Gratiolet's study of the brain of the Hottentot Venus, "who was no idiot": simple on the surface, with perfectly symmetrical hemispheres, characters so similar to those of the lower animals or of a white man affected by an arrest of cerebral growth. Marshall had further elaborated on this thesis, arguing that the Negro's brain is superior to that of the Bushman, but cannot reach the level of the White's one. Among Europeans, "other circumstances being alike", the size of the brain would bear a general relationship to the mental power, "although apparent exceptions to the rule sometimes occur".[69]

In 1870, Alfred Russell Wallace collected some of his essays, previously published in various periodicals. One of them, a quite famous one, in 1864 had addressed the origin of human races and the antiquity of man, then reprinted "with a few important alterations and additions".[70] On these issues, the Darwinian theory was open to a multitude of striking deductions. For instance, a precious clue to the origin of races resulted, according to Wallace, from the view that man's physical structure had ceased to be affected by natural selection insofar as his social, moral, and intellectual faculties had begun to develop. Before this process, however, the human species would still be subjected to physical variations induced by greater extremes of climate, changes in food and conflicts with enemies: "Thus might have arisen those striking characteristics

68 Ibid., pp. 53-54.

69 Maudsley, *The Physiology and the Pathology of the Mind* (cit. note 46), pp. 52-55.

70 Alfred R. Wallace, *Contributions to the Theory of Natural Selection. A Series of Essays* (London: Macmullan and Co., 1870), pp. 303-331; for the first version see "The Origin of Human Races and the Antiquity of Man deduced from the Theory of «Natural Selection»," *Journal of the Anthropological Society of London*, 1864, 2:clviii-clxxxvii.

and special modifications which still distinguish the chief races of mankind."[71] In the meantime, natural selection would soon give pride of place to the mind: it would most powerfully influence speech, the arts, division of labor, restraint of the appetites, social feelings etc. Wallace felt thus enabled to strike a balance between the opposite views of quarrelsome anthropologists: man might have been once homogeneous, but only in a distant time in history, when he had not yet acquired "that wonderfully developed brain", which now, even in his lowest forms, raises him far above the highest brutes. He became a veritable man, though, when his superior faculties fully developed, and in this sense one may fairly believe in the existence of many originally distinct human races. Due to the antiquity of man, both monogenists and polygenists, so to speak, were right.

The last essay in the 1870 miscellany is a further development on a few sentences at the end of a review of Lyell's works published the year before by Wallace, who had tried to give a better explanation of the limits of natural selection as applied to man.[72] Among other things, there he noted some facts that almost made him doubt for a moment whether the size of the brain represented a direct indicator of mental power. It was indeed rather surprising that "savages" had sometimes shown a brain that was just a little less bulky, or even more voluminous, than Europeans, while their faculties were very little above those of animals. Even more extraordinarily, the few extant remains of pre-historic men – Neanderthal, Engis, Les Eyzies skulls – did not exhibit any diminution in the size of their brain case. So, "the idea is suggested of a "surplusage" of power; of an instrument beyond the needs of its possessor". In other words, the rudiments of all superior powers and feeling would exist in primitive races in a latent state, such that moral and intellectual faculties may occasionally appear:

> We see, then, that whether we compare the savage with the higher developments of man, or with the brutes around him, we are alike driven to the conclusion that in his large and well-developed brain he possesses an organ quite disproportionate to his actual requirements – an organ that seems prepared in advance, only to be fully utilized as he progresses in civilization. A brain slightly larger than that of the gorilla would, according to the evidence before us, fully have sufficed for the limited mental development of the savage".[73]

71 Wallace, *Contributions* (cit. note 70), p. 320.

72 Ibid., pp. 332-371. See his (anonymous) "Sir Charles Lyell on Geological Climates and the Origin of Species," *Quarterly Review*, 1869, *126*:359-394.

73 Wallace, *Contributions* (cit. note 70), p. 343.

Implying a close connection between needs and the development of the organs, the laws of evolution could not therefore account for that oversized brain, which seemed instead to prove the existence of "some power", distinct from that which had boosted the development of the lower animals. In his 1869 review, Wallace had even alluded to a "Higher, Overruling Intelligence" that forced Darwin to differ "grievously" from him – as he wrote in a letter – not seeing any necessity for an additional cause in regard to man and hoping he did not murdered too completely "your own and my child".[74]

Shortly thereafter, Darwin repeated with some emphasis how he did not understand Wallace's assertion that natural selection could have endowed the savage with a brain not much higher than that of an ape. Little was known yet about the functions of the organ, but it was plausible that, as mental faculties gradually developed, brain size grew; as intellectual powers further developed, "the most intricate channels of intercommunication" should connect its various parts together, and each of them would perhaps tend to be less fit to respond to sensations or associations in a definite and uniform manner. The continued use and advancement of language, for instance, would have influenced the mind in such a way as to enable it to engage in long trains of thought, and this intimate connection between the brain and the faculty of speech was evident in some odd cases of cerebral disease. According to Darwin, the belief in the existence of some close relationship between brain size and the development of intellectual faculties was supported by a comparison of the skulls of savage and civilized races, of ancient and modern people, and by analogy with the whole vertebrate series. In support of this belief, Davis, Broca, and Prichard were quoted. Darwin himself had shown that the brains of domestic rabbits were considerably reduced in bulk, compared to those of wild rabbits or hare. Having been closely confined for generations, they had had fewer opportunities to use their instincts, senses, voluntary movements, and intellect.[75]

He had no doubts that the various races, if carefully compared and measured, greatly differed from each other in many physical traits (even the brain convolutions), in constitution, acclimatization, and liability to diseases. Their mental characters seemed likewise very dissimilar. Having voiced such truth, Darwin moved on to thoroughly examine the arguments in favor of classing human races as different species, and then the opposing ones. The naturalists who had welcomed the principle of evolution – "this is now admitted by the greater number of rising men" – could only think that they must have

74 Darwin to Wallace, 14 April 1869, <http://www.darwinproject.ac.uk/letter/DCP-LETT-6706.xml>.

75 Charles Darwin, *The Descent of Man, and Selection in Relation to Sex* (London: John Murray, 1871), I, pp. 137-138, 37, 56-57, 145-146.

descended from a single primitive stock, which probably appeared in Africa. Everyone knows about Darwin's monogenism, as well as his choice to identify the mechanism that could explain racial divergence in sexual selection, an evolutionary factor so important as to deserve most of the pages of *Descent of Man*.[76] Incidentally, in its second part, man was qualified to be "more courageous, pugnacious and energetic than woman", with a brain absolutely larger, not being fully clarified yet "whether relatively to the larger size of his body, in comparison with that of woman".[77]

In the human species, the ennobling influence of natural selection benefits the brain above all other organs. So Ernst Haeckel – a daring interpreter of Darwinism in Germany – taught when he introduced his monistic system. It is the most perfect intellect (*Verstand*), not the best revolver, that enables one to win in the long run, and the cerebral qualities that assist in the victory would be transmitted to the descendants. Moreover, the strongest lever for every progressive development of the organism and, above all, of the brain, had consisted in the progress of an articulate language. That way, in Asia apes (*Menschenaffen*) had slowly turned into real men (*Echte Menschen oder sprechende Menschen*), through the intermediate stage of the *Affenmenschen*, and the brain, or mental activity, had risen among the more civilized nations to a degree which wild savages could not imagine. Despite all the efforts of retrograde forces, Haeckel was confident that natural selection would foster mankind's adaptability – which meant limitless brain changes – toward the utmost perfection.[78]

In 1878, such a widely recognized authority as Rudolf Virchow said something about the obscurities that still hindered the path,– just when he was dueling with his ex-student Haeckel on the theory of evolution and the freedom of science – at a session of the Berliner Gesellschaft für Anthropologie, Ethnologie und Urgeschichte. There, he read a letter and a manuscript sent by the Russian scientific traveller Nikolai von Miclouho-Maclay from New Guinea. Virchow remarked that, for instance, Johann Caesar Godeffroy – a trader and the founder of a naturalistic museum in Hamburg – had not gone beyond the collection of human skulls and skeletons, or of photographs. The anatomical investigation of distant races, which only European governments could boost by founding *Beobachtungsstationen* in different continents, had to be extended to make real progress. A comparative encephalology (*eine vergleichende*

76 Ibid., I, pp. 216-236.

77 Ibid., II, pp. 316-317.

78 Ernst Haeckel, *Natürliche Schöpfungsgeschichte. Gemeinverständliche wissenschaftliche Vorträge über die Entwickelungslehre im Allgemeinen und dejenige von Darwin, Goethe und Lamarck im Besonderen...Zweite, verbesserte und vermehrte Auflage* (Berlin: Georg Reimer, 1870), p. 156.

Encephalologie) was an urgent desideratum for science. Von Miclouho-Maclay was right when he viewed craniology as such with some suspicion. Virchow knew that the other materials available at the time were too scarce to do anything but work temporarily on skulls, albeit with the hope that, as soon as possible, studies on brains would be connected to the existing knowledge.[79]

That same year, speaking at the Congrès international des sciences anthropologiques held in Paris, Virchow reiterated that urgency, and formulated the proposal – approved by the participants with a unanimous vote – that the European countries should establish anthropology laboratories in their respective colonies:

> Or, les personnes qui ont suivi aujourd'hui les démonstrations importantes de M. Broca sur le cerveau et sur les diverses applications de l'encéphalologie aux études anthropologiques, seront convaincues qu'il est surtout nécessaire de faire les études anthropologiques dans les pays memes où vivent les peuples que l'on étudie. Il est impossible de réunir les matériaux nécessaires pour ces recherches en Europe. On nc peut pas avoir des cerveaux assez bien conservés pour les étudier comme il convient. Elles ne sont possible que dans les colonies.[80]

Nonetheless, opening a *Festschrift* for Adolf Bastian in 1896, Virchow had to take note that the brain, from the point of view of racial anatomy, was still *terra ignota*.[81]

4 Antinomies and Paradoxes

Paris, 19th October 1876: at a session of the Société d'anthropologie, twenty of its members founded the Société d'autopsie mutuelle, on the initiative of the physician Auguste Coudereau. They agreed that the intellectual and moral future of humanity depended entirely on the most exact knowledge of the cerebral functions and the localization of the various faculties. The eleven articles of their statute insisted on the need to fill the serious gaps in the psychophysiological field, where the few notions already available resulted from autopsies

79 Nikolai von Miclouho-Maclay (and Rudolf Virchow), "Anthropologische Notizen, gesammelt auf einer Reise in West-Mikronesien und Nord-Melanesien im Jahre 1876," *Zeitschrift für Ethnologie*, 1878, *10*:101-119, p. 119.

80 Rudolf Virchow, "Création de laboratoires d'anthropologie dans les colonies," in *Congrès international des sciences anthropologiques* (cit. note 7), pp. 222-224.

81 Rudolf Virchow, "Rassenbildung und Erblichkeit," in *Adolf Bastian als Festgruss zu seinem 70. Geburtstage 26. Juni 1896 gewidmet von seinen Freunden und Verehrern* (Berlin: Dietrich Reimer, 1896), pp. 3-43, p. 13.

practiced exclusively on hospitalized pathological cases, of whose life everything was unknown. To be fruitful, the observations had to move toward brains that belonged to individuals of the educated classes, such as scientists, writers, industrialists, politicians, etc. Therefore, post-mortem examinations had to be brought out of hospital walls, in the interest of public health and the longevity of future generations. The proponents of the new Société predicted that numerous prejudices would hinder such practice for a long time, though they were not discouraged by that realistic feeling. It was a matter of setting the example first, by signing a sort of living will for knowledge and prevention:

> Je soussigné, désire et veux qu'après ma mort il soit procédé a mon autopsie, afin que la découverte des vices de conformation ou des maladies héréditaires à laquelle elle pourrait donner lieu, puisse servir de guide dans l'emploi des moyens propres à en combattre le dévéloppement chez mes descendants. Je désire en outre que mon corps soit utilisé au profit de l'idée scientifique que j'ai poursuivie pendant ma vie. Dans ce but, je lègue mon cadavre, et notamment mon cerveau et mon crâne, au laboratoire d'anthropologie, où il sera utilisé de la façon qui semblera convenable.[82]

Those twenty *libre penseurs* belonged to the same generation, the one born around 1830, had opposed the Second Empire, and in the new climate of the Third Republic shared the will to affirm secular and materialistic principles. Their scientific and political commitment nourished each other, enabling the Société to stay open until the eve of the First World War, reaching up to about one hundred members over time. In 1878, the first autopsy was performed by Broca, assisted by Coudereau and Henri Thulié, on the corpse of Louis Asseline, a radical and materialist politician, a friend of Émile Zola and editor of Diderot's works, "un des conférenciers les plus féconds et les plus recherchés", who had died at forty-nine. Among other things, the report spoke of a brain larger than the average at that age, but with thick and coarse convolutions, despite Asseline's intelligence was of great finesse. Another sign of inferiority lay in the very deep parieto-occipital fissure, a trait that Broca had detected in the hemisphere of a gorilla and was often visible in women. Therefore the case deserved much attention even though the anterior parts of the brain were so

82 "Statuts de la Société d'autopsie mutuelle," *Revue scientifique de la France et de l'étranger*, 1876-1877, *11*:527-528. See Nélia Dias, "La Société d'autopsie mutuelle et le dévouement absolu aux progrès de l'anthropologie," *Gradhiva*, 1991, *n.10*: 26-36; Jennifer Michael Hecht, *Scientific Modernity, Atheism, and Anthropology in France* (New York: Columbia University Press, 2003), pp. 6-40; Hagner, *Geniale Gehirne* (cit. note 12), pp. 166-176.

developed as to hint to outstanding intellectual faculties, and the skullcap was extremely thin, with not fully ossified sutures: "Faut-il attribuer cette particularité au travail et au grossissement incessant du cerveau? Il serait audacieux de l'affirmer, il est bon cependant de poser la question".[83]

Other autopsies and reports came soon, for instance those of Coudereau, the demographer Adolphe Bertillon, who had taught medical geography at the École d'anthropologie, and of Léon Gambetta, one of the most prominent figures of the Third Republic and a superb speaker, whose third convolution – the motor center of language – was found to be extremely developed.[84] Whereas at first, until 1893, ideological motivations had prevailed, from then on the president Jean-Baptiste Vincent Laborde, professor of biological anthropology, assisted by a new steering group composed of professional anthropologists (Georges Hervé, Charles Letourneau, Henri Thulié), decided to make some statutory changes. The emphasis was now less on hereditary concerns than on anatomical aspects. In her essay, Nélia Dias estimates that, despite all the strenuous efforts to determine the correlation between the physical and the mental, the outcome of autopsies was not up to expectations.[85]

The idea that intellectual work could influence the volume and shape of the head had already a long and tortuous history, which Alexandre Lacassagne – then professeur agregé at the Val-de-Grâce military hospital – briefly retraced

83 Henri Thulié, "Sur l'autopsie de Louis Asseline, membre de la Société d'Anthropologie de Paris," *Bulletin de la Société d'Anthropologie de Paris*, 1878, *1*:161-167, p. 165. A brief discussion followed, in which someone was surprised to hear of a slight protuberance identified on the right side of the brain, while Asseline was known as a skilled lecturer: should not he have it on the left side, in the area of Broca? The answer reported another small convexity on the left parietal, and reminded that the right cerebral side often served as a substitute for the left one (p. 167).

84 Chudzinski Théophile, Mathias Duval, "Description morphologique du cerveau de Gambetta," *Bulletins et mémoires de la Société d'anthropologie de Paris*, III série, 1886, *9*: 129-152. The detailed verbal description was accompanied by eight drawings. Chudzinsky prudently warned that "ce ne sont là [...] que des pièces, des documents, dont une interpretation générale ne pourra être tentée que lorsque ces documents seront devenus plus nombreux. L'idée directrice de ces études sera naturellement la suivante: étant donnée une série d'encéphales ayant appartenu à des sujets caractérisés par le développement evident, incontestable de certaines facultés, rechercher si ces cerveaux présentent, dans leurs circonvolutions, des caractères particuliers, également évidents et incontestables, de sorte qu'on puisse dire que tel trait de la morphologie des circonvolutions est en rapport avec le développement de telle faculté cérébrale. Nous sommes encore loin de pouvoir tenter de nombreuses déterminations de ce genre." (p. 150).

85 Dias, "La Société d'autopsie mutuelle" (cit. note 82), pp. 33 and 35. Dias also points out that one of the merits of those materialists consisted in associating the seriousness of their scientific interests with the pleasures of the table, during periodic and sometimes bizarre collective dinners, opportunities for sociability (p. 36).

DIAMÈTRES	DOCTEURS en médecine.	SOLDATS sachant lire	SOLDATS illettrés.	DÉTENUS	DIFFÉRENCES
Longitudinal.	85^{mm},29	81^{mm},97	79^{mm},13	81^{mm},40	En faveur des docteurs : 4^{mm},56
Antérieur (bi-frontal).	48^{mm},91	43^{mm},65	42^{mm},35	41^{mm},62	— — 6^{mm},37
Postérieur (bi-occipital).	52^{mm},58	49^{mm},06	50^{mm},27	49^{mm},96	— — 2^{mm},82

FIGURE 3.10 Comparative measures taken on the heads of different categories of individuals by means of the "conformateur", in Alexandre Lacassagne, Cliquet, "De l'influence du travail intellectuel sur le volume et la forme de la tête," *Annales d'hygiène publique et de médecine légale*, II série, 1878, *1*:50-72, p. 62.

in 1878, paying tribute to "l'illustre Gall" for paving the way to some successors. The most outstanding of these were Parchappe and Broca, the results of whose research still had to be proven. For this purpose the future criminologist Lacassagne, in his hospital and with an assistant, had applied the *conformateur* – an instrument used by hat-makers to take the exact shape of the head – to almost 300 military personnel (medical doctors, literate or illiterate soldiers), a fairly homogeneous sample of individuals from the same milieu, with the advantage of eliminating the variables of stature and age. Furthermore, 91 inmates, refractory to every authority, were also subjected to the procedure. Their longitudinal, anterior transversal and posterior transversal diameters had been measured: in order that the differences between the four categories could be perceived at a glance, the averages were presented in a *tableau*: for each type of diameter, the doctors were ahead of the other categories, in particular regarding the frontal region, hence much more developed in those who had continued to cultivate their intelligence.

The two zealous measurers were far from claiming they had improved craniometry and anthropology, or from wanting to deduce anything 'philosophical' from their findings. However, the figures produced by the *conformateur* also had a hygienic implication. If it were true that intellectual work increased the front of the skull, the locus of the higher faculties, then education not only made man better, it also enhanced his brain, and through the generations such development would have eventually become a racial character. Lacassagne and Cliquet's report was presented at a session of the Académie de médecine, and discussed by colleagues who doubted the general value of such results. All of them conceded that there was indeed a relationship between the cranial capacity and the development of intelligence, but it was also known that small heads were sometimes associated with first-rate individuals. How could one forget that Gratiolet had foregrounded this circumstance, focusing instead on the complexity of convolutions as a decisive factor?[86]

Numerous observations – made by Lélut, Parchappe, Wagner, Broca – had confirmed the importance of brain weight, to which, however, it would have been misleading to attribute an absolute value. In 1878, this was the belief of Samuel Pozzi, a former student of Broca destined for a brilliant career as a gynecologist and a lover of famous women. First, individual variations – depending on age, sex, diseases, height, etc. – demanded large numbers, and if the organ had been weighed on a sufficient scale of Europeans, not as much had been done for distant populations, since the operation should be performed on fresh brains, and not on those preserved in alcohol. Various causes of error, therefore, awaited the unprejudiced observer. The density of cerebral matter, as well as the volume and richness of convolutions, were probably increased by intellectual activity. For instance, the brain of an Australian superior to his fellow men might be heavier and more wrinkled than that of a mediocre Parisian.[87]

The study of the brain of idiots – in Pozzi's opinion – raised considerable problems, showing that sometimes individuals devoid of any intelligence had a cerebral weight equal to or above the average. Not to mention the comparative analysis of brains of human races, which would not offer variations in weight corresponding to the enormous difference that exists in intellectual development. Pozzi recalled the statistics that had been published a few years earlier by Crochley Clapham, clinical assistant at the West Riding Asylum, on 1200 patients subjected to post-mortems. Its data supported the truth of Wagner's conclusion that superiority of size could not be regarded as a constant

86 Alexandre Lacassagne, Cliquet, "De l'influence du travail intellectuel sur le volume et la forme de la tête," *Annales d'hygiène publique et de médecine légale*, II série, 1878, *1*:50-72. Almost twenty years earlier, the surgeon, illustrator, and painter Émile-Théophile Blanchard had already referred to the hat-makers, "juges compétent en pareille matière", who knew that headgears sold to country men were smaller than those manufactures for city dwellers: see his *Note sur la conformation particulière de la tête observé dans le Limousin* (Limoges: Chapoluaud Frères, 1859), p. 23. However, Broca undertook to prove the risks inherent in making use of the *conformateur*, an ingenious tool "excellent dans la chapellerie", but to be excluded in anthropology: see his "Sur la fausseté des résultats céphalométriques obtenus à l'aide du conformateur des chapeliers," *Bulletins et mémoires de la Société d'Anthropologie de Paris*, III série, 1879, *2*:101-106.

87 Samuel Pozzi, "Du poids du cerveau suivant les races et suivant les individus," *Revue d'anthropologie*, 1878, II série, *1*:277-285.

accompaniment to intellectual primacy.[88] A further illustration of that truth was proposed, in 1878, by Clapham's weighing of the brains of 16 Chinese, 4 Pelew Islanders, and one Bengalese constable, "with much care, eliminating as far as possible all elements of fallacy". Fifteen of them had been victims of the fury of the great Typhoon, which raged Hong-Kong in September 1874, and all of them belonged to the "Coolie" or lowest grade of Chinese society. Of course, he could say nothing of the amount of intellect displayed by them during life, but, judging by their fellows of the same order, one was justified in supposing it rather poor, at least as far as performance is concerned. The average weight of their brain was nonetheless surprising, as if the Chinese males were Mandarins, and the Pelew Islanders held high posts in the government of their kingdom:

> The only explanation which I can offer of the large size of the brain in these cases is that it was essentially musculomotor in function, – the Chinese especially being very liberally endowed by nature with muscular tissue, much more so indeed than any European nation; – that it was in no wise an index of the amount of intelligence possessed is, I think, sufficiently clear.[89]

In fact, upon scanning the brains more attentively, the solution of the conundrum was at once rendered apparent: the primary convolutions were too well

88 Crochley Clapham, "The Weight of the Brain in the Insane," *West Riding Lunatic Asylum Medical Reports*, 1873, *3*:285-298; 1876, *6*:11-26. 500 cases were also curiously tabulated according to their "religious persuasion", noting that "not only have the Roman Catholics heavier brains than the Protestant Dissenters, and these again than those of the Church party, but also that the Cerebellum, Pons, and Medulla, as compared with the entire brain, are proportionately larger in the Church of England cases than in the Roman Catholics." Clapham did not see any reason to doubt that the same relation would hold good "in the case of those enjoying their liberty" (pp. 23-25). It is worth mentioning that for a while James Crichton-Brown turned the West Riding Lunatic Asylum, under his direction, into a research institute and promoted the keeping of rigorous medical records. His first contribution to *Brain*, the journal he co-founded in 1879, referred to the introduction of a more complete method of examining the brain: see his "On the Weight of the Brain and its Component Parts in the Insane," *Brain. A Journal of Neurology*, 1879, *1*:504-518; 1880, 2:42-67.

89 Crochley Clapham, "On the Brainweights of some Chinese and Pelew Islanders," *The Journal of the Anthropological Institute of Great Britain and Ireland*, 1878, *7*:89-94, p. 92. More generally, on the attitude towards Asian populations see Michael Keevak, *Becoming Yellow: A Short History of Racial Thinking* (Princeton: Princeton University Press, 2011).

defined, there being a marked deficiency in the number and depth of the secondary gyri and an almost "Simian symmetry" of the hemispheres.[90]

In attempting to explain the antinomies and paradoxes generated by weighing brains, Pozzi also resorted to the dual function of the organ, *idéo-moteur* and *musculo-moteur*. Experimental physiology was then demonstrating that the application of electricity to certain cerebral regions caused spasms in corresponding groups of muscles, and that their destruction was followed by paralysis. A savage spends his whole life hunting and waging war: this continuous and extreme physical exercise would justify the relatively considerable weight of his brain. One could similarly understand the mediocre weight of a savant's brain. But no illusion was allowed: the puzzling question of brain weight was still open and perhaps for a long time, although the histological knowledge would have most probably provided new and precious elements.[91]

Contrariwise, in the words of the polygraph Gustave Le Bon, still far from the international fame that his *Psychologie des foules* would have earned him, all was clear enough: inequalities in cerebral development did indubitably correspond to diversity of feeling and intelligence. Just a century before – he argued – most philosophers had assumed that all men were endowed with equal organization and intelligence, while every difference was nothing but an effect of education. The subsequent, careful study of human races would have revealed the inanity of that belief and proved that scientific and intellectual differences were indeed profound and innate. Le Bon's long essay of 1879 was awarded by the prix Godard of the Société d'Anthropologie, not without reservations from the jury about the defiant tone of its thesis.[92] Especially original

90 In a different context, but during the same period, "Ape-like peculiarities" were abundantly found by A. J. Parker while analyzing the cerebral convolutions of the Negro brain, and communicated to the Academy of Natural Sciences of Philadelphia. See the minutes of the sessions: "Cerebral Convolutions of the Negro Brain," *Proceedings of the Academy of Natural Sciences of Philadelphia*, 1878, *30*:11-15; "Simian Characters in Negro Brains," ibid., 339-340.

91 Pozzi, "Du poids du cerveau" (cit. note 87), pp. 284-285.

92 See Charles Letourneau, "Rapport sur le prix Godard, année 1879," *Bulletins de la Société d'Anthropologie de Paris*, III série, 1879, 2:373-386. In 1878, Le Bon had already had a polemical exchange with Clémence Royer on cerebral sexual differences: see *Congrès international des sciences anthropologiques* (cit. note 7), pp. 118-119. Broca himself, president of the Congress, had made a point: "L'un des facteurs de la difference du poids entre le cerveau de l'homme et celui de la femme vient en grande partie de l'éducation que les civilisés se donnent à eux-mêmes et qu'ils ne donnent pas à leurs femmes. Quant à l'influence de l'éducation sur le volume du cerveau, nous la connaissons et nous l'avons démontrée par des études faites sur des lettrés et sur des illettrés. [...] Que l'instruction se perfectionne et se generalize dans un État et on verra le poids moyen du cerveau augmenter." (p. 119).

was its methodological intent, consisting in the declaration that the average value obtained by adding "des grandeurs quelconques" and dividing the sum by their number was an imaginary abstraction, which did not have any match in reality. Quetelet and Bertillon had already criticized that primitive procedure, in particular the Belgian statistician who elaborated the graphic method of the binomial curves. Le Bon proposed to adopt both a different curve system and a new, simple, and portable instrument called *compas de coordonnée*, especially useful for travelers to measure skulls or heads. The advantage of his system was explained as follows:

> La propriété que possédent ces nouvelles, de traduire immédiatement par leur aspect la variation en centième des éléments existant au sein d'un groupe donné, me les a fait designer sous le nome de *courbes centesimales* ou *courbes de série*. Au lieu de confondre, comme le font les moyennes, des résultats très-différents, ells donnent au contraire un tableau très-fidèle des moindres nuances que présentent les éléments d'un groupe dont on veut connaître les facteurs. N'y eût-il qu'un individu sur cent d'une capacité cranienne ou d'une taille donnée, la courbe l'indiquera immédiatement, alors que les moyennes n'eussent même pas permis de soupçonner sa présence.[93]

Such mathematical practice was so effective that it could highlight, for example, that the superiority of a race consisted much more in its greater or lesser proportion of voluminous skulls than in the difference of its average cranial capacity.

Le Bon condensed the outcome of his meticulous research in thirteen points. To sum them up, the variations in the volume of the skull (and brain) would be much higher than they look when one merely observes the averages. The cranial capacity can vary to twice its size, and differences in excess of 600 cm^3 are found in the upper races, a gap that exceeds 800 cm^3 in subjects of different races. The influence of height on brain weight would be minimal, while body weight has a greater effect, and the sex factor is even more conspicuous. That the distance in the cranial volume between male and female would increase through the scale of civilization is noteworthy, "en sorte qu'au point de vue de la masse du cerveau, et par suite de l'intelligence, la femme tend à se

93 Gustave Le Bon, "Recherches anatomiques et mathématiques sur les lois des variations du volume du cerveau," *Revue d'Anthropologie*, II série, 1879, 2:27-104, p. 36.

différencier de plus en plus de l'homme."[94] In the most intelligent groups, like the Parisians of the time – "où le rôle de la femme est nul" – there would be a noticeable proportion of the female population whose skull approached in volume to that of the gorilla more than that of the more developed male one. Measurements taken on 1200 heads of Parisians and on a number of *paysans* were ranked as folllows, as to the volume of the heads: "1° savants et lettrés, 2° bourgeois parisien, 3° nobles d'anciennes familles, 4° domestiques, 5° paysans."[95]

As a consequence of all these data, Le Bon believed that the shape, structure and volume of the brain were closely related to the development of intelligence, the third of these factors being the most important. Not only the average cranial capacity of the superior races exceeded that of the inferior ones, but also their true superiority resulted above all from their containing a greater number of large skulls. The difference in cranial capacity between individuals of the same race varied widely from one race to another and increased in the higher races, which would demonstrate its tendency to grow more and more, in spite of any egalitarian tenet.

Anatomists and physiologists had sometimes believed, and sometimes doubted, that the mass and the weight of the brain were in direct relation to its function as an organ of mental activities. It was with this plain statement that in 1880 Theodor Ludwig Wilhelm von Bischoff began his attempt at making the issue less intricate.[96] He had retired at the end of a long career spent teaching anatomy in Heidelberg, Giessen, and from 1854 in Munich. He had graduated in Heidelberg with a thesis *De Nervi Accessorii Willisii Anatomia et Physiologia*, and, while he devoted himself to embryology by carrying out a series of developmental studies on mammals, he had also cultivated a constant interest in the nervous system. It is worth recalling that, when Tiedemann died

---

94 Ibid., p. 102. Some time after, Émile Durkheim quoted approvingly Le Bon's statement about the differences between man and woman: see his *De la division du travail social* (Paris: Félix Alcan, 1893), p. 62.

95 Le Bon, "Recherches anatomiques et mathématiques" (cit. note 94), p. 104. A decade or so later, Le Bon took up the subject in a large volume, stating how it was useless to employ anatomical features – as had been the case for a long time – for racial classification. Only gross divisions could come from the color of the skin or hair, the shape or volume of the skull. Psychology could better define the boundaries between the races, in character and intelligence, waiting for the time, probably far away, when the progress in brain studies would discover the real explanation of racial differences. See his *Les premières civilisations* (Paris: Flammarion, 1889), and an advance excerpt: Id., "Psychologie. L'influence de la race dans l'histoire," *Revue scientifique* (*Revue rose*), III série, 1888, *15*:525-532.

96 Theodor L. W. von Bischoff, *Das Hirngewicht des Menschen. Eine Studie* (Bonn: P. Neusser, 1880).

in 1861, his son-in law Bischoff read the eulogy and disclosed the results of the autopsy that the old 'egalitarian' anatomist had prescribed in his will, at the service of science. His brain had seemed strangely light (1254 grams) for a presumed general average of 1387 so Wagner had taken advantage of this to reiterate his skepticism about a direct link between weight and intelligence. That same interest had driven Bischoff, in 1868, to explore the convolutions of men and apes, sure as he was of an unbridgeable distance between them. Quite understandably, he harbored a clear distrust of the Darwinian theory.[97]

In his 1880 work, Bischoff processed data from 906 brains – weighed with their membranes – which had given an average weight of 1362 grams in males (aged between 17 and 80 years), and of 1219 in females (aged between 15 and 80 years). Although rather wide, the sample was made up only of inhabitants of Bavaria who had died in hospitals or in prison, many of them of tuberculosis, which had the effect of decreasing their encephalic mass; consequently, they belonged to a social class usually tending to produce the smallest contingent of heavy brains. 150 pages of tables impressively display, on a case-by-case basis, brain weight, body weight, age, height, and cause of death. A chapter of Bischoff's text takes into consideration the influence exerted by race and nationality, with a chronicle of the attempts made in this direction.

First of all, earlier methods for filling skulls with various materials were rejected, due to the technical difficulties they presented and to the noticeable difference between the real brain weight and the *Schädelinnenraum*. Therefore, a perfect substitute for direct weight would never be at hand: Bischoff agreed with Virchow about the urgency of a *vergleichende Encephalologie*, and moreover he thought that in part the procedures adopted up to then had discouraged scholars from finding and investigating brains of other races. It seemed to him unbelievable that American anatomists and physicians had not committed themselves to studying the brains of Blacks or natives, despite having plenty of material available. Again, only a few *Negergehirne* had been weighed in the past and, unfortunately, in multifarious ways and with differing weight units; the situation was not better with respect to other races. The question of the potential difference in the brain weight of various European

97 Theodor L. W. von Bischoff, *Gedächtnissrede auf Friedrich Tiedemann. Vorgetragen in der öffentlichen Sitzung der k. Akademie der Wissenschaften am 28. November 1861, als am allerhöchsten Geburtstage Sr. Majestät des Königs Maximilian II. Von Bayern* (Münich: Verlag der k. Akademie, 1861); Id., "Die Grosshirnwindungen des Menschen mit Berücksichtigung ihrer Entwicklung bei dem Fötus und ihrer Anordnung bei den Affen," *Abhandlungen der mathematisch-physikalischen Classe der königlich bayerischen Akademie der Wissenschaften*, 1870, *10*:387-497. For the discussion around Tiedemann's brain see Hagner, *Geniale Gehirne* (cit. note 12), pp. 156-159.

TABELLE I

über

gewicht, Körpergewicht, Alter, Grösse und Todesursache

von 906 Leichen

geordnet nach dem Hirngewicht.

1. Männer 559,

von 17–80 Jahren.

Hirngewicht in Grms.	Körpergewicht in Grms.	Alter	Grösse Ctm.	Todesursache	Bemerkungen
1018	45000	66	162	Marasmus	
1039	35110	78	171	Marasmus	
1069	48900	40	164	Pleuritis	
1075	55110	35	169	Tuberculosis	
1077	42800	30	160	Tuberculosis	
1095	43600	36	169	Phtisis	
1105	44800	39	168	Tuberculosis	Sträfling
1113	53000	58	166	Pneumonia	
1120	40950	70	158	Marasmus	
1133	45780	49	170	Morb. Brigthii.	
1150	49320	65	167	Tuberculosis	
1152	54000	74	172	Atrophia cerebri	
1153	62870	?	?	Decapitatus	Mörd.Langgut
1158	59930	23	161	Pneumonia	
1162	40000	62	148	Marasmus	
1162	43000	59	155	Morb. Brigth.	
1163	49650	67	157	Oedem. cerebri	
1166	53300	38	?	Pyaemia	
20110	867120				

FIGURE 3.11
First page of the first table in Theodor L. W. von Bischoff, *Das Hirngewicht des Menschen. Eine Studie* (Bonn 1880).

nations had also been raised and discussed, but neither the prerequisites nor the results of the previous endeavors were reputed to be satisfactory by Bischoff.[98]

In his *Schlussbemerkungen* he honestly apologized for not having made himself great progress in the field, but still hoped that his *Studie* could supply some useful elements for further elaboration. He did not miss the opportunity to counter the thesis of those who foresaw a future of equality between men and women, in case the latter would reach emancipation. As a matter of fact, even imagining a possible improvement in the achievements of both sexes, the hiatus between them would have remained intact. Significantly, in his fight to preserve the old gender-specific and hierarchical role patterns, in 1872 Bischoff had hindered women's access to medical studies, as it would contravene their

98 Bischoff, *Das Hirngewicht des Menschen* (cit. note 96), pp. 65-91.

nature, both physical and moral.[99] No surprise, then, that he took a radical position about the impossible advancement of the inferior races, which, coming into contact with the more developed ones, would predictably tend to become extinct, due to their limited capacities: he doubted whether education and crossbreeding could help elevate their condition.[100]

Recently admitted to the Société d'Anthropologie and soon appointed its assistant secretary, the young Georges Hervé reviewed Bischoff's *Studie* in 1881, taking stock of the matter, and bringing it back to its essential points. So, in short, the absolute brain weight is defined as extremely variable "suivant les individus", with an identifiable upper limit in the well-known case of Cuvier, although Bischoff reported a brain of 1925 grams in a perfectly healthy worker, and Karl Asmund Rudolphi had written of a man died in 1819, whose brain apparently reached the unlikely weight of 2222 grams. A lower limit compatible with the integrity of intelligence had seemed to set at 1018 grams. Such a wide variation required large data sets, so as to flood, somehow, the individual into the mass, but a glance at the statistics then available showed how incompletely this indispensable expedient had been applied. An absolute lower weight for the female brain seemed established in all races, with a distance that tended to decrease in the inferior ones. The relationship between encephalic weight and body weight had been rather neglected in the past, due to the difficulties and contradictions that it entailed, and one should not forget that body weight was a further factor that acted synergistically to modify the brain weight. In its development, the brain obeyed the general law of organ growth, while Hervé complained that it was much more problematic to formulate laws about changes in brain mass in connection with the different of physical and intellectual attitudes of the human groups:

> Sur cette question, nous ne possédons malheureusement que des données tout à fait insuffisants. Il n'est pas facile, en effet, même pour les

99 Theodor L. W. von Bischoff, *Das Studium und die Ausübung der Medicin durch Frauen* (München: Literarisch-artistische Anstalt Th. Riedel, 1872). Differences in the structure of skull and brain between man and woman would be so characteristic and profound that they should not be overlooked in the judgment of mental differences. Bischoff chiefly emphasized the significant fact that, according to all observers without exception, in all races and peoples the absolute weight of the male brain exceeded the female one of 130-160 grams. He also tried to demonstrate that the relative weight, more favorable to the woman, had no meaning at all. Needless to say, these so-called anatomical *Thatsache* were put in correspondence with conventional evaluations, both psychological and moral (see pp. 15-20).

100 Bischoff, *Das Hirngewicht des Menschen* (cit. note 96), pp. 170-171.

> races européennes, de rèunir en grand nombre les matériaux qu'exige cette étude, et cette difficulté devient un obstacle presque insurmontable quand il s'agit des races exotiques; c'est tout au plus s'il est possible alors de se procurer quelques rares encèphales dont la masse est notablement rèduite par un séjours prolongé dans les liquides conservateurs. A dèfaut de pesées directes, on s'est adressé à la capacité crânienne, et l'on a essayé d'estimer d'après elle le poids probable de l'encèphale dans les différentes races.[101]

In his review Hervé discussed, in addition to Bischoff's book, an essay that had an identical title, by the Italian anthropologist Giustiniano Nicolucci, whose monogenist treatise *Delle razze umane* dated back to 1857-1858, containing a classification of the human species by morphological and linguistic criteria. Since 1863, when he was elected to the first Parliament of the recently proclaimed Kingdom of Italy, he began to wonder from which races the peninsula had been inhabited in prehistoric times. During the Stone Age – he hypothesized – brachycephalic populations, of Turanian origin, had preceded migratory waves of dolichocephalic Arians in the Bronze Age. Living relics of those old lineages could still be found in Liguria and Piedmont, but their skulls also appeared further south, and testified to a wide settlement, never crushed by subsequent events, not even by the dominion of Rome. The ancient authors had already celebrated their tenacity and strength, their love for stability and order: excellent people who, having become a hegemon in Italy, had been able to promote and achieve national unity.[102]

In 1881 Nicolucci thought it appropriate to collect the results of the numerous researches carried out on brain weight by a few colleagues of different nationalities. By adding the weights of 4875 healthy brains detected by them, the average was 1331 grams for men and 1189 for women. Nonetheless – Hervé objected in his review – the figures thus amassed came from very different weighing methods, and therefore the value of his calculation was inevitably played down. In any case, Nicolucci had simply offered a sort of minimum common denominator for the on-going discourse. In short: as the brain grows bigger, so the intelligence expands and increases, both in the individual and in the races; the female brain is always inferior in weight to the masculine; the encephalic

101 Georges Hervé, "Du poids de l'encéphale," *Revue d'Anthropologie*, II série, 1881, *10/4*:681-698, p. 690.

102 Giustiniano Nicolucci, "La stirpe ligure in Italia ne' tempi antichi, e ne' moderni. Memoria letta nella tornata del 6 ottobre 1863," *Società Reale di Napoli. Rendiconto dell'Accademia delle scienze fisiche e matematiche*, 1863, 2:225-234.

hierarchy must be measured by the extension of the cortex and by the complication of its folds and furrows; the average weight of the brains of illustrious men surpasses that of ordinary men; despite the scarcity of knowledge about the brain weight of various races, it could still be stated that the Negro has almost one hectogram less than that of Europeans.[103]

Meanwhile, some Italian alienists gave news of their researches on asylum populations. In 1881, over 300 in-patients who had died in the mental hospital of Reggio Emilia – one of the most renowned back then – allowed Giuseppe Amadei, at the beginning of his career, to focus on the correspondence between height and brain weight, which did not however concern intellectual power, but only those cerebral centers responsible for nutrition, sensation and movement.[104] From the psychiatric institute of the University of Turin, Enrico Morselli insisted on the "little" (*poca cosa*) that was known around the volume, shape, weight, density, permeability, calorific and electrical conductivity of the brain. The obstacles on the way were both the great individual variability within a still uncertain range, and the misleading subjectivity of the observers, which produced inconsistent results. Accurate and repeated observations were needed, that is moving from the "sentimental" to the "experimental" method: the "arid work of multiplied weighing" would be compensated by the goals achieved. Morselli then described his methods of calculation, which he had applied the previous year on a few dozen cases at the asylum of Macerata, discovering that the specific weight of the brain of the insane was on average higher than that of sane individuals.[105]

The dean of Italian psychiatry, Andrea Verga, gave two lectures on the subject in the Ospitale Maggiore of Milan in 1882-1883, reminding his audience that physiologists had begun to regard the palpation of heads or skulls as unscientific, and had preferred to directly focused on the brain, with a view to determining its weight and volume. In doing so, they had somehow revived

103 Giustiniano Nicolucci, "Sul peso del cervello dell'uomo. Studii," *Società Reale di Napoli. Rendiconto dell'Accademia delle scienze fisiche e matematiche*, 1881, *20*:22-30, 40-59.

104 Giuseppe Amadei, "Studj sulle variazioni del peso cerebrale. Il peso del cervello in rapporto alla statura," *Archivio italiano per le malattie nervose e più particolarmente per le alienazioni mentali*, 1881, *18*:199-206.

105 Enrico Morselli, "Il peso specifico dell'encefalo negli alienati. Studio critico e sperimentale," *Rivista sperimentale di Freniatria e di Medicina legale*, 1882, *8*: 58-82, 206-247. In the following years, Morselli did not stop collecting data, expanded his statistical sample, and presented new insights: see his "Studi di Antropologia patologica sulla pazzia – I. Sul peso comparativo dei due emisferi cerebrali negli alienati," *La Psichiatria, la neuropatologia e le scienze affini*, 1887, *4*:279-301; "Studi di Antropologia patologica sulla pazzia – I. Sul peso dell'encefalo in rapporto ai caratteri craniometrici negli alienati," *Rivista sperimentale di Freniatria e di Medicina legale*, 1887, *13*: 365-388.

Gall's old, brilliant intuition about the organ of intelligence. Unfortunately, "quot cerebra, tot pondera, tot mensurae": observers had not yet reached agreement about how to investigate and understand the organ. Verga himself, in the past, had enjoyed weighing a number of brains ("mi sono divertito anch'io a pesare degli encefali umani"), 16 in Pavia and 63 in Milan; to remember the approximate average weights, he had resorted to a rhymed-based mnemonic device: "Pesa il cervello umano un chilo e un etto, e un etto e mezzo il solo cervelletto" – which was, incidentally, a rather low estimate.[106]

Caution imposed by scientific deontology emerges as much from Verga's words as from a *Guida* to the study of convolutions, an expanded edition of which was published in 1884, after the first version of 1878. In those years, his author, Carlo Giacomini, had started teaching in and directing the anatomical institute at the University of Turin. Among his accomplishments, there was a new process to better preserve brains through drying, in order to eliminate the inconveniences caused by previous ones.[107] What his *Guida* offered was a broad perspective on the variations of brain folds according to age, sex, individual and race, but also an introduction to the best methods for determining absolute and specific weight, volume and shape, extension of cortical surface, thickness and quality of cortical and medullary substance. As long as the cortex had seemed not excitable – Giacomini noted – its twisted morphology had been understated; it was supposed to be composite, but such was the discredit to which Gall and his followers condemned it, that the localization of the articulated language identified by Bouillaud and Broca met some resistance. In the 1870s, however, cerebral physiology had entered a new phase, thanks to the experimental stimulations conducted by Gustav Theodor Fritsch and Eduard Hitzig and by David Ferrier: the convolutions, judged for a long time completely irregular, finally appeared constant in number, direction, mutual relations, genesis and vascularization.

In preparing his *Guida*, Giacomini had used both the existing literature on the subject and a vast collection of Italian brains, coming mainly from normal subjects. He ventured some hypotheses, for example about the scarce significance of sexual differences on the cortical surface, or the precarious findings about racial varieties: although the brain weight of colored races had mostly been reputed inferior to the general average, he advised not to take the superiority of the European brain for granted. As for the kind of relationship between

---

106 Andrea Verga, "Del peso e del volume dell'encefalo," in Id., *Studi anatomici sul cranio e sull'encefalo, psicologici e freniatrici* (Milano: Manini-Wiget, 1896), I, pp. 266-277, p. 269.

107 Carlo Giacomini, "Nuovo processo per la conservazione del cervello," *L'Osservatore. Gazzetta delle cliniche di Torino*, 1878, *14*:421-425, 433-440.

brain development and intellectual functions, the anatomist believed that the elements required to untie the knot were not yet available.[108] His conviction about a still too unsatisfactory state of knowledge led Giacomini to question also the assurance of criminal anthropology – represented by the Austrian Moritz Benedikt and, of course, by his colleague Cesare Lombroso – about a special conformation that the brains of social misfits would unavoidably exhibit.[109]

Among his research interests, for two decades Giacomini also dealt with the general *anatomia del negro*, on which he published a series of memories, his collection coming to include 14 Africans of various origins. He started by complaining that the notions of human anatomy were based almost exclusively on the study of the Caucasian race – from which the ideal and most perfect type had been derived – while the other races had been very incompletely addressed. One could assert that science did not possess a complete description of the structure of a single non-white individual. In 1878 two Abyssinian women, mother and daughter, died at the Cottolengo Hospital in Turin, the first specimens in a small series that would increase in the following decade. Over time, Giacomini analyzed a few of their body parts, except for the nervous system, though planning to deal with it and with the sensory organs. His untimely death in 1898 prevented him from continuing the work he had undertaken.[110]

108 Id., *Guida allo studio delle circonvoluzioni cerebrali dell'uomo* (Torino: Loescher, 1884), II ed. A shorter French version helped the international circulation of the text: see Id., "Variétés des circonvolutions cérébrales chez l'homme," *Archives Italiennes de Biologie*, 1882, *1*:1-64.

109 Benedikt had observed four (instead of three) frontal convolutions in criminal brains, an anomaly that would bring them closer to those of carnivores. Giacomini replied that the proportion between normal and abnormal hemispheres was roughly the same in criminals and non-criminals. See Moritz Benedikt, "Sulla questione dei cervelli dei delinquenti. Lettera aperta al prof. Carlo Giacomini in Torino (con note del prof. Giacomini)," *L'Osservatore. Gazzetta delle cliniche di Torino*, 1883, *19*:141-143, 156-159, 174-176. For his part Lombroso stressed the importance of cerebral stigmata in the third edition (1884) of his *L'uomo delinquente*, and in the fifth one (1886-1897) he devoted a whole chapter to the anomalies of the criminal brain. Since 1885, however, Benedikt became increasingly critical of the Italian school, to the point of questioning, at the Second Congress of Criminal Anthropology held in Paris in 1889, its basic principle of a cerebral localization for criminal propensities. Instead, he evoked the possibility of "troubles moléculaires de la substance cérébrale": see his interventions in *Actes du deuxième Congrès international d'anthropologie criminelle. Biologie et sociologie* (*Paris, août 1889*) (Lyon-Paris: Storck-Masson, 1890), pp. 164, 169, 238-240, 416.

110 Carlo Giacomini, "Annotazioni sull'anatomia del negro," *Giornale dell'Accademia di medicina di Torino*, serie III, 1878, *24*:454-470, 506-519; 1882, *30*:729-805; 1884, *32*:462-483; 1892, *40*:17-64; serie IV, 1897, *3*:649-672; *Atti della R. Accademia delle scienze di Torino*, 1886-87, *22*:465-483. See Francesco Loreti, *Contributo alla storia dello "Studio" anatomico della*

He bequeathed his remains to the Institute where he had been trained and where he had spent all his professional life: the skeleton and brain, the latter prepared with his method, both of them turned into a museum display. It was also deemed necessary to illustrate in minute detail his "precious" brain and pass it on to posterity as an object of teaching and comparison.[111]

One of Giacomini's former students, Giuseppe Sperino, took charge of the task and wrote a wide-ranging report, which clearly stated, from the very beginning, that neither the weight of the brain nor the wealth and complication of the convolutions always correspond to great intellectual performances, and that mediocre individuals could sometimes have brains heavier than the average. This being said, Giacomini's brain was found to be rather heavy (1495 grams), rich in convolutions – some of them quite large – and in secondary sulci. Then, Sperino explored the structure of the two hemispheres and showed them in four plates, signaling above all the duplicity of the fissure of Rolando, a very rare character described for the first time by Giacomini himself, who had identified it in a sort of moron. Since then, 23 similar cases had been recorded, mostly in alcoholic, criminal, and uneducated individuals. When Sperino presented the results of his dissection to the Academy of Medicine of Turin, Lombroso did not miss the opportunity to report that the same anomaly had been seen in some delinquent brain, and to argue that it was not at all contradictory in an extraordinary man like Giacomini, given the kinship between genius and degeneracy.[112] Other peculiarities of his brain concerned the development of the *gyrus supramarginalis* in the left hemisphere and of the *gyrus angularis* in the right one, as already observed in other eminent personalities. Sperino concluded by hoping that Giacomini's example would be followed and that a society of autopsies modeled on the French one could be established in Italy as well.[113]

Born into a Jewish family but disobedient to tradition from a very young age, Lombroso faced the question of anti-Semitism in 1894, defending the idea of emancipation and rejecting the increasingly frequent racist attacks, a disruptive occurrence that deserved the utmost attention from the point of view of

---

Università di Torino. Carlo Giacomini (1840-1898) (Torino: Accademia delle Scienze, 1963), pp. 39-45.

111 See Silvano Montaldo, "The Relics of Two 19th-Century Scientists," in *Savant Relics. Brains and Remains of Scientists*, edited by Marco Beretta, Maria Conforti, and Paolo Mazzarello (Sagamore Beach: Science History Publications, 2016), pp. 183-199.

112 See Cesare Lombroso, *Nuovi studii sul genio. Volume II. Origine e natura dei genii* (Milano-Palermo-Napoli: Sandron, 1902), pp. 177-179.

113 Giuseppe Sperino, *L'encefalo dell'anatomico Carlo Giacomini* (Torino: Unione Tipografico-Editrice, 1900).

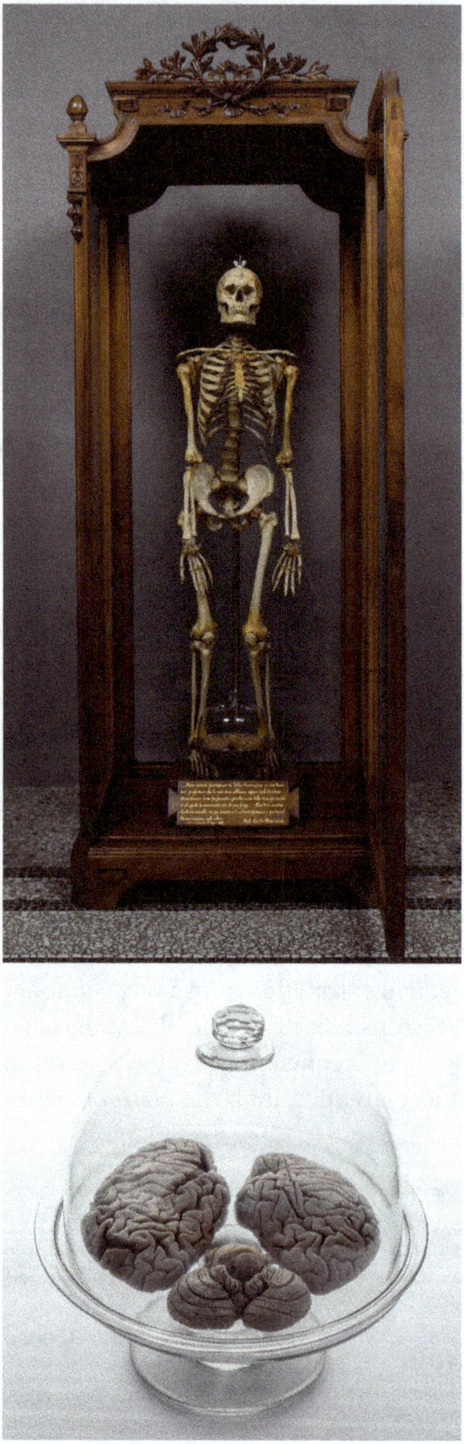

a

b

FIGURE 3.12A-B
Carlo Giacomini's skeleton and brain, Museo di Anatomia Umana "Luigi Rolando", Torino. Photographs by Nadia Pugliese.

socio-anthropological science. On the one hand, anti-Semitism seemed to him an atavistic yet modern phenomenon, recently reinvigorated by nationalist propaganda for political purposes; on the other hand, he reduced Aryanism to a set of false and regressive myths. Regarding the issue of the 'purity' of the Jewish race, Lombroso was certain that for centuries the European population had been a real jumble of "mixed races", which helped promote the growth of its civilization.[114] An advocate of assimilation, he urged his co-religionists to get rid of the ancient, ridiculous rituals and to face the future without the constraints of an anachronistic identity. As maintained by him, three factors might have contributed to determining the propensity of Jews for intellectual professions and commercial speculation: a greater development of their brains, the conditions in which they had been living for a long time, and the habits they had inherited from their ancestors.[115] He also referred favorably to a speech given in 1892 by Felix von Luschan at the 23rd meeting of the Deutsche Gesellschaft für Anthropologie, which had tried to define Jews as a "Rassengemisch" composed of several elements, thanks also to the measurements taken on 60,000 individuals. The Austrian-German anthropologist, whose research on racial subjects would continue very intensely in the following two decades with ambivalent assessments, viewed European societies as prominent due to the great variety of their components.[116]

---

114 A general list of titles compiled for only a decade reached almost a hundred pages: see Joseph Jacobs, *The Jewish Question 1875-1884. Bibliographical Hand-List* (London: Trübner and Co., 1885). The social scientist and folklorist Jacobs was born in Sidney to a Jewish family and often dealt with the topic: see for instance his "On the Racial Characteristics of Modern Jews," *The Journal of the Anthropological Institute of Great Britain and Ireland*, 1886, *15*:23-62, where he supported "the long-standing belief in the substantial purity of the Jewish race" and the assumption that "the vast majority of contemporary Jews are the lineal descendants of the Diaspora of the Roman Empire" (p. 52). The article also briefly mentioned the few available craniometric data, which "seem to show that Jews are predominantly brachycephalic, and are not physically long-headed" (p. 34). However, the scarce statistics that had been produced up to then made it evident "how untrustworthy mere impressions are, even when those of a trained observer" (p. 35).

115 Cesare Lombroso, *L'antisemitismo e le scienze moderne* (Torino-Roma: L. Roux e C., 1894), p. 135.

116 Felix von Luschan, "Die anthropologische Stellung der Juden," *Correspondenz-Blatt der Deutschen Gesellschaft für Anthropologie, Ethnologie und Urgeschichte*, 1892, *3*:94-102 (with two comments by Rudolf Virchow and Moritz Alsberg). Regarding von Luschan's arguments about the *Rassengemisch* see Amos Morris-Reich, "Photography in Economies of Demonstration: The Idea of the Jews as a Mixed-Race People," *Jewish Social Studies: History, Culture, Society*, 2013, *20*:150-183. More in general see John David Smith, "W.E.B. Du Bois, Felix von Luschan, and Racial Reform at the Fin de Siècle," *Amerikastudien/American Studies*, 2002, *47*: 23-38; Peter Ruggendorfer, Hubert D. Szemethy (Hg.), *Felix von Luschan. Leben und Wirken eines Universalgelehrten* (Wien-Köln-Weimar: Böhlau Verlag, 2009).

An appendix to Lombroso's text of 1894 offered a series of anthropometric data on Turin-born Jews: height, hair color, and cephalic index, which indicated an analogy with the characters of the population within which they lived, except for the greater abundance of dolichocephalics, the scarcity of exaggerated cranial capacities, and the higher differentiation of cranial forms, probably due to the succession of ethnic "grafts" (*innesti*) over time. He complained also that a tenacious prejudice of the Jews hindered autopsies or the removal of skulls from their cemeteries; consequently, he could measure only five Jewish skulls from the catacombs of San Calisto in Rome, and compare them with a few others of Phoenician origin, and so detect a very slight "Semitism" of the Jews, even in ancient times.[117]

The brains of Jews had been rather neglected over the course of the nineteenth century. In 1911, the Jewish-American anthropologist Maurice Fishberg highlighted that there were very few observations on record: he knew only of about thirty Jewish brains reported by the Russian Nikolai Vasil'evich Gil'chenko and Richard Weinberg, and by the Czech-Austrian Augustin Weisbach.[118] The same cannot be said about their physical anthropology: in 1902, Fishberg had considered it "one of the most debated questions", due to a comparatively large number of measurements taken on living subjects. Starting from these, some "authorities" had drawn different or opposite conclusions, some of them claiming that the Jews were a pure race (the descendants of the original Semites), others that they showed too variable physical traits to be lumped into a single whole. According to Fishberg, who illustrated his article with tables and graphs, "the problem becomes more complicated the more thoroughly we attempt to analyze it."[119] Although dealing with the question in his 1911 book, he basically set it aside:

> We are accustomed to hear of the great cerebral capacity of the Jew. His friends are always speaking with emphasis of his remarkable brain, while his enemies often speak of the danger the Jew, with his greater cerebral

117 Lombroso, *L'antisemitismo e le scienze moderne* (cit. note 115), pp. 113-121, 141-148.

118 Maurice Fishberg, *The Jews: A Study of Race and Environment* (London-New York: The Walter Scott Publishing Co.- Charles Scribner's Sons, 1911), pp. 57-58. About Gil'chenko and Weinberg see Marina Mogilner, *Homo Imperii. A History of Physical Anthropology in Russia* (Lincoln: University of Nebraska Press, 2013), pp. 257-266. According to Augustine Weisbach, the few available surveys did not support the opinion about a quantitative underdevelopment of the cranium and brain among Jewish, and more certain proof on the connection to race had not been brought forth: see his "Das Hirngewicht der Juden," *Zeitschrift für Demographie und Statistik der Juden*, 1905, *1/3*:5-10.

119 Maurice Fishberg, "Physical Anthropology of the Jews. I. – The Cephalic Index," *American Anthropologist*, 1902, *4*:684-706, p. 706.

> power, may be to his non-Jewish neighbour in Eastern Europe, who has not been endowed with as much brain tissue in his cranial cavity. It is, however, a remarkable fact that measurements of the size of the Jews' head and brain have shown that there is no basis for this belief. Of course it must be remembered in this connection that the size of the brain, and especially of the cranium, has not by far the significance as regards culture and intelligence which some have ascribed to it.[120]

5 Orphans of Broca

Charles Marie Joseph Bra was working as a young alienist at the Asiles de la Seine (Villejuif), when Henri Dagonet, his former teacher and chef de service at the Sainte-Anne hospital, provided him with the material relating to 800 pathological cases and containing the weights of various parts of the encephalon, the type of disease, age, sex, cause of death. Although the extraction and weighing procedure had been uniform, nevertheless Bra confessed in 1882 that

> Notre tâche était cependant assez lourde pour que nous sollicitons l'indulgence du lecteur. L'infinité des chiffres, des moyennes à calculer, la difficulté de ramener les nuances infinies de l'aliénation mentale aux types principaux, de coordonner et présenter les résultats, dans un sujet aussi complexe, de s'en tenir d'une manière definitive au plan préalablement conçus, toutes ces causes réunies nous ont rendu ce travail pénible.

Bra's warning was as follows: firstly, the impossibility, at that time, to measure intelligence should be accepted; secondly, the encephalon represented such a composite organ that the series of weights accumulated till then were of little worth. After working on it, his point of view was decidedly circumspect: "il faut donc ne pas exiger des pesées cérébrales plus que ce qu'elles ne peuvent donner." The anatomical and physiological conditions of composition, structure, quality, which contributed to the almost mysterious function called intelligence and ensured the balance of its processes, were largely independent of the encephalic weight. However, as stated by most authors and sources, intelligence would increase in proportion to encephalic weight and volume. So, the old riddle was getting in the way of the debate again.[121]

120 Fishberg, *The Jews* (cit. note 118), p. 56.

121 Charles Marie Joseph Bra, *Étude sur le poids de l'encéphale dans les maladies mentales* (Paris: Adrien Delahaye et Émile Lecrosnier, 1882), pp. 6-7, 9. *Inter alia*, in his elaboration

How numerous were those who had paid attention to the topic, it can be seen from the brief review written by Charles Debierre, newly appointed Chef de travaux d'anatomie et physiologie at the Université de Lyon. In 1884, he tried to check the data found by his *devanciers* by focusing on two categories of army individuals, "sous-officiers" and illiterate soldiers – same age group, same height – measured with Broca's cephalometric method. Six kinds of curves were investigated, in addition to the total circumference of the skull and the cephalic index: the first group was found to have a more voluminous anterior skull and a lesser rear one, compared to the second group, and the other measurements too were better in the *sous-officiers*: "*tout est donc comme si* effectivement le travail intellectual avait pour résultat d'accroître les dimensions du crâne, mais surtout du crâne anterieur."[122] Although the brain does not exactly fill the skull, due to the thickness of the bones, the abundance of cephalic-rachidian liquid etc., one could still say that its development would determine that of the skull, so the measurement of one could be estimated through that of the other. Debierre claimed that the brain should vibrate in relation to the number of its "éléments vibratoires", like the current produced by a battery, to use a fairly crude similitude. It was a matter of "matière grise pensante", so that a small brain would behave in a superior manner when rich in convolutions.

In comparatively assessing the cerebral weight of human races, the terms did not change: Debierre mentioned Le Bon's belief in a higher frequency of voluminous skulls in the upper races, a phenomenon also found in his own sample of military personnel. More than twenty years before, Broca had discovered that the cranial capacity of the Parisians had grown significantly from the 12th to the 19th century. That the skulls of the Troglodytes – coming from the caves of l'Homme mort or Cro-Magnon – had a great volume, could not certainly falsify the general assumption, since it was only their rear part that prevailed: "c'est une race *occipital*." However complex the issue, due to its multiple factors, Debierre placed emphasis on the "gymnastique intellectuelle individuelle", as done by Broca, whom he quoted like the greatest authority: education did not only make man better, but allowed him to become superior to himself, to enlarge his brain and to perfect its forms. Spreading education meant also improving the breed.[123]

---

of data, Bra found that the weights of the encephalon and of the brain in particular were, during the initial phases of madness, superior to the physiological average ones, and furthermore recorded greater weights in the depressive forms: see his conclusions, pp. 96-99.

122 Charles Debierre, "De l'influence du travail cérébral sur le volume et la forme du crâne," *Bulletin de la Société d'anthropologie de Lyon*, 1884, 3/2:243-264, p. 252.

123 Ibid., p. 259. A discussion followed Bra's speech at the Société d'anthropologie de Lyon: it was objected that the cranial superiority of the *sous-officiers* depended rather on

For the umpteenth time, in 1885 it was Adolphe Bloch who asked whether the intelligence was related to the brain volume: "voice une question qui intéresse au plus haut point les anthropologistes, à en juger par les discussions qu'elle soulève fréquemment au sein de la Sociétés d'anthropologie et par le nombre de travaux qu'elle à déjà inspirés."[124] His purpose, declared from the outset, was to prove that intelligence was not measurable *only* by the volume of the brain and that, in order to find the causes of its variability, research should move in multiple directions. Exceptions to the presumed law of a direct relationship were too frequent and significant to go unnoticed: Bloch did not spare a long list of examples and anecdotes in this regard, also recalling that Broca himself, in his dispute with Gratiolet, had admitted that no "homme éclairé" could give credit to that kind of measure, if isolated from other factors. Very clever individuals with a small brain had been seen, and others quite ordinary with a big one; on the other hand, reputedly lower races had shown considerable cranial amplitude. Was the topographic study of convolutions perhaps more reliable, as an alternative solution? Their complexity reached a peak in the elephant and cetacean, so it must have been independent of intelligence. It surely had a greater meaning in man, and yet it was not enough to give a final answer, and neither was the asymmetry of the hemispheres, which had seemed to correlate with intellectual development.

Bloch believed that one of the most important factors had been overlooked: the quality of the brain cell, that is, its degree of "impressionabilité" (or "excitabilité", "vibration") under the influence of sensory stimuli and of those either produced by internal organs or by the work of the mind. It might be "native" or "acquise": the former was a sign of superior intelligence, the latter could derive from a sustained mental work. An appendix listed numerous and precious documents that had been gathered in the meantime within the *Élements d'anthropologie générale* by Paul Topinard, who had added five other factors to the encephalic volume, to be examined when assessing the degree of intelligence. After all, Broca's brain weighed 1484 grams: "c'est une bonne moyenne,

something innate and inherited. Debierre replied that it would only be so if he had compared individuals of different races, but in the case of the soldiers the racial homogeneity emphasized the cerebral gymnastics: since the function increases the organ – as taught by Lamarck – the brain of the thinker must develop like the dancer's calves. For his part, Lacassagne reported that he had gone through different phases, first believing in the possibility of enlarging the individual brain with intellectual work, then realizing that – instead of having such an immediate influence – the *culture de l'esprit* became manifest only after many generations, accumulating by inheritance (pp. 260-262).

124 Adolphe Bloch, "L'intelligence est-elle en rapport avec le volume du cerveau?," *Revue d'anthropologie*, III série, 1885, *14*:577-619.

mais ce n'est toujours pas un chiffre élevé, si l'on admet que l'intelligence est en rapport avec le volume du cerveau."[125]

On July 9, 1880, Broca had suddenly died of an aneurysm, at the age of 56, and Paul Topinard succeeded him both as director of the École d'Anthropologie and as secretary of the Société d'Anthropologie de Paris. His first contributions to the field had been two detailed surveys of craniological and ethnographical information on Tasmanians and Australians. In 1872, he had drafted an extensive review of literature about Aboriginal physical anthropology, at the request of the Commission permanente pour l'Océanie – in the form of instructions to travelers for the collection of useful materials – and did not miss the opportunity to express his compassion for the native inhabitants:

> Cette malheureuse population indigene, qu'on a reléguée au dernier echelon de l'espèce humaine, comme une transition de l'homme au singe, se voit en certains endroits refoulée jusqu'à 1000 kilomètres des côtes, et diminue rapidement. Il faut donc se hâter et, tandis qu'il en est temps encore, l'étudier et lui assigner sa place dans la succession des êtres.[126]

Topinard's text is a cornucopia of news on physical and psychological characters, on the attitudes and behavior of two primordial ethnic components – one lower than the other – whose *mélange* in variable proportions would look like a scale with two extremes. Australia was therefore home to a range of human forms, not to mention settlers from various parts of the world, who were causing the progressive extinction of local stocks. A reality that induced Topinard to go straight to the crux of the matter, indicating a causal relationship between cerebral typology and civilization:

> Ne faut-il pas admettre que toute race est comme predestiné par l'organisation de son cerveau à un mode spécial de civilization? La vie sociale des Aryens ne ressemble pas plus à celle des Sémites Arabes que celle-ci ne ressemble à la vie sociale des Australiens du *bush*.[127]

In 1885 Topinard, the secretary, presented to the Société d'Anthropologie "trois Australiens vivants" who happened to be in Paris, the survivors of a group of

---

125 Ibid., p. 619.

126 Paul Topinard, "Sur les races indigènes de l'Australie," *Bulletins de la Société d'Anthropologie de Paris*, 1872, II série, 7:211-327, p. 211.

127 Ibid., p. 284.

nine who had been deported three years earlier from two islands off the coasts of North Queensland by the American impresario Robert Cunningham, to be exposed in a tour that had already touched a number of US and European cities.[128] Topinard had just published the *Élements d'anthropologie générale* quoted by Bloch, a huge volume of almost 1200 pages, dedicated to his *maîtres*, Paul Broca and Armand de Quatrefages, and containing the topics covered by his courses at the École d'Anthropologie over the past eight years. It was divided into three parts, the first introductory and general ("la seuil du temple"); the second, methodological, designed for laboratory personnel; the third aimed at explaining to travelers the anthropometric procedures to be applied. The work was representative of his strenuous effort to give stability and autonomy to the discipline, according to the model that the Parisian school had conceived.

Two chapters of the second part of his *Élements* dealt with the encephalic weight, followed by four on the *Cubage de la cavité crânienne*, for a total of over 250 pages. Reinforced by the ideal consent of most naturalists, Topinard maintained that, among the bodily organs, one dominates and somehow sums up man, distinguishing him in the animal kingdom. While the human encephalon differs a lot from that of the anthropoids, within the genus *Homo* it presents only slight differences; and yet these acquire considerable importance for their physiological and social, philosophical and practical consequences, given the marvelous power of the organ. Although the main role of anthropology was to discover the truth, without descending on the utilitarian terrain, nonetheless in this case it could not escape the task. In fact, the hygiene of the brain and the art of directing its growth in the most advantageous way for the individual and society would descend from the laws of variability and development that anthropologists were seeking, with the mission to lead peoples to the greatest possible cerebral action.

The volume of the encephalon would be only one of the organic factors contributing to the production of intelligence. Others had to be scrupulously taken in consideration: 1) the relative volume of the parts making up the total mass; 2) the development of the convolutions; 3) the relationship between the deep parts; 4) the number and complexity – visible under the microscope – of

128 Paul Topinard, "Présentation de trois Australiens vivants," *Bulletins et mémoires de la Société d'Anthropologie de Paris*, III série, 1885, 8:683-698, which was accompanied by two full-page engravings of the woman named Jenny, after photographs taken by Prince Roland Bonaparte. On the whole affair see Roslyn Poignant, *Professional Savages: Captive Lives and Western Spectacle* (New Haven: Yale University Press, 2004) and Stephanie Anderson, "«Three Living Australians» and the Société d'Anthropologie de Paris, 1885," in *Oceania and the Science of Race*, edited by Brownwen Douglas and Chris Ballard (Canberra: ANU E Press, 2008), pp. 229-255.

the nerve cells of the gray substance; 5) the qualities that were still inaccessible to scientific investigation. The volume and weight had the advantage that they could be expressed by means of numbers, which explained their frequent use and popularity. The other factors were descriptive: to understand the convolutions, many of them had to be observed; they were reproduced in drawings or in plaster casts that each observer saw in his own way. Research on the deep structures of the brain and on the histological quality of its fibers and cells was even more arduous. Thus, essentially, weight was still "la question du jours", which deserved to be addressed and resolved.[129] In this regard, Topinard had already spoken of the *registres* left by Broca – regarding a thousand brains, of various types, weighed in twenty years – and the special method he devised, which involved four phases and eighteen weighings for each specimen. A further and modified version of Topinard's text, published in 1888, defined "le grand problème laissé aux travailleurs du vingtième siècle: des divers modes d'activité cérébrale, considérés comme causes principales des variations individuelles du poids de l'encéphale."[130]

Two names were – according to Topinard – behind the idea of weighing brains: those of Camper and Soemmerring, although it was only in the 1830s that such practice began to take hold – with Hamilton, Sims, Tiedemann, Parchappe, Lélut etc. – and became widespread over the following decades. However, this crescendo of industriousness did not escape the "plague" of anthropology, i.e. the diversity of the operating procedures. Worse still, the authors did not always reveal which one they had adopted, so when faced with an unusual average one could not even understand whether the difference depended on the chosen method. Encephala or otherwise brains had been weighed, with or without membranes; with the ventricles drained of their fluid or not. Finally, individual errors had to be taken into account: "il ne faut faire intervenir, lorque c'est possible, que des pesées obtenues par la meme main." Other causes of divergence stemmed from having worked on diseased brains, or brains stored in alcohol or astringent liquids; the state of conservation was definitely influential, since the decomposition would reduce the weight. The disparate units of measurement sometimes made the comparison even more difficult.[131]

129 Paul Topinard, *Éléments d'anthropologie générale* (Paris: Delahaye et Lecrosnier, 1885), pp. 503-510.

130 Paul Topinard, "Le poids du cerveau d'après les registres de Paul Broca," *Revue d'Anthropologie*, II série, 1882, 5:1-30; Id., "Le poids de l'encéphale d'après les registres de Paul Broca," *Mémoires de la Société d'Anthropologie de Paris*, II série, 1888, 3:1-41, p. 32.

131 Topinard, *Éléments* (cit. note 129), pp. 510-515.

Topinard also distinguished the brain weight from the volume of the encephalon, two terms often and incorrectly used as synonyms. To get the exact volume, one had to divide the weight by the density of the cerebral matter, which fluctuated around 1030, but differed in each part. Brains of equal volume did not necessarily have the same weight, and vice versa, even though the extent of their divergence was minimal. Differences caused by age, height and body weight were to be borne in mind, as well as those dependent on individuality, a delicate issue for all anthropological aspects, but particularly for the central nervous system. Numerous diagrams attempted to highlight the complex cluster of relations between the variables involved, not without Topinard dwelling on the *Encéphales d'hommes d'élite* and trying to reach an arrangement of rather chaotic data; the best established cases were no more than 34, therefore "attendons de nouvelles observations et contentons-nous du fait general: les savants ou hommes célèbres ont un cerveau plus volumineux."[132]

Of course, the phenomena of mental alienation, congenital affections such as idiocy, and criminality also shed light on the question. It was thus discovered that the brain reached extraordinary dimensions in the most opposite conditions: in subjects dedicated to work requiring only muscular strength, first-rate savants, giants, and *macrocéphales*. It was the confirmation that brain activity in relation to intellectual function was not the only cause of a hypertrophy of the organ, and that, consequently, "une masse exagérée de substance cérébrale n'est pas une chose qu'il faille ambitionner." An incontestable fact, as documented by many sources, was the less amount of brain in women; settling for such findings would have been deceptive, given the smaller size of the female body. So, after making a series of calculations to take this reality into account, Topinard had to acknowledge that the female brain average weight was still lower that that of the male.

Finally, another factual statement: after decades of observations, too little was known about racial differences, and most of them revolved around the cranial capacity. The few brains of distant peoples that were available were stored in alcohol, and lent themselves only to very rough estimates. Instead, they should be weighed fresh, except that the number of black, yellow, or red individuals who occasionally died in European hospitals was too low. More than in any other research, the quantity of data was essential here, as was a general agreement on the methods.[133] There was still a long way to go:

---

132 Ibid., p. 546. For a paragraph on the cranial capacity of the *hommes célèbres* see pp. 628-629.

133 Ibid., pp. 568-574.

> Le désarroi, le découragement est tel de la parte de certains anthropologistes qu'ils ne songent à rien moins qu'à abandonner l'étude [...] comme illusoire et impossible. C'est un devoir de lutter contre cette tendance. Ce serait une honte pour le dix-neuvième siècle de renoncer à l'étude d'un caractère de premier ordre. [...] L'anthropologie ne peut donc l'abandonner faute d'entente et parce que son étude présente quelque difficulté.[134]

Not very differently, in 1882 a younger exponent of the Parisian school had diagnosed the state of the art at sessions of the Académie des sciences and the Société d'Anthropologie. Léonce Manouvrier bluntly said that all the attempts made to isolate the various physiological influences acting on the weight of the encephalon had proved unsuccessful. His presentations tried to explain the relationships between the elements involved – height, body weight, weight of the encephalon – by means of a mathematical reasoning on proportions, and announced that he had devoted lengthy researches to the matter. In his opinion, many mistakes had been committed in handling data about those variables, such as to make someone proclaim, for instance, that women would tend to differentiate themselves more and more from men in cerebral mass and intelligence. Similarly, an incorrect analysis had led the same author to conclude that the superiority of one race over the other consisted in a greater frequency of voluminous skulls.

In contrast with Gustave Le Bon, who was to become his polemical target, Manouvrier submitted the well-known fact that, in terms of average absolute cranial capacity, some inferior races such as Polynesians, Patagonians, and Northern Amerindians overtook Parisians. Factors such as body size and weight were not able to produce figures indicating how they affected brain weight. During the heated discussion that followed, Le Bon objected that Manouvrier's reflections were merely theoretical and worthless, in the presence of established facts, and so elicited a strong reaction. Their clash, masked by purely technical aspects, was in fact radically ideological.[135]

Manouvrier had served as an army volunteer during the Franco-German war, received his medical training in Paris, where he became a pupil of Broca,

134 Ibid., pp. 631-632.

135 Léonce Manouvrier, "Sur l'interpretation du poids de l'encéphale et ses applications," *Comptes rendus hebdomadaires des séances de l'Academie des Sciences*, 1882, *94*:143-145; Id., "Sur la valeur de la taille et du poids du corps comme terme de comparaison entre la masse de l'encéphale et la masse du corps," *Bulletin de la Société d'Anthropologie de Paris*, III série, 1882, *5*: 85-105. Samuel Pozzi also took part in the discussion, with the argument already proposed in 1878 – on which Manouvrier agreed – about the role played by the brain in motility.

then *préparateur* at the laboratory of anthropology at the École des hautes études. His medical thesis was discussed and published in 1882, first of all expressing gratitude to Broca, under whose supervision he had begun research on the skull and brain, to be carried forward after the "perte irreparable" of his mentor. Concluding his first demanding work, Manouvrier emphasized that the weight of the encephalon could not be proportional to the intensity of intelligence, due to the multiplicity and complexity of the functions performed by the organ, which included locomotion and nutrition. Before dealing with the issue of that possible proportionality, it was hence indispensable to evaluate the relationship between the brain weight and that of the body parts that most closely represent the intensity of the two aforesaid functions, namely the femur and the jaw. And this he did in his doctoral thesis.[136]

In 1888 a second and more extensive part of Manouvrier's *Recherches* was published: by his own admission, they could help only to a small extent to dispel the thick darkness still reigning over the relationships of weight and form with the functions of the brain, an issue that had been in a too stationary state for so many years. A first inference, deriving from the Manouvriers's approach, explained the gap in brain weight between the sexes. The anatomical and psychological reasons currently given in support of their intellectual inequality seemed insufficient to him. Rather, he was pleased to present results that corroborated his egalitarian opinion about the perfection of the thinking organ and did not hesitate to insist on it:

> Je me serais empressé de l'abandonner si une preuve positive était venue m'en démontrer la fausseté; mais je devais aussi m'y attacher avec plus de confiance à mesure que les faits m'encourageaient. De plus, la question de l'égalité intellectuelle des sexes, dans l'espèce humaine, m'a toujours semblé si importante à divers points de vue que l'espoir de l'élucider davantage a été le motif le plus puissant qui m'ait poussé à poursuivre la tâche difficile et si souvent ingrate que je m'étais imposée.[137]

Psychology would have benefited greatly from the comparative and analytical study of the quantity of encephalic matter: in such a delicate science, which

136 Léonce Manouvrier, "Recherches d'anatomie comparative et d'anatomie philosophique sur les caractères du crâne et du cerveau," *Bulletin de la Société Zoologique de France*, 1882, 7:113-229.

137 Léonce Manouvrier, "Recherches d'anatomie comparative et d'anatomie philosophique sur les caractères du crâne et du cerveau. Sur l'interprétation de la quantité dans l'encéphale et dans le cerveau en particulier," *Mémoires de la Société d'Anthropologie de Paris*, II série, 1888, *3*:137-326, p. 144.

had recently left its metaphysical phase for a positive one, anatomical insight was a precious resource. Nonetheless, the uncertainty caused by the weak series of data, and various other causes of error, should lead to a great circumspection; furthermore, in the current state of knowledge, it was rare that individual cases could be compared fruitfully.

Manouvrier recalled how measuring the cranial capacity offered greater accuracy, independent as it was from the subject's age, last illness, and date of death. The ease of transport and preservation of the skulls made it possible to acquire notions about brain development in ancient times and in races that lived in distant regions. Since the cranial capacity did not exactly correspond to the encephalic volume, he had measured it in individuals of different races whose encephalic weight was known, then divided the former for the latter, and vice versa, thus obtaining two quotients that indicated the variations in their relationship: the "équivalent pondéral de la capacité cranienne" and the "équivalent cubique de l'encéphale", respectively. To roughly evaluate the weight of the brain starting from the cranial capacity (measured with the Broca method), one only had to multiply the latter by 0.87. This kind of information was displayed on two tables in few *individus de race jaune* and of *nègres*.[138] To learn something from the lists of figures provided by several authors in the past decades, Manouvrier considered it essential first to judge the value of the series from which the averages had been calculated. Many authors had made hasty inferences from three or four cases, if not just one, and this bad habit explained the hotchpotch of figures:

> C'est ainsi que les nègres furent considérés tantôt comme ayant un petit crâne et tantôt comme possédant un grand cerveau. On comprit plus tard que tous les nègres ne se ressemblent point, que l'on avait discuté à vide, c'est-à-dire sur des variations individuelles, et que le hasard pouvait jouer un grand rôle dans les compositions des faibles series.[139]

The records compiled by Broca contained a large number of ethnic series, but most of them were still too short to ensure solid averages. Moreover, Manouvrier deplored that, in this regard, Le Bon's "déductions psycho-philosophiques" had nothing to do with scientific truth.

In the conclusions of his long essay, Manouvrier made some rather radical points. The theory of the progressive intellectual differentiation of races and

138 Ibid., pp. 156-160.

139 Ibid., pp. 216-217.

individuals had been wrongly founded on the absolute volume of the brain. The superiority of a race in intelligence was provable neither by the average cranial capacity, nor by the relative number of voluminous skulls, nor by the width of the gap between the maximum and the minimum values. All these data remained devoid of significance until they were associated with precise notions about the organic mass: "au point de vue physiologique, tout ce fatras n'a servi qu'à encombrer la science." And even "l'évaluation quantitative de l'intelligence est, du reste, un problème absolument parlant irrésoluble, même en théorie."[140]

A few years later, Manouvrier had the opportunity to dissect the brain of a three-year-old child who was part of a group of Fuegians and had died of bronchopneumonia while performing in a Parisian "barnum". A decidedly high weight (1305 grams) was found, exceptional for the age of the subject, a fact confirmed by the considerable cranial capacity in other individuals of the same group who were measured. Also the overall morphology as well as its details revealed a substantial equivalence with the white brain, a circumstance that was not amazing at all:

> La *moyenne* intellectuelle des européens et la *moyenne* cérébrale ne sont pas si élévées absolument qu'on doive s'étonner de voir ces moyennes atteintes par un sauvage relativement bien doué pour sa race. [...] Physiologiquement, il n'y a pas lieu de considérer la vie sauvage des chasseurs et pécheurs comme exigeant un fonctionnement cérébral moins compliqué que celui de la pluspart des européens. Aussi bien que nos paysans et ouvriers incultes, les fuégiens savent exécuter des travaux difficiles pour subvenir à leur existence; ils savent exprimer leurs idées dans une langue très riche.[141]

Manouvrier went on to illustrate their technological abilities, social relationships, traditions, and propensity to be educated. He did not doubt that most of the savage races were cerebrally less equipped than most Whites, but this judgment only applied to the average, not to individuals: "c'est avec les aptitudes cérébrales reçues de ses ancêtres qu'un jeune nègre apprend l'anglais ou l'arithmètique." Appropriate external conditions could enhance attitudes that until then had only been applied to elementary actions.[142]

140 Ibid., pp. 289, 322.

141 Léonce Manouvrier, "Le cerveau d'un Fuégien," *Bulletins de la Société d'Anthropologie de Paris*, 1894, IV série, 5:595-614, pp. 610-611.

142 Ibid., p. 613.

Manouvrier happened to discuss, again, the "insoluble" question by contributing to the article "Cerveau" in the *Dictionnaire de Physiologie* edited by Charles Richet.[143] The main obstacle to reaching agreement among the observers was represented by its utmost complexity. Histology had finally clarified that the brain was composed of more or less numerous cellular elements, connected by extensions, which, in turn, were differently branched. In the animal species, these elementary organs constituted more or less complex connected systems. All this structural variety responded to the number and intricacy of the sensitive parts of the body, of its motor functions, and to the array of operations responsible for intelligence. In spite of this, Manouvrier specified that the quantitative aspect did not lose importance: the numerical growth of the cells, of their extensions and ramifications probably translated, *coeteris paribus*, into an increase in the weight and volume of the whole. It was still true, however, that the research done in the previous half century had led to inconsistent results, and he referred to the attempts he had made himself, since 1882, to counteract the causes of error.

Santiago Ramón y Cajal had recently pointed out that the cells of the brain and marrow are in intimate relationship with the amount of muscular, glandular, and sympathetic elements that they must influence. For instance, the enormous development of the optical lobes in birds and reptiles, or of olfactory lobes in dogs, stemmed from the extraordinary richness of retinal ganglion cells or the bipolar cells of the olfactory mucosa. Ramón y Cajal thought that such facts could explain the frequent cases of difference in intellectual power between individuals who had brains of the same weight and volume.

Analytically, Manouvrier reviewed the vast set of aspects involved in the relationship between intelligence and brain weight: believing that the latter could measure the former would be absurd, even though there must exist a relationship between the two entities. He added that the anatomical-physiological conditions vary greatly in their actual power, depending on whether they are enhanced or not by external circumstances or not. The social factor counts beside and perhaps more than the others:

> Un simple ouvrier, un pauvre campagnard auront beau acquérir un haut degré de correspondance dans leur vulgaire milieu, leur correspondance aura beau être étendue en variété et complexité dans ce milieu, on ne

143 Léonce Manouvrier, "§ II – Morphologie générale – Anatomie comparée – Rapports du poids et de la forme du cerveau avec l'intelligence," in *Dictionnaire de physiologie par Charles Richet. Tome II B-C* (Paris: Félix Alcan, 1897), pp. 670-745. The first paragraph of the article (*Historique*, pp. 547- 670) is signed by Jules Soury.

> saurait demander à leur intelligence de suppléer au défaut d'instruction, tandis qu'un homme très médiocrement doué cérébralement pourra acquérir en quelques années une foule de relations internes d'un ordre plus élevé dont l'acquisition première a été le fruit de longs siècles de travail et des efforts [...] De plus, à mesure que s'accroissent les progrès scientifiques, industriels, artistiques, les individus qui ne sont point placés dans les conditions sociologiques nécessaires pour la connaissance et l'utilisation de ces progrès deviennent d'autant plus arriérés et plus incapables de contribuer à l'avancement des sciences, des arts et de l'industrie. [...] Des campagnards et même des ouvriers citadins, admirablement doués, risquent fort d'être considérés comme des imbéciles si on les examine en dehors de leur milieu, tandis que le premier cuistre venu, pour peu qu'il ait été frotté de science ou de littérature et qu'il s'adonne aux «travaux de l'esprit», sera facilement rangé parmi les intelligents et ne manquera pas de se classer lui-même comme «intellectuel».[144]

Exercise was therefore the first prerequisite for cerebral perfection, and only exercise could be the original cause of the superiority of brain weight in distinct individuals, although objective data on this particular phenomenon occurring in adult age were still non-existent, and a method for collecting them was hard to imagine. The acquired element, therefore, seemed to take a revenge on the innate one.

Once again, Manouvrier also denied that the lower weight – in absolute terms – of the female brain could be traced back to an alleged intellectual inferiority. Accordingly, since superiority belonged to woman in relative terms, neither sex could claim for itself a pre-eminence based on quantitative brain development.[145] As to issue of women, Manouvrier had repeatedly challenged the prevailing belief, and in 1881 he had already taken distance from his former mentor Broca, recently deceased. In 1884, he happened to defend the right of women to enter the medical profession, and in 1889 to speak at the Congrès français et international du droits des femmes, saying among other things, that "l'analyse anatomique, dans laquelle tant d'auteurs s'imaginaient trouver des preuves de l'infériorité intellectuelle de la femme, tourne, au contraire, à la confusion de l'orgueil masculine."[146] A few years later, Giovanni Mingazzini

144 Ibid., pp. 694-695.

145 Ibid., pp. 699-701. See Hecht, *Scientific Modernity* (cit. note 79), pp. 217-227.

146 Léonce Manouvrier, "Variétés: l'internat en mèdecine des femmes," *Revue scientifique*, 1884, *3*:592-597; Id., "Indications anatomique et physiologiques relatives aux attributions naturelles de la femme," in *Congrès français et international du droits de femmes* (Paris: F.

– director of the anatomo-pathological laboratory of the asylum in Rome – listed the divergent opinions on the differences to be attributed, in the two sexes, to the Sylvian fissure, *the sulcus interparietalis*, and the development of the frontal lobe with respect to the parietal. Unlike Manouvrier, his researches had anyhow persuaded him that the existence of two distinct types of brain was undeniable.[147]

6 "A Literature by Itself"

By the end of the century, others were wondering about the potential plasticity, so to speak, of the brain. Taking a cue from a memoir upon the measurements made on Cambridge students and read by John Venn – statistician and secretary of the Anthropological Society of Great Britain –, its president Francis Galton reproduced figures of "Head Products" and shortly offered some conclusions. Although it was known that in the general population the brain ceased to grow after the age of nineteen, or even earlier, University students represented an exception to the rule. Those of them who got "high honours" had considerably larger brains than others at the age of nineteen, but not to the same extent at twenty-five: their predominance, in few years, diminished to one-half of what it used to be. Presumably brilliant subjects as a class were both more precocious and more gifted throughout. One had to look upon good success in studies as a fortunate combination of these two helpful conditions.[148] In 1893, a quite similar comparison was hazarded by a famous work by the *Sozial-Anthropologe* Otto Ammon on natural selection in mankind. Through a plethora of data and figures, *Kopf-Formen* of various categories – students of gymnasium and reformatory, soldiers, *Gelehrten und Freunden der Wissenschaft* – showed significant differences in size, the more beneficial of which belonged, of course, to the "superior" ones.[149]

Dentu, 1889), pp. 41-51, p. 49. At the beginning of the new century Manouvrier again expressed his support of the feminist cause: see, for instance, his "Conclusions générales sur l'anthropologie des sexes et applications sociales," *Revue de l'École d'anthropologie de Paris*, 1903, *13*:405-423.

147 Giovanni Mingazzini, *Il cervello in relazione con i fenomeni psichici* (*Studio sulla morfologia degli emisferi cerebrali dell'uomo*) (Torino: Bocca, 1895), pp. 56-70.

148 Francis Galton, "On Head Growth in Students at the University of Cambridge", *The Journal of the Anthropological Institute of Great Britain and Ireland*, 1889, *18*:155-156.

149 Otto Ammon, *Die natürliche Auslese beim Menschen. Auf Grund der Ergebnisse der anthropologischen Untersuchungen der Wehrpflichtigen in Baden und anderer Materialien* (Jena: Gustav Fischer, 1893), pp. 190-269.

Despite recurring bewilderment about the worth of the results obtained from cerebrometric practice, racial brains did not stop being weighed. For example, in 1894, a military doctor serving in German East Africa had sent twelve *Negergehirne* to the Berlin Anatomisches Institut – members of the Zulu, Sudanese, Kenyan and Swahili ethnic groups, most of whom had died of smallpox or dysentery. Another two similar specimens were added later, so that Wilhelm von Waldeyer, director of Institut, could study a fair number of such rare bodily objects. His anatomical eye focused first on detecting their morphological peculiarities, in the lobes, gyri, and sulci, which turned out to be relevant. Then, their average weight was calculated as quite low at 1148 grams, much lower than the 1331 grams recorded by Sanford B. Hunt in brains of North American blacks during the Civil War. These must have absorbed much "white blood" over the centuries, a process that might explain such a remarkable difference. In any case, Waldeyer still believed that the racial brain differences were a very interesting anthropological problem, worthy of further and wider investigations.[150]

In 1895, Henry H. Donaldson, professor of neurology at the University of Chicago, noted how the description of racial differences in the nervous system formed already "a literature by itself", and the traits of widely separated classes in the same community were almost as different as those of unlike races.[151] A few years earlier he had analyzed the brain of Laura Bridgman, the famous blind, deaf and mute woman who had learnt to speak and attained significant mental abilities. The chief modifications found in her brain were caused by growth arrest, due to a pathological destruction of her sense organs occurred at the age of two, so Donaldson was spurred to arrange a program for the study of the nervous system from birth to maturity.[152] In his work on brain growth, he put forward an attempted interpretation of the disparities on which many others had been tinkering for decades:

> One great difference in races will be found to lie in the extent of growth and organization in the nervous system after birth, and especially after puberty. Should it turn out on further examination that some of the lower

150 Wilhelm von Waldeyer, "Über einige anthropologisch bemerkenswerthe Befunde an Negergehirnen," *Sitzungsberichte der Königlich Preussischen Akademie der Wissenschaften zu Berlin*, 1894, pp. 1213-1221.

151 Henry H. Donaldson, *The Growth of the Brain: A Study of the Nervous System in Relation to Education* (London-New York: Walter Scott and Charles Scribner's Sons, 1895), p. 349.

152 Henry H. Donaldson, "Anatomical observations on the brain and several sense organs of the blind deaf mute, Laura Dewey Bridgman," *American Journal of Psychology*, 1891, 3-4:293-342, 248-294.

> races lose their capacity for later training after adolescence, we should look with interest for the changes in the cerebral cortex in order to determine whether growth there practically ceased at puberty; for, by contrast, Venn, studying the size of the head in Cambridge students, found it on the average greatest and growing for the longest time in the group of most successful men.[153]

In other words,

> The provisional picture, therefore, to be formed of the brains belonging to those races least capable mentally is that of one in which the number of cell elements is approximately similar to that in the most capable races; but many of these elements being but partially developed, the organization of the brain is less perfect, though the size is not thereby greatly reduced.[154]

Those where the years when Alfred Binet started to direct the Laboratoire de psychologie physiologique at the Sorbonne and happened to spend a few weeks in Bucharest, to hold a series of lessons. One of his listeners was a twenty-year-old Romanian, Nicolae (then Nicolas) Vaschide, who wrote enthusiastic reports on the event for the local newspapers, and Binet proposed to him to complete his studies in Paris. They soon started to carry out joint research and wrote about twenty articles in *L'Année psychologique,* founded in 1894 by Binet and Henri Beaunis. Before their collaboration abruptly broke off for obscure reasons in 1899, they anonymously published a long *Historique* of literature on the relationships between intelligence and head size and shape. Their own work was mainly conducted *sur le vivant*, leaving out by choice the weight, volume, and shape of the encephalon as well as skull cubage. Not much material was available in this regard, just very few crude measurements.

Underneath that kind of study was Gall's doctrine, which Binet and Vaschide considered historically meaningful, albeit alien to the experimental method. They regarded a second step, the *Recherches sur l'encéphale* published by Parchappe in 1836, with a critical attitude. Compared to Gall, he had been

153 Donaldson, *The Growth of the Brain* (cit. note 151), pp. 349-350.

154 Donaldson, *The Growth of the Brain* (cit. note 151), p. 173. A few years later Jacques Loeb – Donaldson's colleague physiologist at the University of Chicago, educated in Germany but naturalized American – approved for a variety of reasons his conviction that "the absolute mass of the brain cannot be the principal factor in determining intelligence": see his *Comparative Physiology of the Brain and Comparative Psychology* (New York, G.P. Putnam's Sons, 1900), p. 254.

able to pay more attention to detail, with measurements of diameters, curves, and circumferences: even sixty years later, the documents then collected by the psychiatrist of Rouen still seemed worthy of being used. Especially, the two authors agreed with him "pour repousser l'idée d'une proportionnalité entre le volume de la tête et la puissance intellectuelle".[155]Another of his merits had been that to measure heads of men of letters and sciences, as well as those of manual workers or of *aliénés*, whose head volume turned out to exceed that of the most intelligent subjects. Therefore, the quantitative correspondence between head and intelligence was not an absolute law, and prudence had seemed to Parchappe the best policy on the subject, with the approval of Binet and Vaschide.

The third step could only be associated with Broca, whose *De l'influence de l'education sur le volume et la forme de la tête* (1872) was reproduced there in full, both for the relevance of his cephalometric observations and for the role assigned by him to education, which would have the power not only to improve man, but also to enlarge and perfect his brain. Some years later, Lacassagne and Cliquet had taken on the subject again to evaluate Parchappe's and Broca's results and to extend the examination with their *conformateur* to a larger number of individuals. An excerpt from their work was also re-proposed, along with the figures that Le Bon had recently collected on the head circumference of different social categories. With reference to Donaldson's and Manouvrier's contributions, Binet and Vaschide had to conclude that, in general and despite many doubts, "cette relation entre la quantité cérébrale et le degrée d'intelligence paraît aoujourd'hui admise par tous les anthropologistes". Moreover, as for the head shape, the development of the frontal region was, all in all, a particular sign of intelligence. Finally, they still recommended craniometry, easier to practice, to the detriment of encephalometry, at least for the time being.[156]

Cephalometric investigations on living people were conducted at the turn of the century by two anthropologists who would have left their mark, although very differently, on the development of anthropology in the United States. Both of them had migrated from Europe: Aleš Hrdlička and his family in 1881, Czech by birth and barely a teenager, studying and then practicing medicine in the Middletown asylum for the mentally ill; the German Franz Boas in

155 [Alfred Binet, Nicolas Vaschide], "Historique des recherches sur les rapports de l'intelligence avec la grandeur et la forme de la tête," *L'Année psychologique*, 1898, 5:245-298, p. 256. See Michel Huteau, "Un météore de la psychologie français: Nicolae Vaschide (1874-1907), *Bulletin de psychologie*, 2008/2, *n. 494*:173-199.

156 [Binet, Vaschide], "Historique des recherches" (cit. note 155), pp. 296-298.

1887, as a young man, after taking part in a geographical expedition to northern Canada to do field work about the indigenous cultures.

In 1896, Aleš Hrdlička had also spent six months in Paris to study under Manouvrier, a training that led him to start doing research at the Washington Museum of Natural History (Smithsonian Institution) from 1903 to the end of his life. In 1899, he reported on his investigations of one thousand children – white and black, male and female – at the New York Juvenile Asylum, and of another one hundred cases at the New York Colored Orphan Asylum. They were examined, "as thoroughly as possible without offense to their modesty", by applying twelve different measurement, from their height to the width of the forehead. In addition, the average pressure and traction force in each hand were assessed.

The principal aim was to learn as much as possible about the physical state of "abnormal" subjects, mostly coming from very poor classes of people, incorrigible or even criminal people. If they were found to correspond physically to their social and moral state, then one had to consider them as generally handicapped in the struggle for life, and the asylum would be no more than a correctional institute, to compensate natural defects and give prolonged care, without any hope of them being rehabilitated into society as individuals fit to adapt to the inevitable difficulties. If, on the other hand, their strength and constitution did not differ much from average ordinary children, the community could expect a proper educational course to be sufficient to elevate or reform the inmates, and no money would be wasted to such purpose. Such dilemma was not easily settled and received a complex response, when the cases were grouped into categories. No generalization was allowed and similar studies should be extended: Hrdlička recommended that the State Board of Charities should give support to this "broad and promising field" by securing the services of proper and unprejudiced investigators.

Inter alia, he found the three main diameters of the heads of black children to be slightly bigger than those of white ones, just in connection – it was maintained – with an averagely greater height. In contrast, the bin-auricular diameter, which measures the width of the head at about the level of the brain base, was always smaller in the colored children, whose skull gave proof of being absolutely narrower at that location. Hrdlička refrained from speculating about such differences, and, if anything, he also maintained that

> The mental system can not be looked at as a mere reflection of the state of the body, or the reverse; the brain can apparently have properties which are not perceptible in the external parts of the individual.[157]

The top views of skulls exhibited great racial differences in form, so that its more or less elongated conformation had been proved to be a good means of characterizing human varieties. For his part, in 1899 Franz Boas tried to demonstrate, through an array of mathematically-processed data on Indians of various tribes, that the seemingly simple cephalic index – used for half a century, after being introduced by Anders Retzius – was influenced by causes other than the length and breadth of head. Also stature, height and breadth of face were in correlation with an index that could not express, on closer inspection, any important anatomic relation. The most convenient means for gauging cranial size would be given by the circumferences, which should then be included in all anthropometrical schedules designed to investigate racial characters.[158]

In 1896, Boas had already identified the "limitations" of the comparative method that connoted evolutionary anthropology. He refused the idea of a "one grand system according to which mankind has developed everywhere", and the corollary that "all the occurring variations are no more than minor details in this grand uniform evolution". Since the historical growth of human cultures had probably followed "a variety of courses", investigations should be directed at the "processes" by which certain stages of cultures had been developing, when exposed to specific environmental conditions and psychological factors at work. Another method was therefore safer, in many respects, i.e. "the much ridiculed historical method", to be applied on small geographical territories, with the awareness that "the solid work is still before us".[159]

At the very end of the century, the young American economist and sociologist William Z. Ripley published a successful and massive book that had grown out of a series of Lowell Institute Lectures held three years earlier. Based on cephalic index – variable between the limits of 73 and 87 – and other

157 Aleš Hrdlička, *Anthropological Investigations on One Thousand White and Colored Children of Both Sexes the Inmates of the New York Juvenile Asylum, with Additional Notes on One Hundred Colored Children of the New York Colored Orphan Asylum* (New York-Albany: Wynkoop Hallenbeck Crawford Co., [1899]), p. 68. Measures were taken of height, sitting height, arm expanse, weight, depth and width of the chest, maximum circumference of the head, greatest length and width of the head, height of the head, bin-auricular diameter, smallest width of the forehead.

158 Franz Boas, "The Cephalic Index," *American Anthropologist*, 1899, *1/3*:448-461.

159 Franz Boas, "The Limitations of the Comparative Method of Anthropology," *Science*, new series, 1896, *4*:901-908, pp. 904-905, 908.

morphological criteria, he classified Europeans into three main races – Teutonic, Alpine, Mediterranean. An important point he noted was that the shape and size of the head seemed to bear no direct relation to intellectual power or level of civilization: Europe offered the best demonstration of this finding. In his study of the proportions of the head, Ripley declared to be interested in measuring "merely race, and not intelligence in any sense", as if he wished the overcoming of a controversial obsession that had lasted too long.[160]

160 William Z. Ripley, *The Races of Europe. A Sociological Study* (*Lowell Institute Lectures*) (New York: D. Appleton & Co., 1899), pp. 41-43.

CHAPTER 4

Twentieth-century Epilogue

1 Resilience Despite Everything

Although directed to building a new view of cultures, Boas did not neglect physical anthropology. In 1908, at the request of the US Immigration Commission, he examined and measured thousands of immigrants and their descendants in New York. His final report was submitted in December 1911, and published in 1912. His efforts were based on the common belief that body shape was the most stable character of any given race. In the previous few decades, however, indications had already been found showing that a more favorable environment could improve physical development, but no evidence had been collected as to an actual change in racial type. His research, which lasted nearly three years, demonstrated unquestionable modifications in the rate of development as well as physical alterations that could only be explained by the influence of new living conditions: "the adaptability of the immigrant seems to be very much greater than we had a right to suppose before our investigations were instituted."[1] Yet there were limits to such investigations – only a few European types had been tested, largely confined to the male sex, none in numbers large enough – and the greater part of the work remained to be done.

Height, weight, length and width of the head, width of the face across the zygomatic arches, color of hair, eyes, and skin had been studied and measures. At the beginning of his 573-page report – filled with figures, tables, and diagrams – Boas communicated an impressive result with due emphasis:

> In most of the European types that have been investigated the head form, which has always been considered one of the most stable and permanent characteristics of human races, undergoes far-reaching changes coincident with the transfer of the people from European to American soil. For

1 Franz Boas, *Change in Bodily Form of Descendants of Immigrants* (New York: Columbia University Press, 1912), p. 2; Franz Boas, "Change in the Bodily Form of Descendants of Immigrants," *American Anthropologist*, 1912, *14*: 530-562. Two sets of authors have reanalyzed and discussed his findings: see Clarence C. Gravlee, H. Russell Bernard, William R. Leonard, "Boas's *Changes in Bodily Form*: The Immigrant Study, Cranial Plasticity, and Boas's Physical Anthropology," and Corey S. Sparks, Richard L. Jantz, "Changing Times, Changing Faces: Franz Boas's Immigrant Study in Modern Perspective," *American Anthropologist*, 2003, *105*:326-337.

© KONINKLIJKE BRILL NV, LEIDEN, 2020 | DOI:10.1163/9789004431881_006

> instance, the east-European Hebrew, who has a very round head, becomes more long-headed; the south Italian, who in Italy has an exceedingly long head, becomes more short-headed; so that in this country both approach a uniform type, as far as the roundness of the head is concerned.[2]

The cephalic index, frequently used to nail populations or races to a sort of natural immutability, was thus converted into an instrument capable of demonstrating how a new environment could alter characters that had long been deemed permanent.

Boas's conclusion that all the evidence was in favor of a great instability or plasticity of human types was attacked by a number of critics, whose attempts at disproving it in various ways he promptly and diffusely counteracted. The daily press and magazines had given wide resonance to his work, not without exaggerations, such as attributing to him the claim of having discovered the origin of a new American race. Boas pointed out that he had only presented the results of his research and some plausible explanations, without committing himself to theories that could not be proven. Finally, he replied to the charge of believing that his conclusion would completely destroy the value of anthropometry, and in particular of the cephalic index. Quite the opposite: he reassured that the anthropometric method would continue to be a most important means of elucidating the early history of mankind and the effect of social and geographical environment upon man.

"What wonder if civilized man – Boas said in some 1910-1911 lectures – considers himself a being of higher order as compared to primitive man, if he claims that the white race represents a type higher than all others!"[3]. Before putting the stamp of eternal inferiority upon whole races of man, he asked to take a break and try to understand on what grounds such opinions were built. Great, undeniable differences existed in the physical characteristics of human races: yet the mystery to be unraveled was how they related to mental faculties and aptitudes. Boas quoted Karl Pearson's conviction that the onus of proof should be left to those who *a priori* regarded such an association as probable, and he also approved Galton's successor personal propensity to see a weak relationship between intelligence and the size or shape of the head, after studying it.

---

2 Boas, *Change in Bodily Form* (cit. note 1), p. 5.

3 Franz Boas, *The Mind of Primitive Man. A Coursef Lectures Delivered before the Lowell Institute, Boston, Mass., and the National University of Mexico, 1910-1911* (New York: The Macmillan Company, 1911), p. 2.

As for the brain, sufficient data were available to establish the fact that its weight in white peoples was larger than in most other races. Moreover, the assumption that a greater dimension corresponded to an increase in faculty would seem plausible, as many had tried to prove by weighing the brains of eminent men. However, Boas advised against overestimating the force of this argument. First of all, a few of those extraordinary brains had been found to be unusually small. Besides, most of the brain weights in the general series had been obtained in anatomical institutes, where individuals were poorly developed on account of unfavorable life conditions, while the eminent men represented a privileged class. It was not certain, therefore, whether the observed difference in brain size was entirely due to their higher ability.

Boas did not miss the opportunity to enumerate a few additional restrictions. The brain of women is lighter that that of men, nevertheless their faculty could not be deemed of an inferior character, although qualitatively different. From this fact, it might also be conjectured that the smaller brains in males of other races worked as well as those of the larger ones of the white race. The real reason for a lack of close correlation between brain weight and mental faculties was not far to seek: cerebral functioning depended upon the nerve cells and fibers, which did not constitute the whole brain mass. Consequently, if that close correlation existed, it had to be looked for "rather in the morphological traits of the brain than in its size".[4] As others had recently held, real racial differences were exceedingly hard to discover, due to the great individual variability within each single race. Brain weights were so variable that a considerable overlapping occurred, and that even the average size of the white brains were numerously represented among other races:

> We find that 50 per cent of all whites have a capacity of the skull greater that 1550 cc., while 27 per cent of the negroes and 32 per cent of the Melanesians have capacities above this value. If we were to assume a direct relation between size of brain and ability, – which, as we have seen before, is not admissible, – we might, at most, anticipate a lack of men of high genius, but should not expect any great lack of faculty among the great mass of negroes living among the whites.[5]

So, even Boas, in the work that would turn into a manifesto of the new American anthropology, fiddled with a question that in the first decade of the century had continued to intrigue quite a few people. No wonder that a long

4 Ibid., p. 28.
5 Ibid., p. 35.

questionnaire sent to the participants in the Universal Races Congress, held at the University of London in July 1911, began as follows:

> 1. (*a*) To what extent is it legitimate to argue from differences in physical characteristics to differences in mental characteristics? (*b*) Do you consider that the physical and mental characteristics observable in a particular race are (1) permanent, (2) modifiable only through ages of environmental pressure, or (3) do you consider that marked changes in popular education, in public sentiment, and in environment generally, may, apart form intermarriage, materially transform physical and especially mental characteristics in a generation or two?[6]

Among others, Gustaf Retzius inaugurated the new century by presenting the results of his research both on the brain of the Russian mathematician Sonja Kovalevski and on the weight of Swedish brains. It is well known that he had followed in his father's Anders footsteps, both of them collaborating with artists, photographers, and printers to produce images that became an essential part of their publications.[7] Plates with full size brain photographs and drawn diagrams accompanied his contribution on Kovalevski's brain, where he found nothing truly special, except for a Sylvian fissure shortened to the rear (*Verkürzung nach hinten*), so that the posterior part of the supramarginal gyrus was widened and pushed out like an operculum over the fissure, especially in the right hemisphere. Two years earlier, Retzius had observed the same peculiarity in the brain of the Finnish astronomer and mathematician Hugo Gyldén, who had directed the Stockholm Observatorium for a quarter of a century. He also mentioned that the neurologist Paul Julius Möbius preferred to localize the mathematical organ in a part of the third frontal convolution, inspired by the phrenological doctrine that he had just rediscovered. He had got to know Gall's works fairly late and just by chance and, to his astonishment, he had seen hidden treasures in them. In 1905 he also verified the correctness of his view on

6 Gustav Spiller (ed.), *Papers on Inter-Racial Problems Communicated to the First Universal Races Congress held at the University of London July, 26-29, 1911* (London: P. S. King & Son, 1911, pp. xiv-xv. On that singular event see Helen Tilley, "Racial Science, Geopolitics, and Empires," *Isis*, 2014, *105*:773-781, and the contributions to a Forum published by the *Radical History Review*, 2015, *92*:99-152.

7 See Eva Åhrén, "Figuring it out: Visualizations in the work of Swedish anatomists Anders and Gustaf Retzius, 1829-1921," *Nuncius. Journal of the Material and Visual History of Science*, 2017, *32*:166-211. By the way, at the end of the 1880s, Gustaf Retzius had tried to found in Stockholm a group inspired by the Parisian Société d'autopsie mutuelle, without any success.

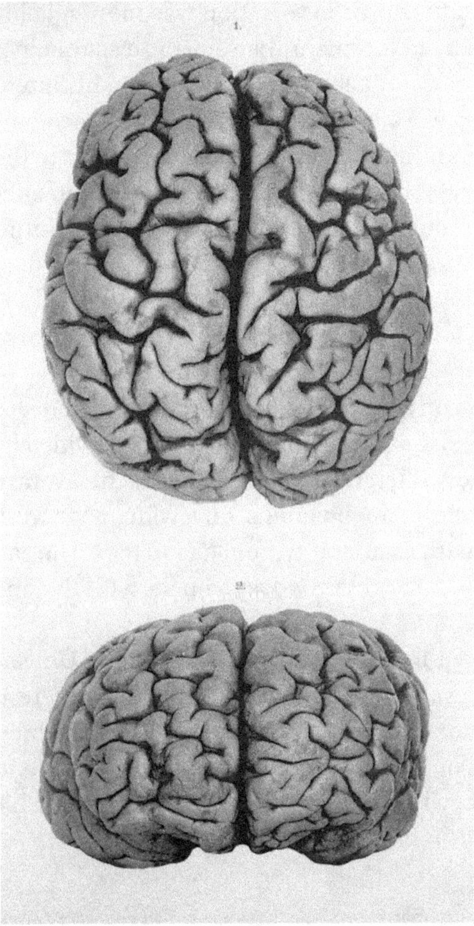

FIGURE 4.1
Tafel I, in Gustaf Retzius, "Das Gehirn des Mathematikers Sonja Kovalevski," *Biologische Untersuchungen*, 1900, 9:1-16.

the skull of his grandfather, the mathematician August Ferdinand Möbius, whose remains had been unearthed.[8]

8 Paul Julius Moebius, *Ueber die Anlage zur Mathematik* (Leipzig: Barth, 1900). The *Anhang* – one third of the book – was *Ueber Franz Joseph Gall* (pp. 195-331); Id., "Ueber den Schädel eines Mathematikern," in *Ausgewählte Werke* (Leipzig: Barth, 1905), Bd. 7, pp. 211-221. Moebius' radical anti-feminism caused him to reduce the value of Kovalevski's activity: see pp. 82-83. It cannot be forgotten that the very year in which he dealt with the cerebral seat of the mathematical faculty, his famous *Ueber den physiologischen Schwachsinn des Weibes* came out (Halle: Marhold, 1900), which could not fail to proclaim the innate inferiority of the female brain ("schlechter entwickelt"). A woman who utterly fulfills her maternal task could never have anything like a male brain (p. 15). A little later Moebius again insisted on a 'minor' female brain, albeit suited to her limited needs and activities: see his *Geschlecht und Kopfgrösse* (Halle: Marhold, 1903), especially pp. 42-47.

Nevertheless, Retzius believed that his own localization was more reliable and, above all, he stressed the usual need for "mehr thatsachen zu sammeln".[9] As for the weight of the Swedish brains, for a long time it had been his intention to ascertain it, but only in 1887 work began by weighing 150 specimens (100 male, 50 females), little by little, and submitting a provisional report to the Society of Swedish Physicians in 1898. Their number, however, was not sufficient to calculate a credible average value, and therefore the procedure continued to amass up to 700 brains (450 male, 250 female), of subjects who had died in Stockholm hospitals. Retzius claimed that, of the previously published examinations of brain weights, only a few had been carried out on equally large scale. The outcome of all his labour was the compilation of twelve tables of weight data – displaying information on age, height, illness and cause of death – which were then compared with those of European populations provided by other researchers. In the end, minor differences were detected: the average weight of Swedish male brains was close to 1400 grams, on a wide range from 1118 to 1743; 1248 grams for females, with a variability from 940 to 1553. The gap between the two sexes (151 grams) was in quite good agreement with the one found elsewhere in Europe.[10]

It was 1901 when a medical student in his fourth year at Columbia University was invited to examine the corpse of the anarchist Leon Czolgosz, President McKinley's killer. Edward Anthony Spitzka performed the autopsy under the supervision of Carlos D. MacDonald, professor of mental diseases and medical jurisprudence in New York, who, in illustrating the case, praised his young assistant with these words:

---

9 Gustaf Retzius, "Das Gehirn des Mathematikers Sonja Kovalevski," *Biologische Untersuchungen*, 1900, *9*:1-16, p. 16; Id., "Das Gehirn des Astronomen Hugo Gyldéns,", ibid., 1898, *8*:1-22 (with six plates).

10 Gustaf Retzius, "Ueber das Hirngewicht der Schweden," *Biologische Untersuchungen*, 1900, *9*:51-68, with twelve tables of data (pp. 56-67). In his Huxley Lecture, read at the Royal Anthropological Institute of London in 1909, Retzius chose to talk about the North European race. He began with a defense of craniology, since it had become latterly fashionable to diminish its importance. The study of the cranium shape had led anthropologists, in the previous decades, to look at the race that was spread over the north of the continent and formed, due to emigration, an increasingly large element in the white population of North America. The lecture was also a tribute to Blumenbach – praised as the founder of the discipline – and to his father Anders, who had shown in the 1840s that the Caucasian variety was far from uniformity, divided into several races with different shapes of the skull, susceptible to measurement. Retzius pointed out that since then the cephalic index had been playing an essential role in anthropology. See Id., *The So-Called North European Race of Mankind. A Review of, and Views on, the Development of Some Anthropological Questions* (*The Huxley Lecture for 1909*) (London: Royal Anthropological Institute, 1909).

> In view of its great importance both to medical science and to medical jurisprudence, the writer regards it as fortunate that the State was able to secure the services of so able a brain anatomist and skilled operator and draughtsman as Mr. Spitzka, to make the post-mortem examination.[11]

As the offender had been judged mentally healthy and therefore sentenced to death, his brain did present no marked peculiarity of shape or size: "it was firm to the touch, and no portion of it, despite most careful examination, felt softened or indurated." A plaster cast of Czolgosz's head was made, photographed and shown, along with some drawings by Spitzka himself. Quite amazing for his young age and status, in those early years of the century he published a large number of accurate articles related to brain anatomy, which had mostly to do with "accomplished" men or criminals. An "unverblümter Rassist" who urged that more time be also devoted to racial brains, since "cerebral physiognomy" could well reflect the difference between stupidity and intelligence.[12]

In 1902, Spitzka complained that a "meager amount of material" had been so far accumulated: few and often unsatisfactory descriptions, a limit particularly to be regretted in the case of populations rapidly becoming extinct. Notwithstanding this, assuming that brains of different races showed typical differences in cerebral surface and morphology was entirely reasonable. The establishment of a systematic "anthropological encephalometry" was still to be attained, and he meanwhile contributed a report about three Eskimo brains, from Smith Sound, the sea passage between Greenland and Canada. That hyperborean race – most travelers had told – was sharp-witted, exhibited remarkable aptitudes and considerable intellectual power. In spite of the crudeness of delineation and imperfection in detail, their ivory sculptures of animal and human forms displayed a remarkable quality. Like those of the Chinese, "their drawings have but one defect, being faulty in perspective".[13] In the past, only

---

11 Carlos F. MacDonald, "The Trial, Execution, Autopsy and Mental Status of Leon F. Czolgosz, Alias Fred Nieman, the Assassin of President McKinley. With a Report of the Post-Mortem Examination by Edward Anthony Spitzka," *The Journal of Mental Pathology*, 1901-1902, *1*:179-209, p. 186. A short report of the case was given by Spitzka's father, the neurologist who happened to serve, twenty years earlier, as an expert witness at the trial of Charles J. Guiteau, accused of the assassination of President James A. Garfield: Edward Charles Spitzka, "The Czolgosz Case," *The Philadelphia Medical Journal*, 1901, 8:693-695. See Duane Haines, "Spitzka and Spitzka on the brains of the assassins of presidents," *Journal of the History of the Neurosciences*, 1995, *4*:236-266.

12 Michael Hagner, *Geniale Gehirne. Zur Geschichte der Elitegehirnforschung* (Göttingen: Wallstein Verlag, 2004), p. 227.

13 Edward Anthony Spitzka, "Contributions to the Encephalic Anatomy of the Races. First Paper: – Three Eskimo Brains, from Smith's Sound," *American Journal of Anatomy*, 1902-1903, *2*: 25-71, 28.

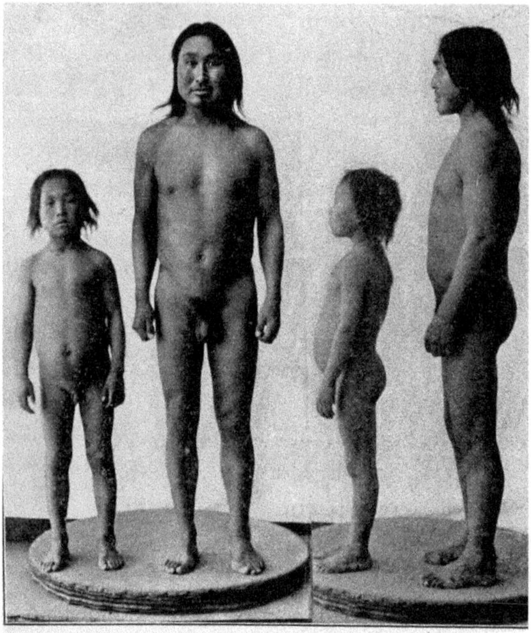

PLATE I.—Kishu and Menee (photographed on their admission to Bellevue Hospital).

a

FIGURE 4.2A-B
Photographic portrait of Kishu and his son Menee; Plate IV: Kishu's hemispheres, in Aleš Hrdlička, *An Eskimo Brain* (New York 1901).

four Eskimo brains had been described: three by Théophile Chudzinski – a Polish-born assistant of Broca, who had found in 1881 a considerable volume of their hemispheres combined with a great simplicity of the convolutions. More recently, Hrdlička made meticulous observations, with photographic views, on the brain of Kishu, an Eskimo chieftain who had died of tuberculosis at Bellevue Hospital in New York. As a whole, it was heavier and larger than the average brain of white men of similar stature, with a pronounced and rather more complex gyration.[14] Spitzka was similarly persuaded as to his three Eskimo brains, after illustrating their peculiarities:

> The hitherto popular notion that the typical Eskimo skull exhibits a low order of intelligence, and is characterized by a small brain-capacity has been generally refuted, and the fallacy of this idea becomes more

14 Théophile Chudzinski, "Sur les trois encéphales des Esquimaux morts de la variole, du 13 au 16 janvier 1881, dans le service de M. Andrieux à l'hôpital Saint-Louis," *Bulletins de la Société d'Anthropologie de Paris*, 1881, III série, *4*:312-318; Aleš Hrdlička, *An Eskimo Brain* (New York: The Knickerbocker Press, 1901), pp. 33-35.

PLATE IV

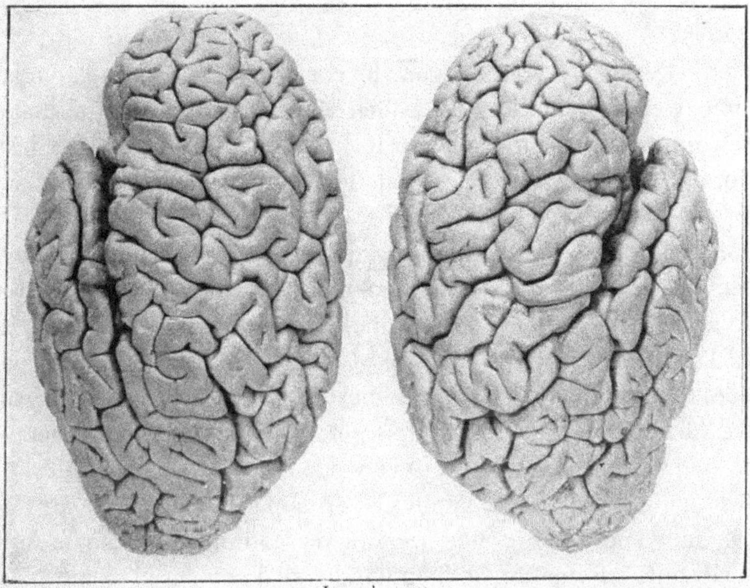

Lateral aspect

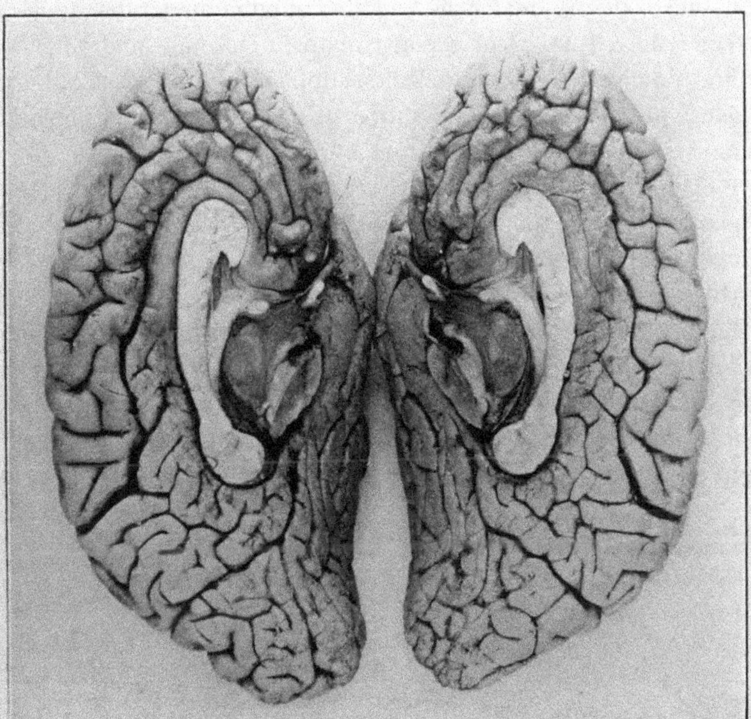

b

> apparent with the demonstration of so highly developed a brain as these specimens have shown the Eskimo to possess.[15]

He gave an equally positive opinion when he commented on a few data relating to the brain weight of the Japanese, which compared favorably with that of Europeans of similar height, and superior to other races of short stature: facts that were "of not a little significance in relation to the learning, industry and aptitudes of this progressive race."[16]

In 1908, Spitzka stressed again the "pressing need" for fruitful encephalometric research in the exotic races, "so rapidly becoming impure or even extinct": because of the on-going disappearance of Amerindian, African, Australian, and Eskimo populations anatomical knowledge had to be recorded without delay. He considered he had been lucky at having been able to pursue comparative studies upon the brains of some of them, and lastly on natives of the Andaman and Nicobar Islands, of very bad reputation as a most primitive race. Belonging to the Negrito family, the Andamanese were of dwarf stature but well made and proportioned, wearing no clothing and living in huts. Spitzka added some information on their handicraft and focused on morphology: small and round skull, prognathous face, with several simian affinities. The brain of one of them, a male who had died of pulmonary tuberculosis in 1905, had been removed and immersed in a mixture of formalin and water by a medical officer on duty at Port Blair. Its dissection, performed at the Jefferson Medical College of Philadelphia almost three years later, revealed a weight of only 1113 grams, with an estimated 1200/1250 grams in the fresh state. Broad and short, it had frontal lobes less massive than in Whites, and Spitzka listed a series of measures without comments, just comparing them with those taken on the Nicobarese specimen.[17]

Despite his repeated calls to multiply efforts on racial brains, Spitzka stopped dealing with them and preferred to investigate those of individuals who had shown special qualities during life. In 1906, for instance, at a meeting of the American Philosophical Society he read a study of the brains of six "eminent scientists and scholars" and, two years later, he published a report about them. After Paris – where the Société d'autopsie mutuelle had begun its activity in 1876 – the American Anthropometric Society (AAS) was founded in Philadelphia in 1889, with a similar ambition but a rather different operating style,

15 Spitzka, "Contributions" (cit. note 13), p. 66.

16 Edward Anthony Spitzka, "The Brain-Weight of the Japanese," *Science*, 1903, *18/455*: 371-373.

17 Edward Anthony Spitzka, "Preliminary Note on the Brains of Natives of the Andaman and Nicobar Islands," *Proceedings of the American Philosophical Society*, 1908, *47*: 51-58.

while Burt Green Wilder, an anatomist at Cornell, started collecting brains of "educated and orderly persons".[18]

When Spitzka spoke in front of his associates, the AAS loot amounted just to eight specimens – two of them in poor conditions – unfortunately deprived of Walt Whitman's brain that, "together with the jar in which it had been placed, was said to have been dropped on the floor by a careless assistant."[19] Since 1902, he had been working on them and now he delivered some results, but first he listed 137 "notable individuals", only four of them women, whose brains had been weighed or examined over time. A hodgepodge of data, which shows, on one hand, the extension of the obsessive brain hunt that had taken place in the last few decades and, on the other hand, their loose heterogeneity.[20] Spitzka's summary sounds vaguely despondent:

> It were unwarranted to propose conclusions of wide significance upon so little material and only brief comments are offered here upon the most notable findings in these brains. In general the cerebral surface shows complex development with intricate fissuration and a bold contour of the numerous gyres. In some brains one or another region preponderates over other regions in the degree of development.[21]

In particular, compared with ordinary men, those brains seemed to have larger callosa, "an index in somatic terms of how we may distinguish the brain of the genius or talented man from that of persons of only ordinary abilities."[22]

How could intelligence, understood as a complex set of psychic qualities, be measured? Since for many reasons it was impossible to do it directly, *en elle-même*, one solution might consist in evaluating the corresponding spatial phenomena, i.e. the body modifications that vary with it. In 1903-1904, this was the

---

18 See Hagner, *Geniale Gehirne* (cit. note 12), pp. 225-226. It should also be added that in 1903 the Société d'autopsie mutuelle revised its statute, elected a new *bureau* – composed of Henri Thulié (president), Georges Papillault (secretary) and Georges Hervé (treasurer) – and launched an ambitious recruitment campaign, with a more comprehensive questionnaire to be filled in by aspiring members: see "Notes et informations. Société d'Autopsie," *Revue scientifique*, V série, 1905, *3*:572-573.

19 Edward Anthony Spitzka, "A Study of the Brains of Six Eminent Scientists and Scholars belonging to the American Anthropometric Society together with a Description of the Skull of Professor E. D. Cope," *Transactions of the American Philosophical Society*, 1907, *21*:175-308, p. 176. See also Brian Burrell, "The Strange Fate of Whitman's Brain," *Walt Whitman Quarterly Review*, 2003, *20*:107-133.

20 Spitzka, "A Study of the Brains" (cit. note 19), pp. 177-206.

21 Ibid, p. 298.

22 Ibid, p. 303.

underlying intention of Nicolas Vaschide, now *Chef de laboratoire* of experimental psychology at the École des hautes études, and Madeleine Pelletier, who would soon be the first French woman to graduate in psychiatry. Both barely thirty years old, they were driven by a reforming impulse. For seven years, he had been visiting schools in Paris and in the département de la Seine, from the *écoles élémentaires* to the École normale supérieure and the École Polytechnique, looking for physical signs of intelligence in thousands of students. At first sight the usefulness of such research could have seemed questionable: would disclosing a psychic diagnosis to someone only pique the vanity of the superior subject and induce discouragement in the inferior? If it was risky to let them know their position in the intellectual hierarchy, yet such knowledge was indispensable to educators.

The aspiration of the two collaborators was therefore decidedly pedagogical, although conceived and pursued in the psycho-physiological field. All degrees of learning suffered from *psittacisme* – they complained – because it gave an enormous space to memory at the expense of intelligence. Above all, they were interested in finding a rigorous differential evaluation, according to the principles and practices that psychotechnics was then trying to define:

> S'il était possible de reconnaître dès l'enfance, au moyen de signes physiques spéciaux, les intelligences supérieures, on pourrait pousser plus loin leur education, les preparer plus spécialement à la haute culture, afin que, devenus adultes, ils soient une élite intellectuelle capable de promouvoir la société dans toutes les branches de son activité. [...] Si l'on arrive un jour à comprendre cette verité si simple qu'il faut diriger les intelligences supérieures vers les modes d'activité difficiles et les intelligences inférieures vers les travaux plus faciles, le bonheur et le progrès de l'ensemble seraient augmenté.[23]

Whatever the power of the social factor in increasing individual capacities, Vaschide and Pelletier excluded that *intelligences d'élite* could be grown on any cerebral terrain, because an undeniable innate component helps to determine their production. So, in measuring the heads of 140 students of Villejuif, they were not concerned with education, but with potential intelligence, and furthermore convinced that the most reliable object of investigation was not the brain – too susceptible to be changed by age and health condition – but the skull. That their subjects were at different stages of cranial development did

23 Nicolas Vaschide, Madeleine Pelletier, "Recherches expérimentales sur les signes physiques de l'intelligence," *Revue de Philosophie*, 1902-1903, 3:796-831, pp. 805-806.

not seem to invalidate the test. The game consisted in comparing the information given by the teachers on the intelligence of each individual student, the results of some tests and data about the body: stature, bi-acromial diameter, muscle strength, head diameters and heights, and consequent cephalic volume. The outcomes were arranged in a large number of *tableaux*, and the conclusions assured that, on average, the individuals of superior intelligence had shown to have larger heads; that the greater volume of the head seemed independent of stature and body development; that a wider forehead was likely to occur in smarter subjects.[24]

When Madeleine Pelletier collaborated with Vaschide, she was on her way to become a feminist and socialist militant, after studying under Charles Letourneau and Léonce Manouvrier, both heirs to Broca's school. As an anthropologist, she had to accept the relationship between brain size and intellectual development, although recalcitrant to the idea that women's smaller brains proved their inferiority. Under Manouvrier's direction, in 1900 she had measured 55 Japanese skeletons and skulls and calculated their cranio-femoral, cranio-mandibular, and cranio-cerebral indexes. *Inter alia*, she pointed out that, in relation to the body size, "la femme a une capacité crânienne et par conséquent un poids encéphalique proportionnellement plus élevé que l'homme." Nevertheless, Pelletier seemed to diminish the importance of the relationship between intelligence and encephalic weight:

> Considérer l'intelligence comme un bloc dont on pourrait avoir plus ou moins, serait antipsychologique. L'intelligence, c'est tous nos états de conscience: perceptions, souvenirs, sentiments, idées generales [...] Mais l'intelligence c'est aussi l'énergie du tempérament qui produit cette vivacité et cette rapidité dans la succession des états de conscience, qui caractérise les intelligences élevées. Enfin l'intelligence est aussi, surtout peut-être, cette agregation entre les états de conscience qui fait que l'on perçoit les rapports des choses; cette sorte de chimie mentale dont les réactions nous sont encore inconnues, peut-être est-ce à cela que correspondrait la quantité constant du cerveau?[25]

24 Vaschide, Pelletier, "Recherches expérimentales sur les signes physiques de l'intelligence," *Revue de Philosophie*, 1903-1904, *4*:168-195, p. 195.

25 Madeleine Pelletier, "Recherches sur les indices pondéraux du crâne et des principaux os longs d'une série de squelettes japonais," *Bulletins de la Sociétè d'Anthropologie de* Paris, v série, 1900, *1*:514-529, p. 524. See Charles Sowervine, "Women's Brain, man's Brain: Feminism and Anthropology in Late Nineteenth-Century France," *Women's History Review*, 2003, *12*:289-308.

This doubt did not prevent her, shortly thereafter, from collaborating with Vaschide on the project described above. They confessed, however, to their awareness of a possible vicious circle: not knowing how to measure intelligence, then how could they assert that the brain increases with it? After all, when science has no precise data available – they unconvincingly argued – it must settle for vague ones and work with them, otherwise it would never advance.[26]

Despite all the reservations already expressed at the end of the century about a misguided brain-intelligence equivalence, Manouvrier's curiosity was still attracted by phenomena such as the extraordinary volume and weight (*hypermégalie*) of the brain of Joseph Bouny, a notary in a small town of the Gironde. Struck by his big head, his countryman and friend Broca had persuaded him to bequeath his brain to the Paris Laboratoire d'Anthropologie. Having a reputation as a man of character and great intelligence, he seemed a perfect case to investigate post-mortem. The scale marked 1935 grams for his brain, 575 grams above the average weight of adult French males: an abnormal figure that led Manouvrier to wonder about its causes, believing that Bouny could not be possibly classed as one of the "dégénérés supérieurs". After separately analyzing the two hemispheres – which showed more asymmetries than usual –, and noting the exceptional depth of the sulci, could a clear correlation be finally established between the peculiarities of Bouny's brain and his degree of intelligence? Once again, "dans l'état actuel de nos données scientifiques", the question remained unanswered, the nature of that relationship being only hypothetical.[27]

What Vaschide and Pelletier were doing in Paris at the beginning of the century, Karl Pearson undertook in London. As "a New Year's greeting to Francis Galton" on January 23, 1902, he read a proviso about the data he had obtained from the Cambridge Anthropometric Committee on undergraduate students, and had processed, with a team of assistants, in order to ascertain "which if any, physical characters are sensibly correlated with intellectual ability". Their statistical elaboration focused on the length and breadth of the head, and on the cephalix index. The outcome of this first work is noteworthy for its honest recognition of a poor result:

> Very brilliant men may possibly have a very slightly larger head than their fellows, but taking the general population there is really a very insignificant association between size of head and ability. For practical

26 Vaschide, Pelletier, "Recherches expérimentales" (cit. note 24, p. 801)

27 Léonce Manouvrier, "Considérations sur l'hypermégalie cérébrale," *Revue de l'École d'anthropologie de Paris*, 1902, *12*:391-414.

> purposes it seems impossible, either in the case of exceptionally able men or in the bulk of the population, to pass any judgment from size of head to ability or *vice versa*.[28]

Yet, still in 1906, Pearson presented "more precise and extended investigations" carried out in his Biometric Laboratory, established in 1903. The material dealt with was the measurement of a thousand Cambridge graduates and of more than five thousand school children, girls as well as boys. After describing the work done, the statistician himself hinted that its results might be "of little significance", as they could also be argued, without elaborate statistics, from the impressions of a careful and observant teacher. Science is however – he added – the verification or refutation of impressions and opinions, so that the mainly negative conclusions of his paper were useful to place on a sounder quantitative basis the view that little could be judged as to intelligence from the anthropometric measurements.[29]

The first issue of *Biometrika. A Journal for the Statistical Studies of Biological Problems* was published in 1901, edited by Pearson, Charles B. Davenport, and W. F. Raphael Weldon, "in consultation with Francis Galton", honoring Darwin with a photograph of his statue at the University Museum of Natural History in Oxford, and the motto *Ignoramus, in hoc signo laboremus* underneath it. In 1904, Pearson published his Huxley Lecture on the laws of inheritance in man, centered upon mental and moral characters, in comparison with the inheritance of the physical characters, and an article in collaboration with Marie A.

28 Karl Pearson, "On the Correlation of Intellectual Ability with the Size and Shape of the Head (Preliminary Notice)," *Proceedings of the Royal Society of London*, 1902, *69*:333-342, p. 342. A second part of the notice introduced a non-cephalometric evaluation parameter, given Pearson's fixation with the importance of athletics: "While the intelligent are only *slightly* the more healthy, the athletic are *notably* the more healthy element in the community. Further, the athletic are considerably more intelligent than the non-athletic; they are the more popular and more noisy element; and they tend to quick rather than sullen temper.": Alice Lee, Marie A. Lewenz and Karl Pearson, "On the Correlation of the Mental and Physical Characters in Man. Part II," ibid., 1903, *71*:106-114, p. 114.

29 In essence, the research had produced these findings: "While no characters in school children so far dealt with show very high correlation with intelligence, we may yet say that the intelligent boy is markedly conscientious, is moderately robust, athletic, and popular: he tends rather to quick than to sullen temper. He is more self-conscious and quieter than the dull boy; he has a *slightly* bigger head, and possibly lighter pigmentation than those of more mediocre intelligence. His hair has a larger percentage of curliness." The intelligent girl had similar features, except that "she is less self-conscious than the dull girl, and noisier than the girl of mediocre intelligence." See Karl Pearson, "On the Relationship of Intelligence to Size and Shape of the Head, and to Other Physical and Mental Characters," *Biometrika*, 1906, *5*:105-146, p. 136.

Lewenz on the measurement of internal capacity from cranial measurements. The following year, the periodical hosted a long article by the young American biologist Raymond Pearl, who was then completing his studies in Europe. He recalled first of all that earlier workers had collected, studied and discussed brain-weights to a considerable extent as measures or indices of brain-power. After a time, though, it became evident that the close and definite relationship, which had been supposed to exist, was something of a chimera, although increase in brain-weight and in psychic capacity went in general hand in hand along the taxonomic scale. In Pearl's opinion, dispelling the illusion did not mean belittling the brain-weight, which remained an important anthropological, rather than psychological, character. The human brain was justly to be regarded as the highest product of organic evolution, and only by understanding the correlation between its weight and other bodily characters could different races, or different groups of the same race, be compared with each other. Nevertheless, the method and use of the data were a problem:

> Now, as everyone knows who has even an elementary knowledge of statistics, it is possible to make the same statistical material lead to quite different conclusions by grouping it in different ways, when the tabulations and averages are the only sources from which conclusions may be drawn. As a matter of fact this has happened in work on brain-weights. Different investigators, working in different ways, have arrived at quite different numerical appreciations.[30]

A desire to apply adequate statistical methods to a biometrical problem of peculiar interest led Pearl to undertake his research, having obtained a big Brunsviga arithmometer on a grant from the Carnegie Institution.

As a biostatistician, Pearl did not weigh a single brain but borrowed and reworked the available materials, such as the results presented in 1902 by the Leipzig pathologist Felix Marchand, who had weighed hundreds of brains at the Institut für Pathologie in Marburg from 1885 to 1900, and those produced by Theodor Bischoff in 1880, already discussed in the previous chapter. Other sources from which he drew included the series of Swedish brain-weights recently collected by Gustaf Retzius in Stockholm and those gathered by Jindřich Matiegka in Prague, or the English tables compiled by Boyd and Marshall half a century earlier.[31] From a biometrical point of view, all this raw material was

30 Raymond Pearl, "Biometrical Studies on Man. I. Variation and Correlation in Brain-Weight," *Biometrika*, 1905, 4:13-104, pp. 14-15.

31 Felix Marchand, *Ueber das Hirngewicht des Menschen* (Leipzig: Teubner, 1902); Gustaf Retzius, "Ueber das Gehirngewicht der Schweden," (cit. note 10); Heinrich Matiegka, *Über das Hirngewicht, die Schädelkapazität und die Kopfform, sowie deren Beziehung zur*

peculiarly problematic: its great bulk came from autopsies performed in general hospitals or other public institutions of a similar character, and consequently did not represent random samples of the general population dying outside a hospital. The brain-weight was obviously affected by the age and by the cause of death.

Such inconvenience, of which Pearl was aware, did not prevent him from making some pretty clear points, for instance that there were definite racial types in brain-weight, while each of the series – Swedish, Hessian, Bavarian, Bohemian, and English – exhibited a very fair degree of homogeneity; that brain-weight and skull capacity were sensibly equal; that variation might be considered to follow the normal law of distribution of errors; that the correlation of brain-weight with age, stature, and body weight was in all cases low; that the sexual differences were practically constant in all the races studied, whether considered absolutely or relatively; that there was no evidence that brain-weight was sensibly correlated with intellectual ability. Finally, Pearl could not fail to repeat the usual caveat:

> There is great need for further large and homogeneous collections of brain-weight statistics. When these are available for a considerable number of races it will be possible to pass from intra-racial to inter-racial problems.[32]

That the topic was still considered worthy of interest is shown by the fact that Pearl's article was followed by one written by the anatomist Reginald J. Gladstone, who had been collecting data in the *post-mortem* room of the Middlesex Hospital, London, for over three years. His practice had consisted in measuring the heads of dead patients and afterwards, for each case, in taking out and weighing their brains. With the help of the mathematical treatment provided by John Blakeman, Gladstone's aim was to predict – within the limits of normal variation – the approximate weight of the brain, when in possession of some chief measurements of the head. Blakeman himself, assisted by Alice Lee and Karl Pearson, then carried out a painstaking biometric analysis of his work, collated with that of Pearl.[33]

---

psychischen Tätigkeit des Menschen. 1. Über das Hirngewicht des Menschen (Prag, Verlag der Königl. Böhmischen Gesellschaft der Wissenschaften, 1902).

32 Pearl, "Biometrical Studies on Man" (cit. note 30), p. 83.

33 Reginald J. Gladstone, "A Study of the Relations of the Brain to the Size of the Head," *Biometrika*, 1905, *4*:105-123; John Blakeman, Alice Lee, Karl Pearson, "A Study of the Biometric Constants of English Brain-Weights, and their Relationships to External Physical Measurements,", ibid., pp. 124-160. The last of their conclusions granted that there was no

While in England the most sophisticated statistical tools were applied to the same old problem, a young assistant of Paolo Mantegazza in Florence – first Italian chair of anthropology in 1869 – authored a merciless review of what craniology had accomplished about the problem of intellectual development. Aldobrandino Mochi began by re-proposing the battle of his mentor, who in 1875 had tried to counteract the compulsive tendency of his colleagues to measure skulls with ever-new instruments and to pile up a jumble (*farragine*) of figures that nobody cared to read. Thirty years earlier, Mantegazza had already asked whether all that time-consuming craniological labor was really indispensable.[34] In 1904, Mochi proclaimed that human intelligence is so polymorphic and elusive, linked to such a wide range of environmental conditions, that a single measure would never apply to all its forms; and that no uniform method could ever establish a hierarchical scale of all its manifestations. But, even assuming that a psychological yardstick (*metro psicologico*) were available, it would take a statistically large series of skulls, each one belonging to an individual whose psychic value was known: a task thwarted by the craniological collections of that time. Therefore, a rigorous research on the relationships between skull and intellect had to be given up, at least for the moment. A similar impediment made also studies into the relationship between brain and intelligence unreliable. Although much attention had been focused, time after time, on the cerebral size and shape, development of convolutions, etc., yet the question remained elusive. Too approximate methods were faced with insurmountable difficulties.[35]

From its beginnings, craniology had yearned for a shadow of a numerical expression of intelligence by means of increasingly sophisticated measures and had devoted themselves to the cult of "hierarchical characters", which was still somehow worshipped despite its flaws. Mochi reviewed the past and present methods to which diagnostic value of intelligence had been attributed as a measure of intelligence: general form of the skull, development of its frontal region, proportions between anterior and posterior parts. From two tables – the first containing data taken from the existing literature, the second with

sensible *relative* difference between the brain-weights of man and woman, when proper allowance was made for the relative difference in their size (p. 160).

34 Paolo Mantegazza, "Dei caratteri gerarchici del cranio umano. Studi di critica craniologica," *Archivio per l'Antropologia e la Etnologia*, 1875, *5*:32-81. Criticism and reform of craniology were one of the many leitmotivs in Mantegazza's activity: see his "La riforma craniologica," ibid., 1880, *10*:117-137 and "Di alcune recenti proposte di riforma della craniologia," ivi, 1893, *23*:45-55.

35 Aldobrandino Mochi, "Sui rapporti tra lo sviluppo intellettuale e la morfologia craniense. Tesi per la libera docenza in Antropologia," *Archivio per l'Antropologia e la Etnologia*, 1904, *34*:83-142.

measurements made by Mochi on 686 skulls of various races – it was easy to infer that a greater cranial capacity did not always correspond to a higher psychic evolution. In other words, the brain mass would be linked partly, but not constantly and necessarily, with the development of intelligence, and its volume would obey other factors in addition to the intellectual ones.

Given the aforementioned limits, did anthropology – of which craniology had been *magna pars* in the previous century – admit his failure? Not at all: Mochi charged the anthropologist with the pivotal mission of studying individual and collective characters in their variations through time and space, such a vast task that craniology was reduced to only one of its many chapters. Having relegated all hope of anatomically reading the secrets of the psyche to the realm of scientific utopias, any amount of work could still be done. At that time, Maria Montessori, who obtained her *libera docenza* in anthropology in 1904, was dealing with the relationships between brain volume and intellectual state, by observing and measuring 105 schoolchildren in Rome. The problem posed by Broca almost half a century before – whether education could promote brain enlargement – seemed to her still unsolved, and she merely communicated that she had found in more intelligent children not only a greater development of the head, but also some differential features of the face: a more spacious forehead, a larger zygomatic diameter, a leptorrhine nose. Of course, the influence of social conditions on those parameters had to be scrutinized.[36]

2 Further Views in Conflict

In 1906, a young instructor in anatomy at the University of Michigan bemoaned that "yet no exact measurements of the brain, such as we have of the skull" could be found, despite all the attempts made in the past century to determine the points of separation between the Caucasian and the Negro brains. So, Robert Bennett Bean – a descendant of Thomas Jefferson – had thought it appropriate to remedy the situation by examining 152 brains: 103 from American Negroes and 49 from Caucasians. The work was undertaken at the suggestion of his professor at the Johns Hopkins School of Medicine, Franklin P. Mall, prompted by Hrdlička's assumption that racial differences existed in the brain, of which Bean was fully convinced. Most of those specimens came from the

36 Maria Montessori, "Sui caratteri antropometrici in relazione alle gerarchie intellettuali dei fanciulli nelle scuole. Ricerche di antropologia pedagogica," *Archivio per l'Antropologia e la Etnologia*, 1904, *34*:243-297.

collection at the Anatomical Laboratory of the Johns Hopkins. The perfect preservation of their shape was obtained by injecting the bodies of fresh cadavers with carbolic acid, alcohol and glycerin through the femoral arteries under pressure, leaving the corpse to rest for 12 hours. After removing the brain, firm and solid, it had to be suspended in 10% formalin and sodium chloride. Outline drawings were also made of the brains in their normal position, viewed from above; lateral and mesial outlines were drawn after separating the hemispheres, as well as the lateral border of both.

In particular, a comparison between two brains of males about the same size and weight taken and close in age (37 and 40 years old) – a White and a Black – showed that

> the Caucasian brain conforms more nearly to a circle in its contour in the different planes than does that of the Negro, which is squared at the ends, and flatter on the sides and above, especially along the frontal lobes, thus exhibiting a distinct box-shaped appearance. The shape of the Negro brain is manifested in the mesial outline by the abrupt rise of the contour from the axis at its posterior end, by the nearly straight line over the anterior association center, by the nearly vertical line along the anterior aspect of the frontal lobe, and by the horizontal line along the inferior border if this lobe.[37]

In other words, viewed from the side, the Negro brain appeared to be pressed back, while the Caucasian to be pushed forward, the result being that the frontal lobe of the first brain, i.e. its anterior association center, was considerably smaller. Charts and tables were tasked with demonstrating this smaller dimension, which even Tiedemann – called by Bean "that eminent continental champion of the Negro" – had to acknowledge. Relying on Paul Flechsig's work on the development of the fiber tracts and cortical areas, the American anatomist took it for granted that the posterior association center was "objective", and the anterior "subjective": so, he found a way to revive an old *cliché*:

> The Caucasian – more particularly the Anglo-Saxon, which was derived from the Primitives of Europe, is dominant and domineering, and possessed primarily with determination, will power, self-control, self-government, and all the attributes of the subjective self, with a high development of the ethical and aesthetic faculties. The Negro is in direct contrast by

37 Robert Bennett Bean, "Some Racial Peculiarities of the Negro Brain," *The American Journal of Anatomy*, 1906, 5:353-431, p. 361.

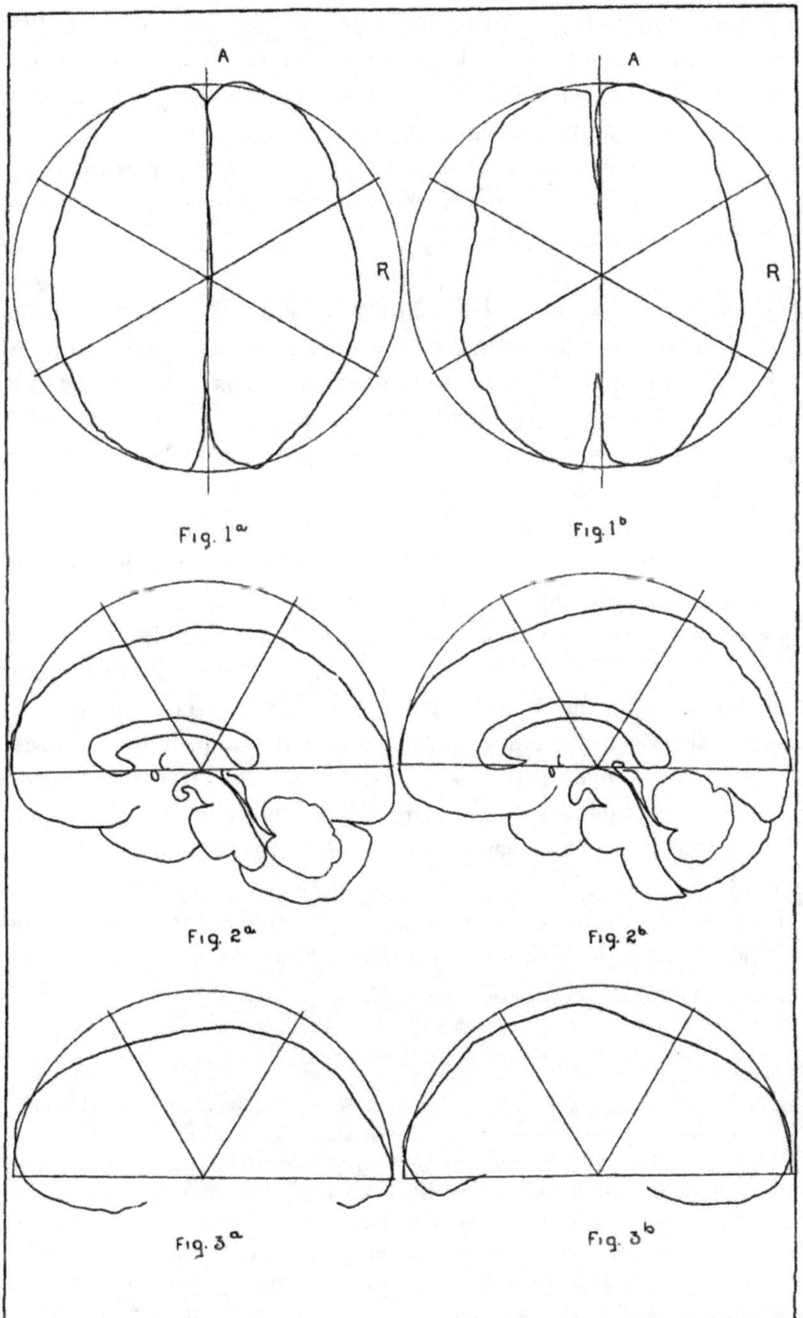

FIGURE 4.3 Comparison between brain outlines of a Caucasian male (fig. 1a-3a) and of a Negro one (1b-3b), in Robert Bennett Bean, "Some Racial Peculiarities of the Negro Brain," *The American Journal of Anatomy*, 1906, 5:353-431, p. 360.

> reason of a certain lack of these powers, and a great development of the objective qualities. The Negro is primarily affectionate, immensely emotional, then sensual and under stimulation passionate. [...] One would naturally expect some such character for the Negro, because all the posterior part of the brain is large, and the whole anterior part is small, this being especially true in regard to the anterior and posterior association centers.[38]

Bean believed he could validate this antithesis of subjective/objective, front/back, White/Black by availing of the authority of Broca, Gratiolet, Topinard, Spitzka. In addition, the smaller size of the Negro's *corpus callosum* had to be reckoned, especially the genu connecting the left and right frontal lobes, while there were so many factors involved in brain weight that it seemed unprofitable to discuss this quite controversial question.

While Bean was testifying against the colored minority, Atlanta University published the proceedings of the Eleventh Conference for the study of the Negro problems, held on May 29, 1906 – a series started ten years earlier by W.E.B. Du Bois, who taught history and economics there.[39] After dealing with various topics, from mortality to housing, from economic and social organization to education and religion, the second cycle took up the subject of the physical conditions of Negroes again, and greatly expanded the range of data, drawing on the most varied sources. It was usually assumed – the anonymous author (Du Bois) said – that there were significant differences between the European and African brain, the latter being inferior. However, Joseph Deniker had pointed out that "the weight of the encephalon varies enormously according to individuals", whereas the alleged greater lightness of the brains of non-European races had not been ascertained by a sufficient number of examples.[40] For his part, the African American sociologist Monroe Nathan Work, at that time

38 Ibid., p. 379.

39 *The Health and Physique of the Negro American. A Social Study made under the direction of Atlanta University by the Eleventh Atlanta Conference* (Atlanta, Georgia: The Atlanta University Press, 1906). Eight pages of a *Bibliography of Negro Health and Physique* precede the text (pp. 6-13), while the back cover contains a quote from Boas: "An unbiased estimate of the anthropological evidence so far brought forward does not permit us to countenance the belief in a racial inferiority which would unfit an individual of the Negro race to take his part in modern civilization. We do not know of any demand made on the human body or mind in modern life that anatomical or ethnological evidence would prove to be beyond the powers of the Negro."

40 Ibid., p. 24. A long passage was taken from the English translation of Deniker's major work: *The Races of Man. An Outline of Anthropology and Ethnography* (London: Walter Scott, 1900), p. 97.

professor at Georgia State Industrial College, was quoted for having raised serious objections to the ruling opinion. First of all, insufficient data had been collected: about 700 hundred brains in all, compared to hundreds of millions of blacks living in Africa and America. Secondly, almost no account had been taken of age, stature, social class, occupation, nutrition, and cause of death: each of which affected both the weight and structure of the brain. Third objection, the differences in the average weight of Negro and white brains were not great enough to warrant the conclusion that, if an equally large number of their specimens had been made available, the result would be the same.

The main target was obviously Bean's work, which had just been published. His generalizations had been based on too few examples, so the images he drew to certify that the size and shape of the front cerebrum were different in the two races could only be a very weak proof. More generally, the statement that white brains had more elaborate convolutions and deeper fissures needed to be refuted as well. These morphological aspects depended on several variables:

> As for example, a brain possessed of an extensive cortex with the elements incompletely associated can be a much folded brain, because in order to apply it to the surface of the cerebrum it must be thrown into many gyri. On the other hand, the associating fibers may be so developed as to increase the central mass, thereby giving a larger surface to which the cortex may be applied and thus tend to increase the cortical folds. These facts, with those from comparative anatomy respecting the fissuration and convolution of the brains of beasts and birds, seem to indicate that there is no certain relation between brain convolution and intelligence.[41]

Finally, the best evidence seemed to indicate that the organization of the central nervous system was continually modified through life, with significant changes caused by age, occupation, nutrition, disease etc. This fact made it very doubtful whether any uniformity as for the finer details of structure could be found in any race. Characters which were said to be distinctive of a particular race were found with more or less frequency in other races: so, what had been fixed as peculiar in the Negro or the white brain was not true of all of them and probably had no special connection with the mental capacity of either race.

41 *The Health and Physique of the Negro American* (cit. note 39), p. 27.

In 1909, Mall, who had asked Bean to undertake the aforementioned comparative analysis of the brains and had even provided the material, nevertheless had something very critical to say about it and set out to dismiss the groundless thesis of some of his predecessors and contemporaries. They had claimed they could determine, almost at a glance, whether or not a given specimen came from a great man, a woman or from a Black. It had been over half a century since Emil Huschke had detected "a *decidedly* greater amount of frontal lobe, fully one per cent (!) in the male rather than in the female."[42] Speaking of the black brain, the German anatomist had emphasized its resemblance with the female and childish ones, not to mention its proximity to the type of apes. A dozen years later, Theodor Meynert confirmed that men had more cerebral substance in front of the central sulcus than women, though this was in conflict with Broca, who had calculated a slightly higher percentage of the frontal lobe in the female brain (43.7 vs. 43.5%). In 1894, Nikolaus Rüdinger, an anatomist in Munich, had studied the brains of twin fetuses of different sex and apparently demonstrated that the development in the male was more advanced than in the female, with a larger frontal lobe. More recently, Wilhelm von Waldeyer judged that Rüdinger was right in the majority of specimens, although this was not always the case, a circumstance that Mall had personally verified.[43] He was anyhow convinced that "with the methods used by the above named investigators it cannot be definitely inferred a marked difference between men and women in the relative amount of brain in front of the central sulcus."[44]

Mall considered all those attempts unsuccessful, as well as others aimed at highlighting the same unequal distribution of brain substance in Blacks or in men of genius. Spitzka's and Bean's works were his polemical targets. Above all

42 Franklin P. Mall, "On Several Anatomical Characters of the Human Brain, said to vary According to Race and Sex, with Especial Reference to the Weight of Frontal Lobes," *The American Journal of Anatomy*, 1909, 9:1-32, p. 1. *Schädel, Hirn und Seele* (1854) was Huschke's book quoted by Mall.

43 The reference was to Nikolaus Rüdinger, "Über die Hirne von Zwillingen," *Verhandlungen der Anatomischen Gesellschaft auf der 8. Versammlung in Strassburg*, 1894, *8*:177: see Wilhelm von Waldeyer, "Über Gehirne menschlicher Zwillings- und Dillingsfrüchte verschiedenen Geschlechtes," *Sitzungsberichte der Königlich Preussischen Akademie der Wissenschaften zu Berlin*, 1907, pp. 114-126.

44 Mall, "On Several Anatomical Characters" (cit. note 42), p. 6. A little later, on the contrary, the neurologist and psychiatrist Giovanni Mingazzini – very active in Rome for a few decades – thought it was sad that the female brain revealed, right from the onset, its inferiority: a more squat and narrow frontal lobe, more symmetrical and poorer in furrows. And as such it would be preserved, like a childish sign, in adulthood: an eloquent warning to the so-called feminists. See his *Il cervello e i fenomeni mentali. Discorso inaugurale* (Roma: Pallotta, 1912), p. 16.

– he remarked – in studies of that kind no one had previously paid any attention to the "personal equation" of the researcher, a factor of such considerable importance that he had decided to make all his measurements without knowing the race or sex of the individuals from which the brain had been taken. As was to be expected, his own figures, which came from observations conducted on eighteen of the brains examined by Bean, dramatically differed from those reported by the latter, and, furthermore they showed no variation in the genu or in the *corpus callosum* due to either race or sex. It was likewise incorrect to state that the frontal lobe of the Negro brain weighed relatively less than that of the White.

Mall also referred to a discussion that had taken place – and that was a cause of so much trouble for anatomists – around the occipital peculiarity called *Affenspalte* by Rüdinger in 1882 and described as belonging to monkeys. At the beginning of the new century, the Australian-British anatomist and ethnologist Grafton Elliot Smith had been attempting to show that what he redefined as "sulcus occipitalis lunatus" was present not only in monkeys and apes, but also in man. His first observations of the human lunate sulcus were made on Egyptian and Sudanese brains that he examined while teaching at the Cairo Governmental School of Medicine. Therefore it would have been easy for Smith – Mall commented – to draw a wrong conclusion regarding this sulcus, but luckily he did not go there.[45]

In his institute in Berlin, Waldeyer hosted Sergio Sergi, a young Italian who had graduated with a thesis on the physiology of the cerebellum and eventually succeeded his father Giuseppe in the chair of anthropology at the University of Rome. While training in Germany (1906-1907), he had the opportunity of studying a group of brains that had belonged to individuals of the Herero ethnic group, coming from the colony of Deutsch-Südwestafrika, about which Waldeyer had already informed the Akademie der Wissenschaften in January 1906. Only two years earlier – it is worth remembering – the German troops had exterminated the Herero population that resisted the colonial conquest of

45 Mall, "On Several Anatomical Characters" (cit. note 42), pp. 20-21. See Nikolaus Rüdinger, "Beitrag zur Anatomie der Affenspalte und der Interparietalfurche beim Menschen nach Race, Geschlecht und Individualität," in *Beiträge zur Anatomie und Embryologie. Als Festgabe Jacob Henle zum 4. April 1882 dargebracht von seinen Schülern* (Bonn: Cohen & Sohn, 1882), pp. 186-198; Grafton Elliot Smith, "The Morphology of the Retrocalcarine Region of the Cortex Cerebri," *Proceedings of the Royal Society of London*, 1904, *73*:59-65; Id. "The so-called «Affenspalte» in the Human (Egyptian) Brain," *Anatomischer Anzeiger*, 1904, *24*:74-83; Id., "Studies in the Morphology of the Human Brain, with Special Reference to that of the Egyptians. No. 1: The Occipital Region," *Records of the Egyptian Government School of Medicine*, 1904, 2:123-173.

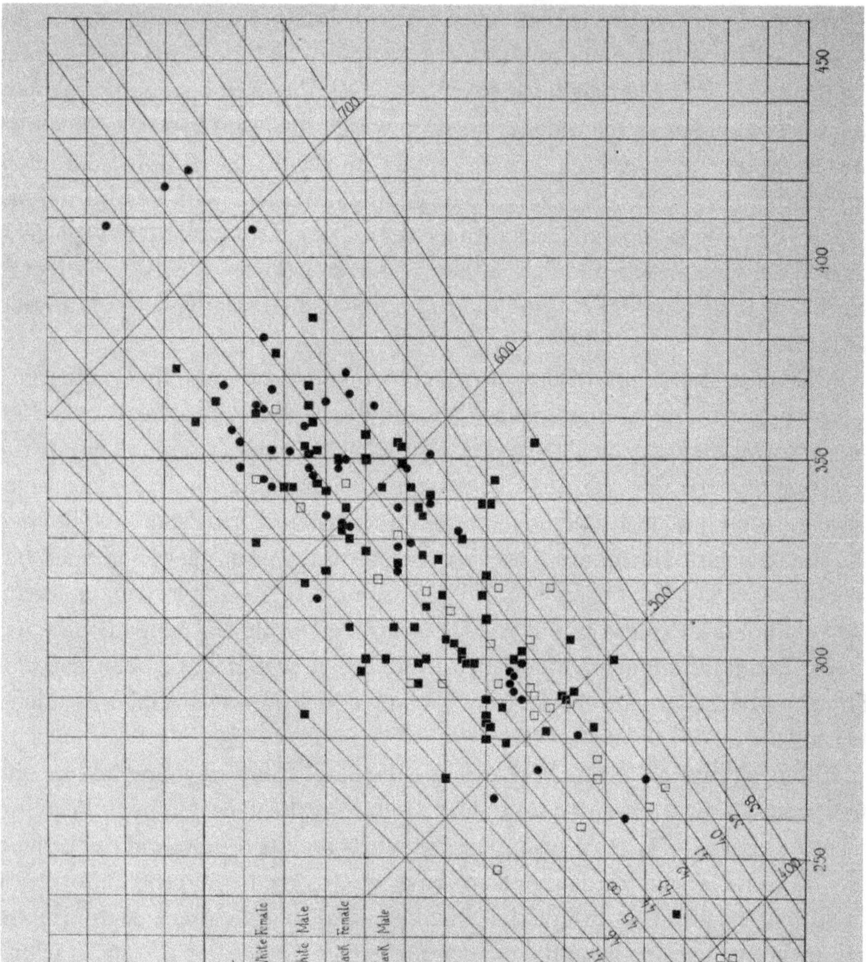

FIG. 2. Showing the relation of the brain substance lying in front (ordinates) of the sulcus centralis to that lying behind it (abscissæ). Each symbol represents a half brain.

The figures are in grams. The long diagonal lines, 38-50, indicate the percentage of the precentral brain weight. The lines marked 400-700 indicate the weights of the hemicerebra. The weights given are reduced to those of the fresh brain.

FIGURE 4.4 Fig. 2, in Franklin P. Mall, "On Several Anatomical Characters of the Human Brain, said to vary According to Race and Sex, with Especial Reference to the Weight of Frontal Lobes," *The American Journal of Anatomy*, 1909, 9:1-32, p. 14.

their territories, the first genocide in the 20th century, as stated by the Whitaker Report (1985).[46]

The full report of Sergi's research was published with a preface by Waldeyer and then awarded a prize by the Académie des Sciences in Paris, but some articles anticipated a synthesis of it. Sergi divided the brain into several regions and meticulously described them one by one: *fissura Silvii*, *sulcus Rolandi*, the frontal lobe, the parieto-occipital region, the temporal lobe, the region of the cingulum, *fissura calcarina*, *fissura parieto-occipitalis*. Overall, the Herero brain differed from the European one in a long series of morphological peculiarities, from the predominance of simple forms in the posterior termination of the Sylvius fissure, the greater division of the frontal sulci, the frequency of a single or slightly split upper temporal sulcus, the constancy of the rhinencephalon-temporal gyrus, the greater division of the callosal-marginal sulcus, to the frequency of the gyrus of the superficial cuneus and that of the gyrus of the superficial posterior lingual cuneus. Regardless of the technicalities, Sergi ruled out the possibility of attributing the primitive nature of the Herero to a smaller extension of the frontal lobe: the value of this index did not seem to be a correlative of the degree of intellectual and social development. To easily follow the analysis of the sulci, twenty-one plates of drawings of each brain and photographs of each skull were presented, albeit taken from Schultze's report.[47] In a further comparative study of the sagittal sulci of the frontal lobe, Sergi found most numerous anastomoses between them at the upper and lower borders in the brain of Japanese, while in Indians they occurred only partially, and

46 Wilhelm von Waldeyer, "Gehirne südwestafricanischer Völker," *Sitzungsberichte der königlich Preussischen Akademie der Wissenschaften*, Erster Halbband, Januar 1906, pp. 3-8. Waldeyer received the anatomic material from the zoologist Leonard Schultze, who happened to be on a collecting trip in Southwest Africa when the outbreak of the fighting hindered his task. Nevertheless, the circumstance made it possible to remove from fresh native corpses parts that facilitated his study of living bodies of prisoners: Leonhard Schultze, *Zoologische und anthropologische Ergebnisse einer Forschungsreise im westlichen und zentralen Südafrika ausgeführt in den Jahren 1903-1905. Erster Band: Systematik und Tiergeographie* (Jena: Gustav Fischer, 1908), p. viii. See Andrew Zimmerman, *Anthropology and Antihumanism in Imperial Germany* (Chicago and London, The University of Chicago Press, 2001), p. 245; Reinhart Kössler, *Namibia and Germany. Negotiating the Past* (Windhoeck, University of Namibia Press, 2015), pp. 275-276.

47 Sergio Sergi, *Cerebra Hererica* (Jena: Fischer, 1909), which shared the Prix Fauvelle 1911 with Elliot Smith and Levy-Valensi as "une oeuvre très importante", "un ensemble de documents de premier ordre" (Raoul Anthony, "Rapport du Prix Fauvelle," *Bulletins et Mémoires de la Société d'anthropologie de Paris*, VI série, 1911, 2:391-393); Sergio Sergi, "Sulla craniologia degli Herero," *Bollettino della R. Accademia medica di Roma*, 1908, *34*:100-116 ; Id., "Sulla morfologia del cervello degli Herero," *Atti della Società Romana di Antropologia*, 1908, *14*: 71-74.

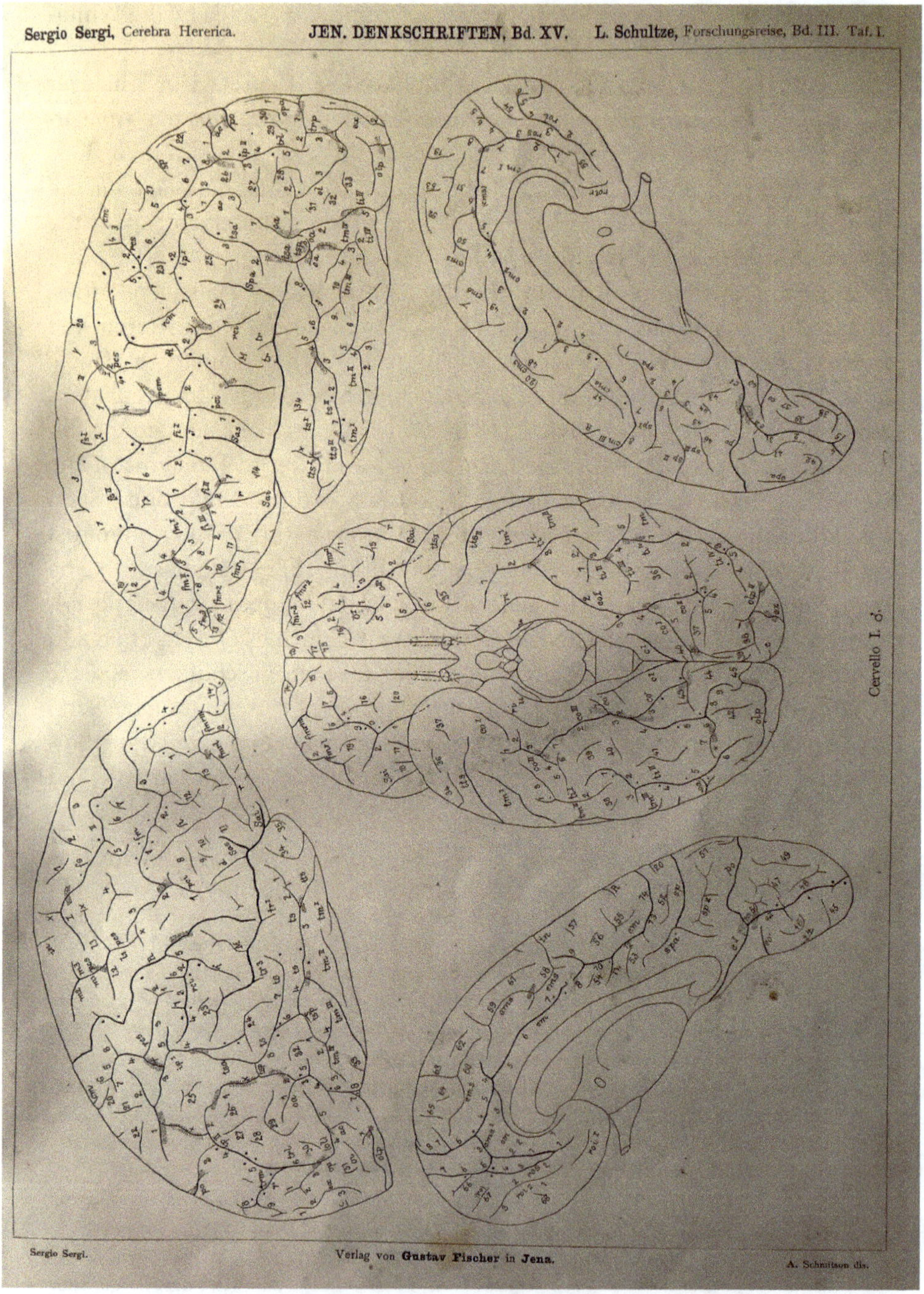

FIGURE 4.5 Drawing of a Herero brain with a detailed analysis of its sulci, in Sergio Sergi, *Cerebra Hererica* (Jena 1909).

in the Herero were entirely lacking: a presumed proof that the frontal lobe of the former was more evolved.[48]

The urgent need for a civil right organization that could fight the tendency to restrain black Americans' right to vote, to end racial discrimination and to improve their poor living conditions in the South and elsewhere, was brought into sharp focus by a famous race riot that took place in Springfield, the capital of Illinois, in 1908. 1909 marked the birth of the National Association for the Advancement of Colored People (NAACP), with a first meeting held on February 19th – the 100th anniversary of President Lincoln's birth – by a group of twenty African Americans and a few white reformers. A National Negro Committee convened a conference that took place in New York a few months later and tried to involve Booker T. Washington, the conservative black leader who, from the Tuskegee Institute in Alabama, had been advocating the progress of his people through education and entrepreneurship. In vain, because he declined the invitation fearing that his participation could jeopardize his work in the South. As a matter of fact, it was the group headed by W.E.B. Du Bois to run the conference that turned out to be rather stormy but somehow managed to adopt a common platform that demanded the Congress and the Executive to enforce the Constitution, to ensure equal educational opportunities and to recognize the right of the Negro to vote anywhere in the country.

William English Walling, a white liberal reformer, signed the preface to its *Proceedings*: he emphasized that no organization existed yet – composed of colored and white people alike – whose mission was the preservation of rights threatened from many quarters. It was highly important to establish a relation with other centers or activities that pursued the same goal. A series of successive, more specific conferences could raise awareness of the situation of the Negro and establish "a basis of fact, reasoned policy and even of science for its future conduct".[49] In his initial address, William Hayes Ward, editor of *The Independent*, noted that, as time went on, a new generation had grown without any memory of the Civil War or the Emancipation Proclamation, of the adoption of the Fourteenth and Fifteenth Amendments. The old wave of sympathy for the cause of the Blacks had subsided, in the South the ruling majority wanted to keep them – considered inferior and incapable of emancipation – in virtual serfdom. The scientific basis of this popular opinion was still voiced in speeches, editorials, books, popular novels and plays. For some years before

48 Sergio Sergi, *Note morfologiche sulla superficie metopica del lobo prefrontale in cervelli di Indiani e di Giapponesi* (Roma: Tip. Pallotta, 1912).

49 W.E.W., "Preface," in *Proceedings of the National Negro Conference 1909. New York, May 31 and June 1* (New York: The Conference, 1909), p. 6.

the Civil War, *Types of Mankind* by Nott and Gliddon (1854) had been the armory from which defenders of slavery could get their weapons and ammunitions. The book had been all the rage, claiming to have the last word in science: nature had given the Negro nine cubic inches less than to the Germanic stock, and from this all the rest irredeemably came. Once *Types of Mankind* had fallen into oblivion, Ward complained that its dogmas were still being repeated more than a half a century later, and that a multitude of people believed them to be true.[50]

Significantly, the conference was opened by science. Professor of Anthropology at Columbia and a member of Boas's expeditions to the Pacific Northwest, Livingston Farrand said that, although the term "race" was "in hopeless disrepute", he would use it "in a general and popular sense". No matter what physical factor might be taken as a criterion, the variations within the groups so defined were so wide as to cause overlapping and make definite conclusions difficult or impossible. Even so relatively stable a character as the shape of the head had exhibited – in Boas's researches – a pronounced variability. So, reaching a racial classification that could be acceptable to most anthropologists seemed an arduous undertaking. Farrand then wondered whether it was possible, from an anthropological standpoint, to classify human groups upon a psychological basis. Differences in mental expressions were undeniable, but he daringly asked "does civilized man represent a higher stage of mental evolution than the savage?" He answered by bravely assuming that "cultural" and "mental" evolution were by no means synonymous. A few differences did exist between the brains of various racial groups, and a series of them taken at random from Africans, compared with an equal number from Europeans, regularly showed a slightly less degree of size and weight in average, but

> this would simply mean that the great mass of the two series so compared would coincide and it would only be in the extreme members of the two groups that any recognizable differences would appear. Stated in another way it appears that the variation within each group is so wide that for nearly every African brain there would be a corresponding European brain so far as size and weight are concerned. This being the case it seems obvious to any candid mind that inferences with regard to the development of groups so treated are extremely dangerous and that inferences with regard to the mental development of the groups so considered are entirely unjustifiable.[51]

50 William Hayes Ward, "Address," ibid., pp. 9-13.

51 Livingston Farrand, "Race Differentiation – Race Characteristics," ibid., p. 17.

That was even truer given the impossibility of ascertaining the correlation that might exist between mental capacity and brain development. Both in the simplest and in the most complex mental processes, differences were hardly measurable. Nor could the thesis be accepted that the ability to inhibit impulses was a mark of high mental development. It had been assumed that the savage or the more primitive individual was affected by a lack of self-control, an element that Farrand deemed conventional to a large extent. Boasian relativism made him say that what is good form for one group was not necessarily so in the other. The direction taken by inhibitive powers depended largely upon training and habit. Similarly, what is right in one group may not be right in the other: after all, the strictness of conformity to ethical standards among savages was quite comparable to that proper of the civilized man. Ethnology gave full evidence to that end. Of course, differences in mental capacity between the races existed, but they were of a much slighter than it would appear from hasty observation.

At the conference, the second scientific speech was entrusted to Burt G. Wilder, who taught Neurology and Vertebrate Zoology at Cornell University and had served during the Civil War as surgeon of the 55th Massachusetts Infantry composed of free men of color and newly freed slaves. In front of the audience gathered in New York, his focus was on the brain of the American Negro: did any of its physical characteristics warrant discrimination against him, as such? In science, the "personal equation" had to be reckoned with, and Wilder claimed his own approximate impartiality, on the one hand "as a believer in the derivation of the human body from some anthropoid stock" and on the other hand for being tempted to testify, thanks to his army and university experiences, that a white man was as worthy as a colored man, "provided he behaves as well".[52] In his opinion, no facts, deductions or arguments could justify withholding any civil or political rights or any educational or industrial opportunities from men of African descent.

At various times and by numerous writers – Wilder recalled – differences had been alleged to exist between African and Caucasian brains. Color did not seem to him a constant feature: since the brain is primarily derived from the same embryonic layer as the skin, one might not unnaturally expect the African brain to be of a darker hue. However, in the Cornell University collection, the darkest brain belonged to a white physician and poet, while that of a Negro was one of the lightest. Furthermore, museum specimens presented many shades of color due to the nature of the preservatives employed. As for the alleged lower degree of complexity of convolutions and fissures, or for the form

52 Burt G. Wilder, "The Brain of the American Negro," ibid., p. 22.

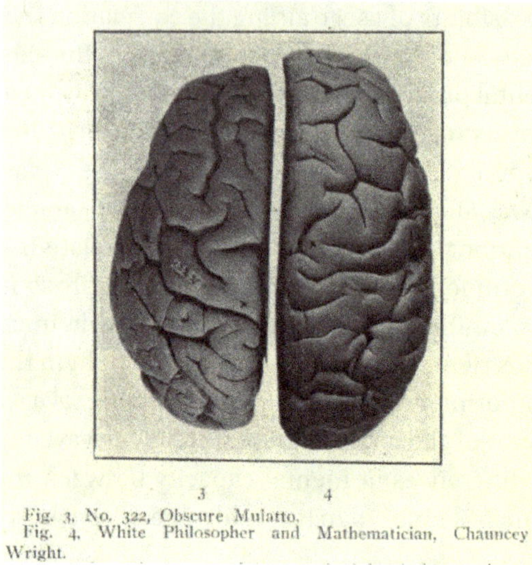

FIGURE 4.6
Comparison between the hemispheres of two brains, in Burt G. Wilder, "The Brain of the American Negro," *Proceedings of the National Negro Conference 1909. New York, May 31 and June 1* (New York 1909), p. 56.

and size of the frontal lobes, Wilder agreed with Mall's skepticism and put forward his own experience: it was risky to assert something certain from observational data on a limited number of cases, which moreover had been found to be always highly variable. As a warning paradox, he reported the morphological resemblance between the brain of the philosopher Chauncey Wright and that of a mulatto.

The last issue considered by Wilder was brain-weight: it was dubious just how much significance should be ascribed to it, whereas past statements and interpretations were not always sound. His conclusion called for maximum caution in making judgments:

> As yet there has been found no constant feature by which the Negro brain may be certainly distinguished from that of a Caucasian [...] For the determination of possible racial peculiarities larger numbers of brains of both races should be examined with impartiality and by more exact methods. Particularly useful would be the brains of persons of African descent who have achieved eminence in any respect. Yet, even if it should appear that certain features or conditions occur more frequently in the Negro, so long as these conditions are not constant, and so long as they sometimes occur with whites, and even with those who are morally and intellectually superior, the greater average frequency in the Negro should not be interpreted to the disadvantage of worthy individuals of that race.[53]

53 Wilder, "The Brain of the American Negro," ibid., p. 40. Eleven tables of figures concerning brain-weight and coming from various sources were added (pp. 40-50), with a testimonial

Wilder hoped that the Afro-American leaders of his time decided to bequeath their brains to some institution that would preserve and study them properly; if that happened, the next generation might find the statistics of brain-weight telling a very different story. Whoever wanted could get a "Form of Bequest of Brain" from him.

Two addresses without a specific topic were given at the Conference. Edwin E. R. Seligman spoke as an economist of Columbia University and as a Jew, a member of a "race" that had both the Jesus type and the Shylock type. The trouble with the Negro was that the ordinary man considered only the Shylock type and had not yet learned to appreciate the Jesus type. Nothing was more tragic than the lot of that American Negro, a cultivated, refined gentleman, "who at the same time is thrown in the caldron and fused with the mass of his unhappy and more unfortunate brethren". However, the scientific man knew no prejudice and could help form an objective view of the situation, which called for patience in recognizing that mankind moves very slowly and that prejudice gives way to science still more slowly; but, on the other hand, he could transmit the confident expectation that in the not too distant future the forces of science and of ethics would grow in their beneficial influence.

The second, short address came from John Dewey, expressing his sympathy with the purpose of the gathering. He mentioned the apparently "disappointing and discouraging" biological doctrine – then in fashion, at the dawn of genetics – that denied the inheritance of acquired characters, but to reverse the first, gloomy impression. Actually the refusal of transmissibility implied that, as far as individuals were concerned, they had "a full, fair and free social opportunity". In other words, the members of a so-called "inferior race" – whose existence Dewey denied – should have the same chances as those of the more favored one. If it had been pointed out that race differences were comparatively slight, it was the opposite for individual differences: all points of skill were represented in every race, which a society should favor in the same way for its own sake.[54]

Not at all discouraged by the criticisms received, Bean persevered in his search for racial characters. Since 1907, he had spent three years in Manila as

on "American Negroes in the Civil War", which argued in favor of their self-restraint and heroism: could white soldiers have behaved any better? Wilder's contribution ended with a rhetorical question: "Shall we now deny civil and political rights, and educational and industrial opportunities, to men merely because they are black, because the average weight of their brains is a little less, and because a certain region of the brain may be more frequently less developed, when two thousand of their fellows, nearly half a century ago, could manifest not merely the highest kind of physical courage, but as high a kind of moral courage, as has been chronicled in the history of the world?" (p. 54)

54 For Seligman's and Dewey's addresses see ibid., pp. 67-73.

assistant professor and director of the anatomical laboratory of the local Medical School. While there, he dealt with studies of Philippine types and a book presented the results in 1910, when he returned to the United States and started working at the Anatomical Laboratory of Tulane University (New Orleans). He took pride in having first established the ear as a differential factor in racial anatomy and a classification criterion, identifying three fundamental units of mankind: Iberian (the European type), Primitive (the Oriental type), and Australoid (the primary Negroid element).[55] The brain was again under Bean's eyes in 1914, when he found some racial peculiarity in the pole of the temporal lobe of the Negro: a smaller size and a different structure than in the White, a larger hippocampus, both in absolute and relative proportion to the weight and size of the entire cerebral hemispheres. As a test of the accuracy of his powers of observation, he selected nine brains at random without knowing their identity, and from the temporal lobes alone he could judge the race correctly in all, except two, which he called white, whereas they had belonged to light-skinned mulattoes.[56]

Later on, Bean analyzed many other organs from the same point of view, until 1932, when, at the headship of the Department of anatomy of the University of Virginia, he summed up his racial knowledge in a book. Despite the time elapsed since his early researches, he did not hesitate to still attribute to the "White Race" a large brain, rich convolutions, deep fissures; corresponding to an active mind, nervous and physical vivacity, strong ambitions and highly developed idealism. In an intermediate position he placed the brain and the mental performances of the "Yellow-Brown Race", and relegated the "Black Race" to the other extreme, below the medium size of the other two: low and narrow frontal lobes, voluminous parietal, protruding occipital. And therefore the psychic correlates were careless, jolly vivacity, emotions and passions of short duration, irrational egoism, weak idealism and co-operative faculties. Blacks loved amusement and sport, were rather artistic in music, had some ability in pictorial and industrial art, but generally lacked steady application. Their poetry was of a lower order, emotional their religion and cursed with superstitious fears.[57] Racial differences were also noted in facial expression,

55 Robert Bennett Bean, *The Racial Anatomy of the Philippine Islanders. Introducing New Methods of Anthropology and Showing their Application to the Filipinos with a Classification of Human Ears and a Scheme for the Heredity of Anatomical Characters in Man* (Philadelphia and London: J. B. Lippincott, 1910).

56 Robert Bennett Bean, "A Racial Peculiarity in the Pole of the Temporal Lobe of the Negro Brain," *The Anatomical Record*, 1914, 8:479-491.

57 Robert Bennett Bean, *The Races of Man. Differentiation and Dispersal of Man* (New York: The University Society, 1932), pp. 94-95, 37-38.

where in the white race a great range of modulations denoted the action of a responsive neuro-mechanism, well-differentiated muscles, and a thin, elastic skin. The coarse bundles of facial muscles, such as were found in the Black Race, together with the great thickness of lips and skin, made this finer mechanism improbable; its neuro-muscular reactions were less controlled and, when the nerve impulses, not so finely graded as in Whites, reached the mimetic muscles, these were set into sudden, strong contractions of a primitive kind.

Writing a first textbook of a young science is a hard and thankless enterprise, and only an urgent need might persuade to undertake it. Thus Rudolf Martin introduced his *Lehrbuch der Anthropologie* in 1914, noting that anthropological works – general and particular – were abundant, but none with the specific character of a manual, which was to codify also the standardization of methods confusedly invented during the previous decades. Martin's academic career began in Zurich and continued in Munich, but with a stint of field research in the Malay Peninsula (1897). His *Lehrbuch* reached as many as 1200 pages, while a posthumous second edition in 1928 grew into three volumes, also with additions by other authors.

The third section of the first volume – more than 400 pages – deals with *Kraniologie*, where the brain is addressed as well. One of the most important, if not the most essential, feature of the human skull was the powerful unfolding (*mächtige Entfaltung*) of the *Neurocranium*, only approximately measured by cranial capacity. A *Tabelle* containing figures of the latter in cm^3, derived from various sources and distributed by continent and population, attribute the highest average to Europe (1450 for men and 1380 for woman), situating it at the border between *Euenkephalie* and *Aristenkephalie*. The *Kulturvölker* of East Asia, the oceanic groups and the Native Americans would come closest to European values. Deep below them, Martin placed the Australian, Vedda and Andaman. Quite differently, brain weight depended on a number of factors – age, body size, muscle mass, nutritional status, and pathological conditions – and admitted of no easy inferences about the intelligence of an individual or a race. Martin quoted Sergio Sergi's contributions, to maintain that there was no noticeable difference in weight of the frontal brain between higher and lower races.[58]

58 Rudolf Martin, *Lehrbuch der Anthropologie in systematischer Darstellung mit besonderer Berücksichtigung der anthropologischen Methoden* (Jena: Gustav Fischer, 1914), pp. 640-652. At the conclusion of his University studies, Martin had travelled through Europe and visited the main anthropological collections, also working as a volunteer assistant at the École d'Anthropologie in Paris. His *Habilitationsschrift* at the Zurich University had been *Zur physischen Anthropologie der Feuerländer* (Braunschweig: Friedrich Vieweg und Sohn, 1893), having had at his disposal the skeletons and some preparations of inner organs

FIGURE 4.7
Primitiveness in a laughing Negro, Robert Bennett Bean, *The Races of Man. Differentiation and Dispersal of Man* (New York 1932), p. 37.

Over time, research on brains that were not of Whites or Blacks had been rather sporadic, partly due to the greater availability of the latter, partly because of the paradigmatic racial antithesis – to be affirmed or, much more rarely, to be denied. In 1916 a young anatomist of the University of Nebraska, J. Jay Keegan, furnished an account of the hemispheres of a North American Indian brain, "thereby filling a gap" since, to his knowledge, no such a description had ever been given.[59] Although a large amount of work had been done upon the Indian tribes from an ethnological point of view, their racial cerebral characters remained still unknown. In particular, Keegan applied the method developed by Grafton Elliot Smith in 1907 by attempting to perform a topographical survey of the human cortex, during the period that also saw Korbinian Brodmann periodically communicate the results of his histological localizations, then merged into a celebrated book.[60]

from five Alkaluf tribesmen (Tierra del Fuego) who had died in Switzerland. Only two of their brains were preserved, already examined in the 1880s by Johannes Seitz, to whose studies he referred (p. 53).

59 J. Jay Keegan, "A Study of a Plains Indian Brain," *The Journal of Comparative Neurology*, 1916, *26*:403-420.

60 See Grafton Elliot Smith, "A New Topographical Survey of the Human Cerebral Cortex, being an Account of the Distribution of the Anatomically Distinct Cortical Areas and their Relationship to the Cerebral Sulci", *Journal of Anatomy and Physiology*, 1907,

The brain examined by Keegan came from the body of a female of the Omaha tribe, fifty years of age, who possessed rather typical Indian features albeit not extreme, since she united French and English blood too in her person. Her intellectual status was far above the average "as superficially judged of Indian people, and above the average of the white race as judged from the high quality of her attainment." She had been trained for medical service with her own people, and was generally recognized by the profession and laymen as a woman of very superior personality.[61] After removing the dura mater, her brain was weighed (1353 grams), and its sulci and gyri were peered into, followed by a meticulous description. The striking difference between the two hemispheres, in almost every fissure or region was noted first: an asymmetry that was in agreement with European brains in general, and a contrast with the supposedly lower type to which the woman belonged: the most convincing evidence of a highly specialized cerebrum.

According to Keegan his study demonstrated that in a race of inferior status all the elements for higher individual development were present. Something similar also had happened the year before, with his graduate thesis, a report of which had been published in 1915, with the help of Charles W. M. Poynter, professor of neurology at the University of Nebraska. The material had consisted of thirteen Negro brains, male and female ones, all obtained from the dissecting material in the local laboratory and chosen for their marked 'negroid' traits, in order to avoid the influence of white blood as far as possible. The report was confined entirely to the macroscopic study of the brain surface and, since the number of specimens was small, the two authors ruled out the possibility of establishing a type of convolution pattern for the Negro. They believed that structural differences in the race, long investigated in the past, did not *ipso facto* indicate inferiority; a quote from Boas certified that such differences were less than the range of variation found in either race considered by itself. Likewise, no fundamental basis had been found for the determination of race, sex, or mental development through the gross weight of the brain, despite a thousand efforts made in the past.

41:237-254; Korbinian Brodmann, *Vergleichende Lokalisationslehre der Grosshirnrinde in ihren Prinzipien dargestellt auf Grund des Zellenbaues* (Leipzig: J. A. Barth, 1909).

61 Keegan, "A Study of a Plains Indian Brain" (cit. note 59), p. 405. He later extended his observation to three brains of full-blood Indians that Hrdlička, Curator of the Division of Physical Anthropology at the US National Museum, had loaned to him: the only ones that had been preserved of that race, together with the Omaha already reported and a specimen in Johns Hopkins. He did not find any discernible difference in them, except for the relative distance from the type of fissuration noted in the most complex brains of the white race: see Id., "The Indian Brain," *American Journal of Physical Anthropology*, 1920, 3:25-62.

The next step had been a study of the convolution pattern, in the attempt to discover particular features for different races. From the few comprehensive attempts available until then, no certainty could be deduced in this regard: although it seemed generally agreed that the Negro brain was less convoluted than the Caucasian, such a character was too subject to individual variation in both races to fix a standard. The various lists of simian or inferior characters, which had been published from time to time, were of doubtful value, as Mall had already stated in 1909. However, after patiently analyzing the thirteen cortices, Poynter and Keegan gave proof of a certain ambiguity: on the one hand, they concluded that the Negro type of brain laid within the limits of individual variations for the Caucasian and did not necessarily denote inferiority or closer relation to the apes, but on the other hand

> perhaps it is safe to say that the balance of evidence sustains the long held idea of the inferiority of the Negro. This inferiority is expressed mainly in the frontal lobe by a flattened anterior association center representing actual deficiency of growth. This growth deficiency is farther evidenced by rostration, by the narrow gyrus frontalis medius and poorly developed sulcus frontalis medius, by the wide gyrus frontalis superior and by the irregularity of the fissures in the gyrus frontalis inferior. General features supposed to have a like significance are low average brain weight and generally greater simplicity of fissuration.[62]

The following year, at a meeting of the American Anthropological Association, Poynter took back that subject and admitted that a review of the literature of cerebral anthropology would give the impression of an unprofitable field. In his opinion relatively little work of a truly scientific character had been done so that it was premature to deny that the brain presented ethnic aspects. Studies had produced records of observable variations and elaborations of extravagant theories based on poor evidence. The issue of brain weight, so widely discussed at one time, could be easily dismissed as devoid of significance. Variations in fissures and convolutions were so numerous in kind and degree that it had proven impossible to determine an irrefutable type for race, sex, or quality of mental attainment. Thus more questions remained than reliable answers: among others, if a race was found to possess a particular fissural pattern, was this due to psychic or environmental influences, or to a definite

62 Charles W. M. Poynter and J. Jay Keegan, "A Study of the American Negro Brain," *The Journal of Comparative Neurology*, 1915, 25:183-212.

morphology produced by evolution? Furthermore, one should not expect to find racial characters in the brain as obvious as pigmentation or prognathism.

Notwithstanding all these self-admitted difficulties, it seemed to Poynter that cerebral anthropology still offered a profitable field of research, especially when the methods lately introduced by Smith to study fissural variations were adopted. For instance, while an exploration of the relative characteristics of the Negro cortex did not suggest a closer relation to the apes, nonetheless they were not as highly developed as those of other races. Essentially, the future of the field was then dependent on functional localization in the cerebral cortex, which involved so many different factors that one could only hope to find a solution to the problem by thoroughly sorting out all the evidence contributed by an interdisciplinary investigation.[63]

3 Innovating Techniques, Popular Science, and Deconstructing Myths

At a meeting of the Royal Academy of Amsterdam, in 1926 Cornelius U. Ariëns Kappers said that the results attained until then in the study of the brains of different races and types were highly disappointing, as based more on general impressions than on careful measurements. Exact information about the cerebral shape among non-European races was very rare, and most authors had kept to descriptions of furrows and convolutions. How to get anthropological data concerning the brain that approached in exactness what was known of other parts of the body? Ariëns Kappers presented several "indices" that seemed to be useful to this purpose. First of all, he recommended that more attention be paid to the fixation and the preservation of the brain, and gave his own precise instructions in this regard. The general relations to be measured were the greatest transverse diameter and the greatest length, the diameter between the triangular opercula and the greatest bitemporal diameter. Many other indices could be best calculated on photographs of the lateral and the mesial side of the hemispheres.[64]

63 Charles W. M. Poynter, "Some Conclusions Based on Studies in Cerebral Anthropology," *American Anthropologist*, 1917, *19*:495-502.

64 Cornelius U. Ariëns Kappers, "Indices for the anthropology of the brain applied to Chinese, dolicho- and brachycephalic Dutch, foetuses and neonati," *Koninklijke Akademie van Wetenschappen te Amsterdam. Proceedings of the Section of Sciences*, 1927, *30*: 81-94. Ariëns Kappers was then Director of the Nederlands Instituut voor Hersenonderzoek, a post that he held from its foundation in 1909 until his death in 1946, making it a prestigious center for the study of comparative brain anatomy. In 1913 he had participated in the 17th

A comparison was thus established between the average figures of 22 Chinese brains and many more Dutch ones. Among the latter, the brachycephalic Dutch were found to be closer to the Chinese than the dolichocephalic one, although with a brain that extended more laterally and less upwards. Ariëns Kappers did not agree with the German anatomist Eugen Kurz, who was inclined to derive the form of the Chinese brain from that of the Orangutan: "One need not go so far, however, as to accept a different anthropoid ancestry for the Chinese and the Caucasians."

In 1928, Ariëns Kappers received an honorary PhD in science from Yale University, and one year later he exhibited his new title on the cover of a voluminous treatise on the evolution of the nervous system, which devoted many pages to the anthropology of the brain. Cephalization, though varying but little in human races, was probably not exactly the same; nevertheless it was quite hard to say if there was a difference in cerebral complexity caused by greater "cognitive" endowments independent of "pragmatic" ones. In this regard, Eugène Dubois – the discoverer of *Pithecanthropus erectus* in 1891 – had related the large cranial capacity of *Homo neanderthalensis* to the muscular apparatus, which was stronger than that of *Homo sapiens* and similar to the Mongoloid one. For his part Otto Hauger – who worked at the Anthropologisches Institut of the University of Breslau, had tried to measure cerebralization – i.e. "die höhere Vergeistigung des Gehirns" – in proportion to the size of the skeleton, detecting a gap between the lower monkeys and the apes, and a greater one between apes and humans. Within humans, it turned out that the two groups he studied, the Australian and the European, differed in cerebralization as an expression of their abilities, and that the former had less brain available for the psychic ones than the European. Consequently, the skeleton of an individual could assumedly be judged from the size of the part of the brain that mediates the psyche.[65]

In his 1928 book, Ariëns Kappers lingered on the application of his evaluation method to the brain of the *Pithecanthropus*, the Piltdown cast – still considered an authentic fossil specimen –, the Neanderthal man and the upper

---

International Congress of Medicine held in London: Id., "Cerebral Localization and the Significance of Sulci," in *17th International Congress of Medicine. Anatomy and Embriology* (London: Frowde, Hodder & Stoughton, 1913), pp. 273-392. See also Michel A. Hoffman, John. I. Johnson, "The C.U. Ariëns Kappers Brain Collection," *Annals of the New York Academy of Sciences*, 2011, *1225 S1*:E64-E84.

65 Eugène Dubois, "On the Significance of the Large Cranial Capacity of Homo Neanderthalensis," *Koninklijke Akademie van Wetenschappen te Amsterdam. Proceedings of the Section of Sciences*, 1923, *23*:1271-1288; Otto Hauger, "Der Gehirnreichtum der Australier und anderer Hominiden, beurteilt nach ihrem Skelet," *Anatomische Hefte*, 1921, *59*:579-617.

Paleolithic races (*Homo sapiens fossilis*) that inhabited Europe during the last glacial and post-glacial periods, the creators of mural paintings and adorned ivory or bone implements. A striking feature of all these *sapiens* types, compared to the Neanderthal, was their higher endrocranial cavity and cast. Most authors considered the Australian aboriginals as closer to *Homo sapiens fossilis* than to Neanderthal: Ariëns Kappers was sure that they formed a very primitive race somatically as well as culturally, apparently able to produce only lower Paleolithic implements, no pottery nor artistic paintings and sculpture. Their sloping frontal lobes and occipital elongation were evident, as well as several other, deeper brain characters. The many pages dedicated to the "Negroid races" by a host of authors were a proof – according to him – of how much knowledge had accumulated about them in the past and still continued to grow, with ever-deepening insights.

Leaving aside the "older tendentious literature on this subject", which had been already variously criticized by others, Ariëns Kappers preferred to rely on the most recent contributions of which he presented a review – adding, of course, that "more researches of these races are necessary". Extensive information was also given about the brains of Eskimo and North-American Indians, Asiatic, Chinese, Japanese, and Indonesians. A final question concerned the taxonomic value of the brain: was it just as great as that of other parts of the body, or was it more subject to individual variations?[66] In his opinion, cerebral morphology was not sufficiently trustworthy for taxonomic studies, and the anthropology of the brain was still far behind the knowledge of other parts of the body: "This is one more reason, however, that we should not lay it aside. If this chapter only has the result that more and better data are gathered, it has not been written in vain."[67]

At any event, in his 1927 article, Ariëns Kappers argued that some factors seemed to attest to a sort of "retardation", especially in the Northern Chinese, and thus to confirm the statement made by Joseph L. Shellshear, who from a European standpoint, had spoken of a "childlike type".[68] After taking part in the First World War as an Australian officer, Shellshear taught Anatomy at

66 Cornelius U. Ariëns Kappers, *The Evolution of the Nervous System in Invertebrates, Vertebrates and Man* (Haarlem: De Erven F. Bohn, 1929), pp. 254-299.

67 Ibid., p. 302.

68 Ariëns Kappers, "Indices for the anthropology of the brain" (cit. note 64), p. 89, 94. Eugen Kurz, "Zwei Chinesengehirne. Ein Beitrag zur Rassenanatomie," *Zeitschrift für Morphologie und Anthropologie*, 1913, *16*:281-328; Id., "Das Chinesengehirn," *Zeitschrift für Anatomie und Entwicklungsgeschichte*, 1924, *72*:199-382. For the Western construction of the oriental races see Michael Keevak, *Becoming Yellow. A Short History of Racial Thinking* (Princeton: Princeton University Press, 2011).

the University of Hong Kong for a dozen years, during which time he carried out fieldwork in prehistory and started assembling his collection of human fossils together with related stone artifacts. In 1926, he proposed that the important part played by the sense of vision in the evolution of man should be investigated by comparing the occipital region in the brain of different races of mankind. Smith had already interpreted the human species as the result of the exploitation, throughout the Tertiary period, of the vast possibilities that reliance upon vision as the guiding sense offered to a mammal whose hands were capable of acquiring skills through action. Manual dexterity had been conquered by developing some cortical mechanisms that correlated information provided by vision and touch.[69]

By identifying the *Affenspalte* in the human brain, Smith had paved the way to a more definite understanding of racial differences in the posterior area of the brain. It had been Shellshear's good fortune – so he told in 1926 – that in Hong Kong he could dispose of a very large number of Chinese brains, both in a fresh condition and in a preserved state: over 400 specimens supported his research, a heterogeneous group that were unlikely to represent a pure race. However, the features of the occipital region differed widely in certain aspects, so as to suggest the existence of a distinctive Mongoloid kind of brain that seemed to agree with a more primitive anthropoid type. A clearly marked *sulcus lunatus* extending well out on the lateral surface was associated with the paramesial, praelunate and transverse occipital sulci arranged in an almost constant pattern. Consequently the Chinese brain appeared to be more primitive than the highly developed European one, but Shellshear limited himself to this fact: "no doubt, as further information of a functional character comes to hand, the meaning of a race retaining such primitive characteristics will become more and more apparent."[70]

Less than ten years later, Smith submitted to the Royal Society a contribution by Shellshear on the brain of a Bushwoman who had already been studied in 1864 by John Marshall, whose daughter had provided the original documents and photographs, while the actual brain had been searched for in vain in the collections of a few London institutions. The progress of knowledge since 1864 enabled to read them in a new way and to highlight that the cortical areas of the brain showed signs of primitiveness in all its four regions, from the occipital to the frontal, where a remarkable resemblance to that of a *Pithecanthropus*

69 Grafton Elliot Smith, *The Evolution of Man: Essays* (London: Oxford University Press, 1924), p. 145.

70 Joseph L. Shellshear, "The Occipital Lobe in the Brain of the Chinese with Special Reference to the Sulcus Lunatus," *Journal of Anatomy*, 1926, 61:1-13, p. 12.

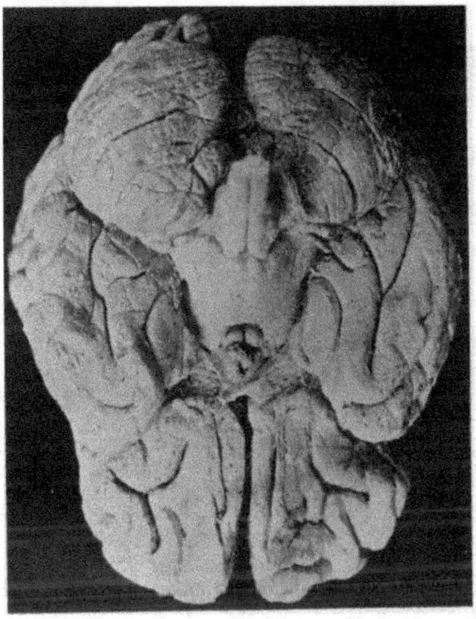

Fig. 1.

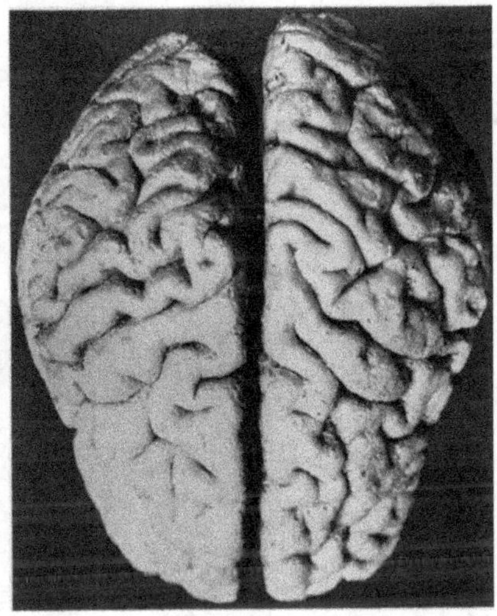

Fig. 3.

FIGURE 4.8 Photographs of the dorsal and ventral surfaces of the brain of a Bushwoman. Plate I, in Joseph L. Shellshear, "The Primitive Features of the Cerebrum, with Special Reference to the Brain of the Bushwoman described by Marshall," *Philosophical Transactions of the Royal Society of London*, Series B, 1934, *223*:1-26.

stood out.[71] Knowledge of the relationship between the anatomical pattern – both macroscopic and microscopic – and functional localization had greatly increased in the recent past, but studies in morphology had not yet reached the desirable level of perfection. Thus Shellshear argued when, in December 1936, he sent a thorough examination of the brain of aboriginal Australians to the Royal Society, wondering "whether the brain could be used as an index of race". Thanks to the labors of numerous researchers, morphological and, to a certain extent, functional significance could be given to many of the sulci; their arrangement lent itself to be used as an index of the relative amount of the expansion of the newly acquired areas in the human brain. Nevertheless, over thirty years earlier, Smith had been right in noting that "we cannot yet read the functional capacities of any given brain by a study of its external appearances": one day, some experts might be able to do it.

In the few previous works on the brain of the aboriginal Australian, certain features had been emphasized as indicative of primitiveness and "lowly

71 Joseph L. Shellshear, "The Primitive Features of the Cerebrum, with Special Reference to the Brain of the Bushwoman described by Marshall," *Philosophical Transactions of the Royal Society of London*, Series B, 1934, *223*:1-26.

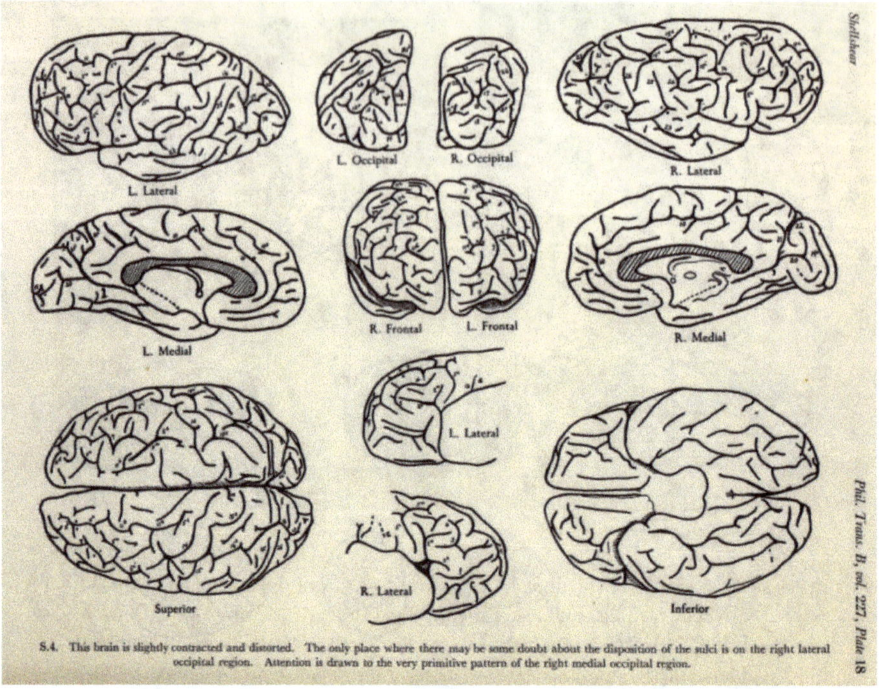

FIGURE 4.9 Drawings of the brain of an Aboriginal Australian. Plate 18, in Joseph L. Shellshear, "The Brain of the Aboriginal Australian. A Study in Cerebral Morphology," *Philosophical Transactions of the Royal Society of London*, Series B, 1937, *227*: 293-409.

estate", and yet Shellshear warned that "the retention of a primitive feature does not necessarily imply lowliness", which indicated a difference in mentality. His study materials consisted of 44 hemispheres in various states of distortion; most of them preserved in Sidney and in Queensland, three brains from Cambridge, one from the Royal College of Surgeons and one from Amsterdam. At the end of the essay, all of them were portrayed in drawings and photos. Detailed descriptions of the sulcal patterns were given, by also comparing them with images of the hemispheres of Chinese brains. Taken together, the aboriginal brain seemed to differ from those of higher human races, although Shellshear was careful to avoid possible misunderstandings:

> whilst many of the features of these brains may point to the Australian aboriginal as an inferior race, it is insisted that the question has no place in a purely morphological paper. Primitiveness, on the other hand, deals with the retention of unspecialized characters in the process of evolution;

> and the actual retention of primitive characters may provide the possibility of greater evolutionary advance.[72]

Shellshear approvingly quoted Mall who, in 1909, had strongly advised to take into account both individual variability and the personal equation of the investigator, which means the inevitable bias as it came to observations and measurements. As for his series of objects, it was only in their occipital region that definite evidence of a lack of development occurred. On the medial surface of both the frontal and parietal lobes, the sulci were not so deeply folded in the Australian as in other races, and further structural differences were observed, similar to those found in some Chinese brains. Except that the study materials were not perfectly preserved, and each conclusion had to be held in abeyance. For the aboriginals, as a whole, "it would appear that the new cortical areas are not developed to the extent that they are in higher races", and the most fully developed of their hemispheres would look ill-formed and underdeveloped as compared with a fully developed Chinese brain.

To the importance of variability the German-Dutch physician and anthropologist Jacob Herman Friedrich Kohlbrugge drew attention in 1935, at the end of his wandering career, spent in Germany, Holland, and the Dutch East Indies. In the first decade of the century he had already written about the sulci in Javanese brains and other races, exhibiting their characters in a series of drawings. His lasting anti-Darwinian and anti-materialistic battle had started even earlier, aimed at distinguishing evolution and *Descendenzlehre*. A 1911 essay on *Gehirn und Kultur* was permeated with this particular attitude, and thoroughly discussed the various and contrasting approaches to the topic.[73] In 1935, while taking stock of the matter, Kohlbrugge debunked the traditional view that the psychic differences perceptible between races should necessarily have a material basis. The earlier observers had been so misled by this view that they constantly saw differences, even when these did not exist, and two mistakes spoiled their reasoning:

---

72 Joseph L. Shellshear, "The Brain of the Aboriginal Australian. A Study in Cerebral Morphology," *Philosophical Transactions of the Royal Society of London*, Series B, 1937, *227*: 293-409, p. 403.

73 Jacob H. F. Kohlbrugge, *Die Gehirnfurchen der Javanen. Eine vergleichend-anatomische Studie* (Amsterdam: Johannes Müller, 1906); Id., "Untersuchungen über Großhirnfurchen der Menschenrassen," *Zeitschrift für Morphologie und Anthropologie*, 1908, *11*:596-609; Id., *Die Gehirnfurchen Malayischer Völker verglichen mit denen der Australier und Europäer. Ein Beitrag zur Evolutionslehre* (Amsterdam; Johannes Müller, 1909); Id., "Gehirn und Kultur," *Biologisches Centralblatt*, 1911, *31*:248-256, 309-316.

> 1° on ne connaissait pas la variabilité du cerveau et l'on partait donc de l'idée préconçue que la forme extérieure du cerveau est fixe; 2° on n'avait pas conscience de la force irresistible qui sort d'une idee preconçue. Le résultat du premier défaut était qu'on ne connaissait pas la construction externe si variable du cerveau des races européennes, parce que personne ne s'était rendu compte de la grandeur de la variabilité de la surface du cerveau des Européens. [...] On se rendait encore moins compte qui'il en est de même, naturellement, pour le cerveau de chaque peuple ou de chaque race.[74]

Unfortunately, the history of biology and anthropology was too scarcely studied, otherwise it would be known that each period is under the influence of a "pensée suggestive", which on the one hand productively illuminates the field of research, but on the other hand completely obscures the horizon.

The desire to seriously delve into the anatomy of the brain in different races could only be satisfied by an international regulation on preparation, measurement and weighing: at that time – Kohlbrugge complained – there was a great diversity of treatment due to which the results of the comparison were falsified from the beginning. Before proceeding with the dissection it was mandatory to record sex, age, height, state of feeding of each corpse. Ariëns Kappers was praised for his method of calculating angles and indices, which allowed to show how the general shape of the brain – brachycephalic or dolicocephalic – determined the direction and the form of the sulci. His insights led to the persuasion that "il y a peu de chance d'établir des différences de races à la structure extérieure ou macroscopique du cerveau."[75]

Nor was the weight a more reliable datum: the average weight of the European brain had been set at 1360 grams, without realizing how worthless this statement was. From three to five different races lived in Europe, just as there was not a Chinese race but a population that owed its origin to a mixture of races: therefore the weight of each of their brains should have been measured, to make some sense of it. It was however a more theoretical than a practical postulate, given the non-existence of a pure race. By and large, Kohlbugge was convinced that weight and intelligence had nothing in common. One just had to think that the fossil skull found at La Chapelle-aux-Saints in 1908 had a volume of over 1600 cm^3, far above the European average. Even worse, a

74 Jacob H. F. Kohlbrugge, "Le cerveau suivant les races," *Bulletins et Mémoires de la Société d'Anthropologie de Paris*, VII série, 1935, 6:61-84, p. 61.

75 Ibid., p. 72.

comparison of fissures and convolutions would never show constant disparities between races:

> En mélangeant les cerveaux des races, personne ne peut distinguer le cerveau des Australiens, etc., de celui des Européens, pas plus que celui de génies des hommes d'intelligence moyenne. Il se peut bien qu'un groupe de cerveaux d'une race, réunis par hasard, semble presenter un rapport réciproque, de sorte qu'on voudrait interpréter les sillons de ce groupe étant des marques distinctives de la race. Si l'on mêle toujours plus de cerveaux dans la comparaison, on s'aperçoit qu'on n'a dessiné que de simples variations fluctuantes.[76]

At the beginning of a lengthy entry for the ninth volume of the *Enciclopedia Italiana*, in 1931 Sergio Sergi prescribed that a complete study of the brain of human races should include three parts: 1. changes in weight and volume; 2. general and particular morphological variations; 3. variations of the minute structure, cytoarchitectonic and myeloarchitectonic. In the first part he quoted the entry on the skull, which he wrote himself for a later volume of the *Enciclopedia*; the third part concerned a still-unwritten chapter on racial cerebrology. So, at the time, the analysis could focus mainly on macroscopic morphology, and more specifically on the arrangement of the sulci. The facts observed until then showed that the differences did not consist in completely distinct varieties, but only in variations in the frequency of forms, which made it impossible to diagnose the race to which one brain belonged. It could only be said that there were traits in a given brain that appeared more frequently in this or that human group. Sergi presented an up-to-date review of knowledge relating on the general form, the insular lobe, the frontal, parietal, temporal, and occipital lobes. Five of the images attached to the text exhibited the variations in the sulci and gyry which were observable in Herero brains, as studied by Sergi at the beginning of the century; other three were borrowed from Kohlbrugge and depicted the cortex of an Australian, a Javanese and a Maori.[77]

The first part of Sergi's entry on cranial capacity describes the various methods for measuring it developed in the past, and emphasizes the individual fluctuations within each ethnic group, much greater than racial variation. Nevertheless the usual hierarchy is confirmed, from the highest point of Europeans to the lowest one of Australians, Bushmen and Andamanese, and

76 Ibid., p. 82.

77 Sergio Sergi, "Il cervello nelle razze umane e negli antropomorfi," in *Enciclopedia Italiana* (Roma, Istituto della Enciclopedia Italiana, 1931), 9, pp. 845-858.

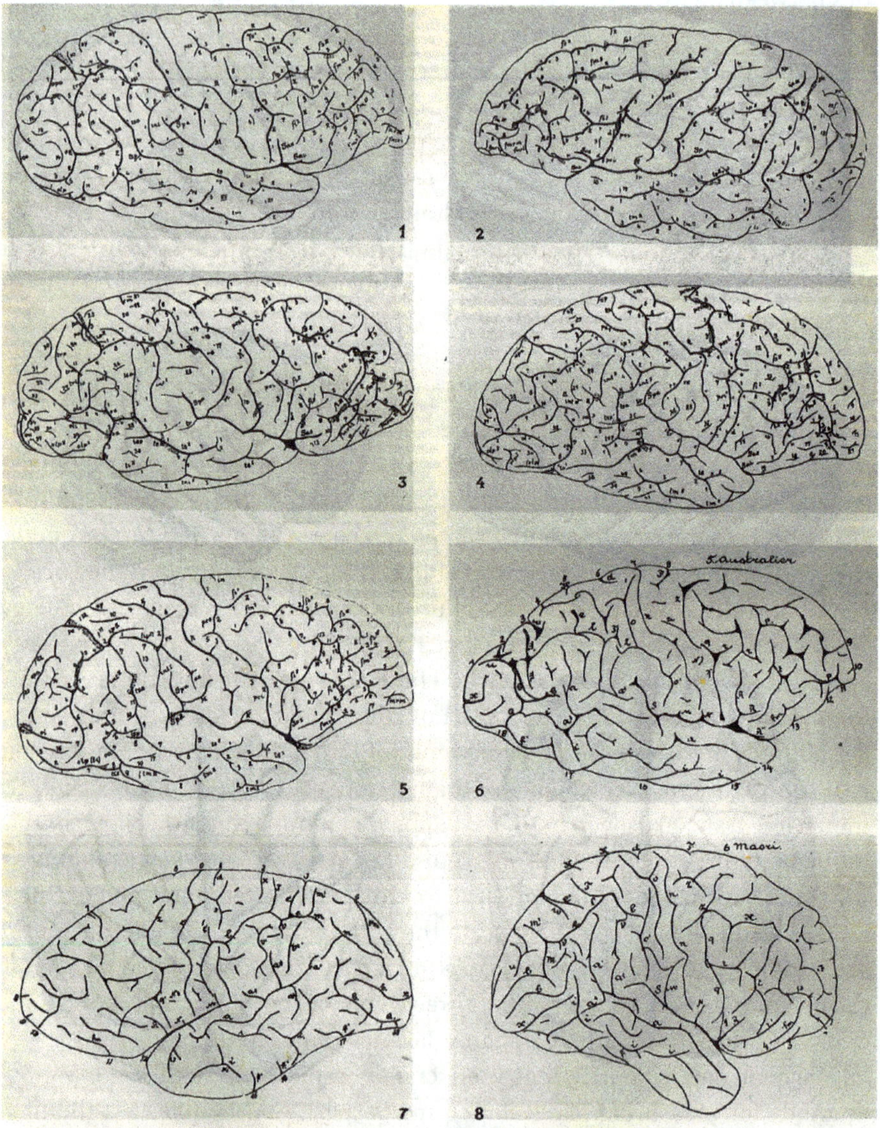

VARIAZIONI DELLA SOLCATURA E DELLA GIRIFICAZIONE IN CERVELLI DI RAZZE UMANE DIVERSE:
1, 2, 3, 4, 5, HERERO (da S. Sergi); 6, AUSTRALIANO; 7, GIAVANESE; 8, MAORI (da Kohlbrugge)

FIGURE 4.10 Variations in sulci and gyri in brains of different races, in Sergio Sergi, "Il cervello nelle razze umane e negli antropomorfi," in *Enciclopedia Italiana* (Roma 1931), 9, pp. 845-858.

the chance that a strong capacity might bear some relation, although neither direct nor simple, to individual psychic potential is defined as "probable". Research carried out in the past tended to prove it, although it was not free from criticism. And perhaps, in conclusion, it would have been appropriate to change the European conception of the cultural level of an ethnic group. The startling capacity of the Eskimo skull might be related to a perfect fitness for extraordinarily difficult living conditions, but Sergi could not tell by which specific adaptations their brains could have specialized. To those who had devalued cranial capacity in favor of the degree of cortical convolution, he pointed out that, *ceteris paribus*, these two elements were in proportional relation.[78]

In 1934, the cortex deserved all the observational care of F. W. Vint, who worked as a Government Pathologist at the Medical Research Laboratory in Nairobi, and was encouraged to study African brains by Henry Laing Gordon, a settler doctor and a leading figure in the eugenic movement that developed in Kenya during the 1930s. Its members were medical professionals and government officers gathered in a Society for the Study of Race Improvement, even though the colonial service was divided between those who pleaded the cause of a remedy to African backwardness through the benefits of social hygiene and others who denied the possibility of improving the status of local populations on the grounds of their innate and hereditary traits, which had to be scientifically proven.[79]

Vint, whose reputation as a serious scientist was then flawless, carried out autopsies in the native hospitals, examining only male adult brains that appeared to be normal. Most of the one hundred cases belonged to the Bantu group and did not include any of the so-called educated classes, rarely seen on the post-mortem table. Brains were weighed without removing the meninges, then suspended in formalin by the basilar arteries: viewed from the side, their upper border had a flattened appearance, while the frontal lobes seen from above showed a rectangular shape; narrower the temporal lobes, blunted and turned inwards the temporal pole. Their average weight was 10.6% (152 grams) lower than that of the European brain. The convolutional pattern – narrowly described sulcus by sulcus – differed somewhat from that of the European, and a lunate sulcus was present in 70% of the sample, conceived as a sign of

78 Sergio Sergi, "Capacità del cranio," in *Enciclopedia Italiana* (Roma, Istituto della Enciclopedia Italiana, 1931), 11, pp. 790-791.

79 See Jock McCulloch, *Colonial Psychiatry and the African Mind* (Cambridge: Cambridge University Press, 1995), pp. 46-49; Chloe Campbell, *Race and Empire: Eugenics in Colonial Kenya* (Manchester: Manchester University Press, 2007), pp. 49-61; Helen Tilley, *Africa as a Living Laboratory: Empire, Development, and the Problem of Scientific Knowledge, 1870-1950* (Chicago and London: University of Chicago Press, 2011), pp. 229-250.

primitiveness. What distinguished the work carried out by Vint was however the microscopic observation of the cortex, which was found to be narrower in almost all its individual laminae. Their depth, compared to the measures already taken by Constantin von Economo on the corresponding areas of European brains, was found to be 14.8% less. The only constant was the number of cells per unit area, which turned out to be the same in the African and European brains, although in the former the pyramidal cells of the supragranular cortex and the Betz cells of the motor area seemed smaller. Three plates complete Vint's article, the first with three views of a "typical Kenya native brain", the other two containing microphotographs of various cortical areas.[80]

Even before Vint's data were published, Gordon presented them at a meeting of the African Circle in London, looking for that "clearer understanding of the East African native, in body and mind" that only science could provide. A post at Mathari Mental Hospital had allowed Gordon to acquire some knowledge concerning the subject of backwardness, an "urgent necessity" to address. Among other things, in a sample of 3444 native brains he had found that their brain did not develop after puberty as it happened instead in the European one, though the reasons for such discrepancy were yet to be discovered. Did they lie in nature or in nurture, or in both? So Gordon resorted to the research undertaken by his "able colleague" Vint: three years of intricate work in his spare time from routine duties that confirmed the supposition about the cortex as the site of a biological deficiency:

> Dr. Vint and I have only cleared a little bush. We believe we have found a promising path to determination of the major causes of the present low biological level of the East African, commonly called backwardness, and to discover remedies for those causes. [...] If cerebral deficiency such as I have indicated is indeed a common ingredient of the material with which we aim to build our East African possession, our efforts and our hopes are vain unless we bring all knowledge to bear which may reduce that ingredient to innocuous proportions.[81]

80 F. W. Vint, "The Brain of the Kenya Native," *Journal of Anatomy*, 1934, *68*/2:216-223. Vint mentioned the English condensation of the book and atlas published by Constantin von Economo in 1926 (*Die Cytoarchitektonik der Hirnrinde des erwachsenen Menschen*): *The Cytoarchitectonics of the Human Cerebral Cortex* (Oxford: Oxford University Press, 1929). Vint had anticipated some of his results: see his "A Preliminary Note on the Cell Content of the Prefrontal Cortex of the East African Native," *East African Medical Journal*, 1932-33, *9*: 30-55.

81 Henry Laing Gordon, "The Mental Capacity of the African," *Journal of the Royal African Society*, 1934, *33*:226-242, p. 240. A letter from the zoologist Lancelot Hogben – who taught Social biology at the London School of Economics – to the Editor of the *Journal* was quite

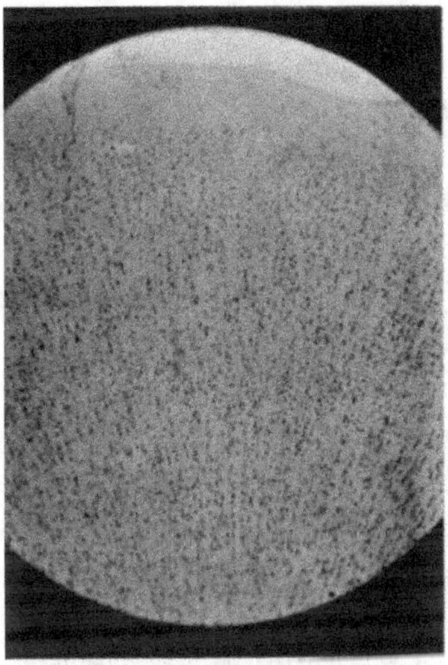

Fig. 4

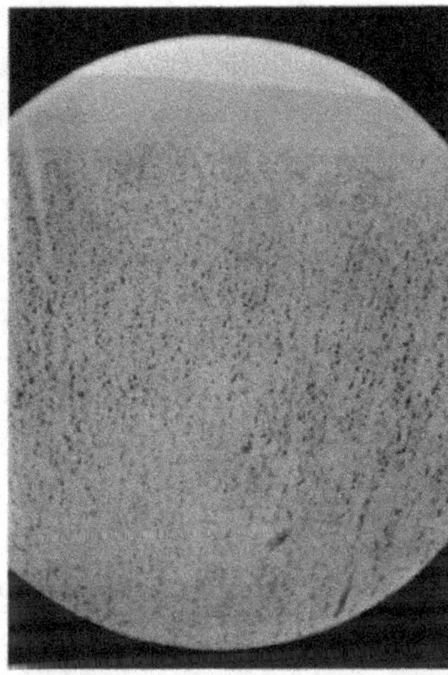

Fig. 5

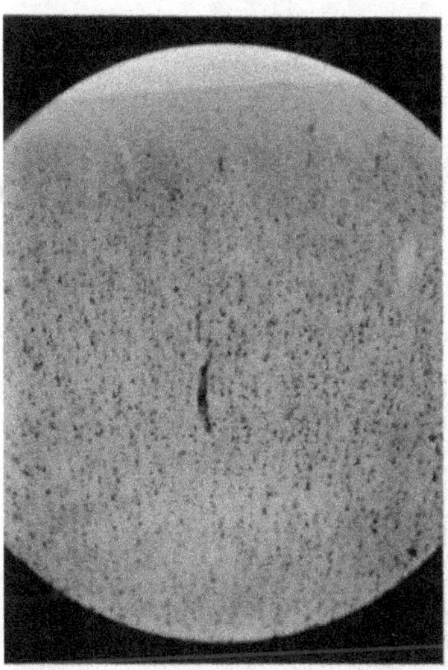

Fig. 6

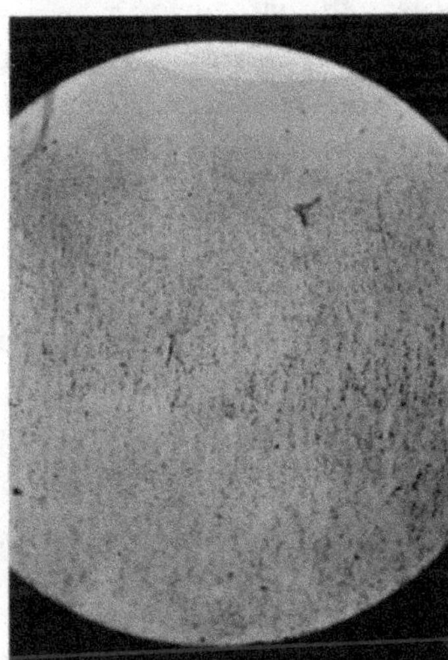

Fig. 7

FIGURE 4.11 Microphotographs of four cortical areas of a Kenya native brain. Plate II, in F. W. Vint, "The Brain of the Kenya Native," *Journal of Anatomy*, 1934, *68*/2:216-223.

Scientific guidance ought to be permanently incorporated – Gordon prescribed – into statesmanship as one of the ideals that distinguished British trusteeship "for weaker peoples". However, his arguments and perspective, expressed to the large audience of the *Times* were not peacefully received. The newspaper gave voice to a series of reactions that were mostly critical, starting from the Kenya-born paleoanthropologist Louis Leakey, who was already working at Olduvai Gorge and considered the study of skulls useful for explaining human evolution, but not intelligence. For his part Julian Huxley, also by virtue of a brief colonial experience reported in *Africa View* (1931), urged to keep in mind the unbelievable human variety by which that part of the continent was populated. Moreover, Europeans often had a greater body mass than that of Africans, which might mean identical brain sizes relative to their body weight. London psychologist Cyril Burt recalled that correlations between skull or brain size and inborn mental capacity had already been dismissed as wrong, and it was up to John B. S. Haldane to insist on the complexity of the relationships between genetic endowment and environment. A letter dated January 8 and sent by Bronislaw Malinowski to the editor of the *Times* remained unpublished to put an end to the series, which goes to prove how hot that issue was in the early 1930s.[82]

A chapter on *Das Hirngewicht* is contained in the fourth volume of an extraordinary editorial enterprise – reprinted multiple times and translated into many languages – that the German-Jewish physician-journalist Fritz Kahn, managed to complete between 1922 and 1931, with the assistance of lots of artists. Lavishly illustrated, his *Das Leben des Menschen* invented a new and popular way of visualizing the human body, strictly combined with technological-industrial modernity. Over time, the multicolored image of the interior mechanical contrivances of the head and trunk, printed as a poster (*Der Mensch*

sarcastic: "The mere facts that Leibniz, who discovered the integral calculus, and Anatole France, one of the most brilliant writers of his age, had heads below the averages of natives, whom Dr. Gordon is seeking to exclude from opportunities of education, is sufficient to remind readers [...] that the modern geneticist attaches the same importance to research as such those which Dr. Gordon is carrying out, as contemporary medicine and physics attach to phrenology and the design of machines to secure perpetual motion." ("To the Editor," *Journal of the Royal African Society*, 1934, *33*:432).

82 Henry Laing Gordon, "The Native Brain – Observations in Kenya – A Comparison with Europeans," *Times*, 1933, December 8; Louis Leakey, "The Native Brain: is Size a True Guide? Reasons for Stunted Growth," ibid, December 13; Cyril Burt, "The Native Brain – Intelligence Tests," ivi, December 15; Julian Huxley, "The Native Brain: Size and Growth," ibid., December 18; John B. S. Haldane, "The Native Brain," ibid., December 19. Malinowski Papers (London School of Economics) contain both his letter and the editor's response.

als Industriepalast), has become iconic: a "historical artifact" of the conjuncture that Weimar Germany was experiencing.[83]

Kahn also devised a few illustrations in setting the question of brain weight. He did not deem surprising to find that, in the animal kingdom, man has no heavier brain in either sense, the absolute and the relative: whale and elephant surpass him in the absolute, small mammals and birds in the relative one. If creatures (*die Geschöpfe*) are ordered in a row by the absolute weight of their brains, man stands next to the larger ones, between the whale and the horse. If, instead, the relative weight is taken as a criterion, then man is placed between the mole and the chimpanzee.[84] By shifting the focus on brain weight within the human species, Kahn recalled what the second volume of *Das Leben des Menschen* had already argued about the skull, namely that its size increases with the development of the species in general, as well as with the social stratum and individual quality. By making use of the *Profilwinkel*, it was easy to see how the forehead – a protective wall for the seat of the higher faculties – was flattened backwards in apes, less in the "Neger", and rose ever more vertically in the "Träger der Kultur". A table highlights the progression of the degree of the facial angle, from the pig, through the apes and four human races, to the European, whose beautiful head with a protuberant forehead verges on 90°: such a face was defined as truly sublime (*erhaben*).[85]

However, Kahn pointed out that the quality of the mental faculties of individuals could not be safely inferred from the weight of their brain. As a matter of fact, Bach, Dante, Raphael had small brains, while the heaviest had belonged to an idiot, whose connective tissue, the glia, had abnormally proliferated, preventing the development of neurons. In any case, the performance of cerebral functions depended more on quality than on quantity, a scientific truth that was visualized by a column of twelve brains (*Vergleichende Oberflächenbetrachtung von Meschenhirnen*) on a scale of graduated weights of 1000 to 2000

83 See Cornelius Borck, "Communicating the Modern Body: Fritz Kahn's Popular Images of Human Physiology as an Industrialized World," *Canadian Journal of Communication*, 2007, *32*: 495-520; Miriam Eilers, "Fritz Kahns *Das Leben des Menschen*. Zur Produktion und Transkription eines populären Werks," *NTM Zeitschrift für Geschichte der Wissenschaft, Technik und Medizin*, 2015, *23*: 1-31; Michael Sappol, *Body Modern. Fritz Kahn, Scientific Illustration, and the Homuncular Subject* (Minneapolis-London: University of Minneapolis Press, 2017).

84 Fritz Kahn, *Das Leben des Menschen. Eine volksthümliche Anatomie, Biologie, Physiologie und Entwicklungsgeschichte des Menschen* (Stuttgart: Kosmos Franckh'sche Verlagshandlung, 1929), IV, pp. 206-207.

85 Fritz Kahn, *Das Leben des Menschen. Eine volksthümliche Anatomie, Biologie, Physiologie und Entwicklungsgeschichte des Menschen* (Stuttgart: Kosmos Franckh'sche Verlagshandlung, 1927), II, pp. 217-218.

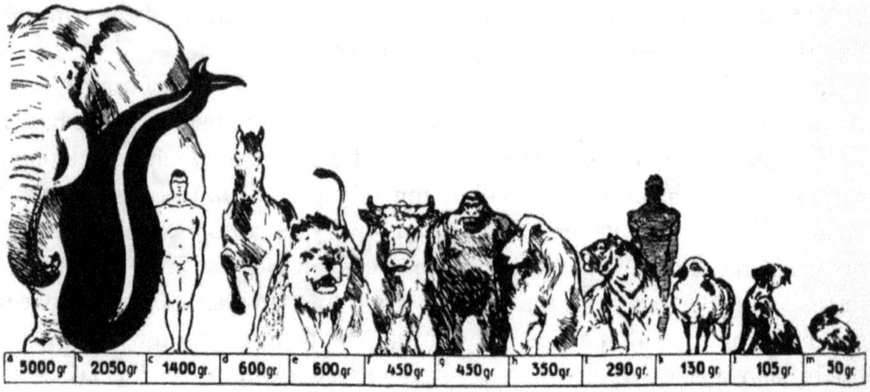

a

b

FIGURE 4.12A-B *Absolute cerebral weight in the animal scale*: Man does not have the largest brain, but he stands between the whale and the horse in terms of brain weight (A), while he stands between the tiger and the sheep in terms of body weight (B). *Relative brain weight*: small animals have more brain mass than large ones in relation to their body weight; in proportion, an adult human should have a brain weight of 1:300 (right), while the real value is considerably higher, namely 1:40 (left). In Fritz Kahn, *Das Leben des Menschen. Eine volksthümliche Anatomie, Biologie, Physiologie und Entwicklungsgeschichte des Menschen* (Stuttgart 1929), p. 207.

grams. The brain of Anatole France – recently dead, one of the most acute and original minds – slightly exceeded the lower limit, and several excellent personalities were below the average weight, set at 1400 grams. Paradoxically, Gall's brain did not reach 1200 grams, while the upper sector of the diagram included only one man of proven superior faculties, the Parisian notary Joseph Bouny.

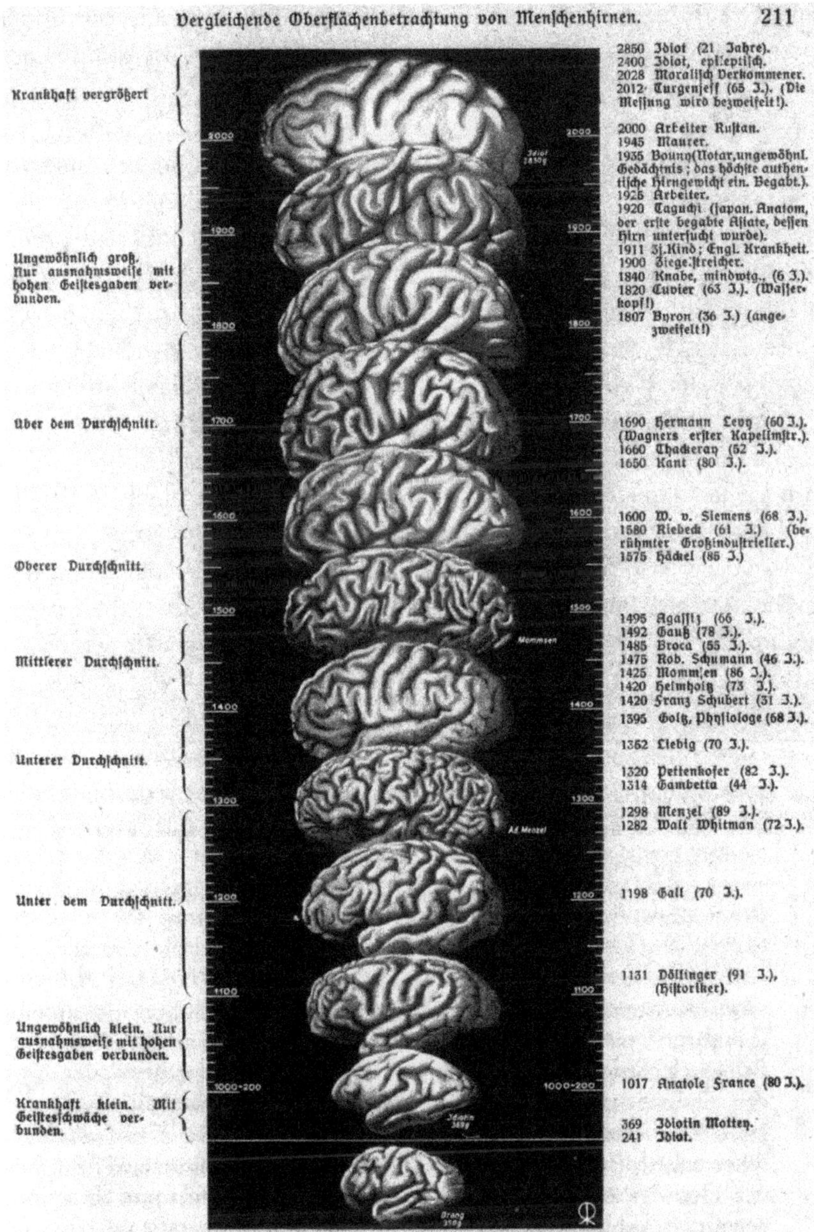

FIGURE 4.13 Lifelike representation of human brains of various sizes, in Fritz Kahn, *Das Leben des Menschen. Eine volksthümliche Anatomie, Biologie, Physiologie und Entwicklungsgeschichte des Menschen* (Stuttgart 1929), p. 211.

Similarly, Kahn noticed that, under the anatomical-racial profile, conditions were not at all simple and clear. First, he reported a lack of useful investigations even about the European peoples. The available data differed so much that they were open to the most arbitrary interpretations and, according to a strange but prevalent law, researchers had always provided better results for their own people than for others. Only a method based on completely new foundations, according to internationally recognized and controlled principles, could supposedly deliver useful data. For the sake of science one should put all the statistics on the racial brain weights together and burn them in a big fire. The still very few measurements taken until then on the so-called inferior (*niederer*) peoples had produced data that were not up to expectations and greatly amazed the investigators. Of all the existing lineages in the world, Australian aborigines and Fuegians were considered as the lowest (*niedersten*), yet Kohlbrugge had denied that an Australian brain could be distinguished from a European one, also vouching that the convolutions of the inhabitants of the Tierra del Fuego were on a par with those of the Europeans. That their brain weight did not show any inferiority was confirmed by Rudolf Martin's *Lehrbuch der Anthropologie*, not to mention the extension of the cortex, which appeared to be greater in the Hottentots than in the Europeans. It was well known that the American Blacks were rapidly reaping the benefits of civilization. In a few years, the number of illiterates among them had dropped from 90 to 20%, they attended high schools, wrote plays, exhibited works of art, composed music, and practiced science. What conclusions – Kahn ironically asked – would come from black anthropologists, who adopted the methods of their white colleagues?

Just like brain weight, the degree of cortical complexity had been assumed as a measure of the level of intelligence, and it had been claimed that the richer in furrows the cortex, the smarter the creature. Kahn cautioned that, in fact, the richness of the sulci and gyri was an expression of a mismatch between the spatial needs of the brain and the available size of the skullcap. Animals possessing a spacious head keep a brain surface as smooth as that of their juvenile stages. Others – like the elephant, the whale and the main predators – that need much space in the skull for big jaws – have to fold their brain into small brain cavities. For instance, the dolphin is a pirate with powerful jaws and a small cranial cavity; consequently his brain is so strongly furrowed that one might think it belongs to a human genius rather than to a cetacean. Conversely, the manatee, which feeds peacefully off marine plants, has a skull with a large cavity and therefore the surface of its brain is smooth. Kahn believed that what distinguished apes and man from all other mammals was not the wealth of the cortical folds, but the fact that, despite a large cranial space, they also had a wrinkled brain.

However, within the human species, no instant conclusion about intellectual faculties could be drawn from the size and folds of the brain. Very intelligent people generally seemed to have a wrinkled cortex, but basically Kahn was skeptical of the emotional reactions of a researcher in front of the brain of a genius, too suggestive and misleading. The anatomist is inclined to find nobility and subtleties in the convolutions of a famous genius, which he would never see in a beggar. In Beethoven's brain the folds could not fail to appear numerous and deep to the observer, and those of Anatole France had been written of as a masterpiece of jewelry. Except that the most tortuous brain was said to have belonged to a poorly skilled chiseller, who had died an alcoholic at the age of 36, and the so-called "poly-gyrie" was not a rare anomaly in the brains of idiots. Kahn did not disavow the fundamental fact of a growing phylogenetic refinement of the cortex, but wanted to draw attention to the danger of generalizing neuroanatomical statements. The publisher, poet and linguist Georg Sauerwein spoke 54 languages and wrote more than 40, but in his brain, despite an earnest search, the anatomist Ludwig Stieda, professor of anatomy at the University of Königsberg, did not detect anything unusual. So he had hoped that the macro-anatomy of the brain would be soon abandoned in favor of microscopic study, at which level actual discoveries were expected to occur.[86]

Kahn could not help commenting on the great scientific and popular interest that, in an age of female emancipation, the question of sexual difference in the brain had raised. He recalled that the average weights significantly varied from 1400 grams in the male to 1250 in the female, with a considerable difference already at birth. The woman, lighter and less muscular, would need less cerebral mass; on the other hand, as a smaller creature she should have a richer cortex, which however she has not. In newborn twins, Waldeyer had found a weight of 175 grams in the male and 165 in the female: Kahn interpreted this gap as a rule, but not as a law. One hundred brains of newborn boys weighed more than just as many of newborn girls, and yet the sex of a single brain could not be safely guessed on the basis of weight and cortical folds.

The same was true for adult brains. Kahn borrowed a visual example from Retzius for the case in question, which showed the brains of two humans whose lives and activities were fairly known, both mentally above average,

86 Kahn mistakenly attributed the research on Sauerwein's brain to Spitzka: ibid., p. 214. See instead Ludwig Stieda, "Über die Bedeutung der Hirnwindungen," *Correspondenz-Blatt der Deutschen Gesellschaft für Anthropologie, Ethnologie und Urgeschichte*, 1906, *37*:137-138; Id., "Das Gehirn eines Sprachkundigen," *Zeitschrift für Morphologie und Anthropologie*, 1907, *11*:83-138.

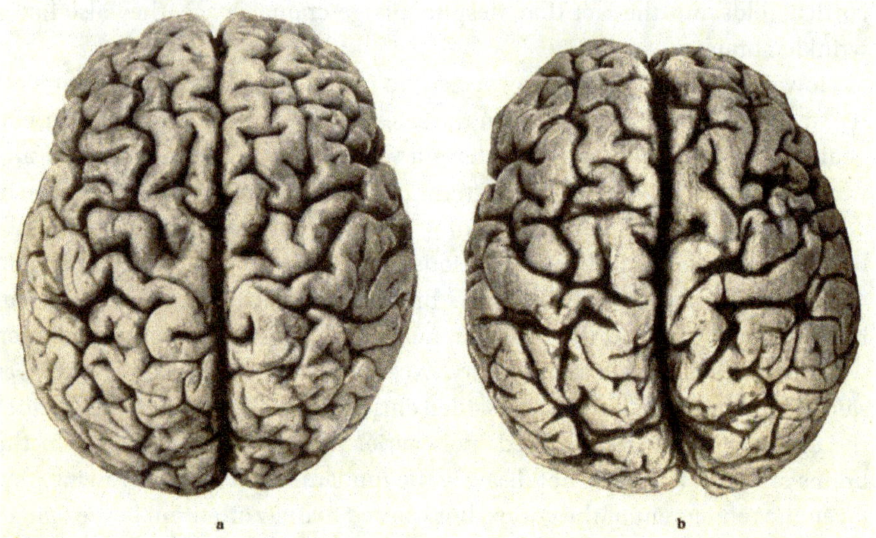

FIGURE 4.14 Brain differences between the sexes: a) the astronomer Hugo Gyldén, b) the mathematician Sonja Kovalevski, in Fritz Kahn, *Das Leben des Menschen. Eine volksthümliche Anatomie, Biologie, Physiologie und Entwicklungsgeschichte des Menschen* (Stuttgart 1929), p. 215.

who had died between their 40s and 60s, both devoted to close scientific fields, whose brains were examined and photographed with the same method. The male brain was of Hugo Gyldén, the female one of Sonja Kovalevski. Hers is evidently smaller and lighter, with a simpler structure of the surface. Kahn believed that the woman used less brain to fulfill her natural function, while the male was none other than a "cerebral creature" (*das Hirngeschöpf*), born with a strong muscular apparatus and thus developing a motor sphere. From a phylogenetic point of view, his senses tend to turn outwards, hunting for prey and thus improving his sensory centers; he must have courage, consideration and a combination of data to fight the enemy and to kill the victim, and that way he must have developed his own system of association. Instead the woman stays at home, raises the flock and feeds the young. His senses and thoughts take place in the hemispheres; her feelings and impulses arise from the deep centers, where every instinct of a living creature is based, which serve the genus and the offspring. *Er denkt, sie fühlt*: so it had been for thousands of centuries, and therefore diversity was established through inheritance. Despite all the modernity celebrated by *Das Leben des Menschen*, here Kahn indulged in reviving an old *cliché*, except that he immediately asked: will it be so in the future? Hard to predict: the ways of evolution are wonderful, the human species is rapidly changing and the brain plastically soft like wax under the artist's fingers. One day the woman might reach the intellect of man and, in general, even surpass him, thanks to that feminine instinct precluded to him, a simple cerebral creature.[87]

In the 1920s, the National Research Council and the Social Sciences Research Council appointed commissions for the study of the American Blacks and organized a Conference on Racial Differences in Washington, in 1928. Franz Boas had approached the thorny issue since the beginnings of his career; then, one of his former students, Melville J. Herskovits – with the help of some assistants – undertook a systematic measurement and a genealogical analysis of black subjects, part of a wider project on *Variability in Race* funded by the National Research Council. In this way, he realized, among other things, that the American Negro descended from racial crossings much more than it had been generally admitted: "a physical type which combined the characteristics of the American and European ancestral population and which was relatively homogeneous had been formed."[88] Ten years later, Herskovits dedicated a book "To the men and women who, in Africa and the New World, have helped me understand their ways of life". The Carnegie Corporation of New York happened to commission the Swedish Karl Gunnar Myrdal to conduct a large survey on the black population of America. In the about twenty partners involved in the task, Herskovits tried to disprove the deep-rooted idea of a Negro without a past, by revealing that the civilizations of Africa, like those of Europe, had contributed to American culture.[89]

Other Boasians were at work, in the middle of the war, to dismantle prejudices and practices that had contributed to its outbreak. For instance, a real mission was that of Ashley Montagu, English by birth but American by adoption, intent on unmasking the nineteenth-century racial typological system, above all by arguing that the advent of genetics had eroded it. A series of his articles, appeared since 1939, were collected in a book that had a great success and numerous reprints after the first edition of 1942. The endless question of the racial brain could not be overlooked: Montagu proposed some quotations from anti-Negro texts and recalled Mall's and Wilder's criticism of the alleged differences in the Negro brain detected by Bean. It was also useful to report

87 Kahn, *Das Leben des Menschen* (cit. note 85), pp. 214-215.

88 Melville J. Herskovits, Vivian K. Cameron, Harriet Smith, "The Physical Form of the Mississippi Negro," *American Journal of Physical Anthropology*, 1931, *16*:193-201, p. 193. Herskovits had already published an anticipation of his research carried on between 1923 and 1927: *The American Negro. A Study in Racial Crossing* (New York: A. A. Knopf, 1928).

89 Melville J. Herskovitz, *The Myth of the Negro Past* (New York-London: Harper & Brothers, 1941).

what G. Levin – from the Bekhterev Institute for Brain Research of Leningrad – had recently shown, namely that the available evidence affords no ground whatsoever for any racial or "inferiority signs" in human brains, whether they be of great men or of savages. Montagu commented:

> Actually, if the Negro brain is somewhat smaller than that of the white, the difference will be found so small that it can hardly be considered in any way significant for the mental functioning of the Negro as compared with that of the white. Within the limits of normal variation, differences in brain size have about as much relation to intelligence and cultural achievement as differences in body size, and as far as the available evidence goes, that is none. [...] The fact is that the external morphology of the human brain, or the characters of size and weight, have little or nothing to do with its functional capacities; these, on the other hand, must be considered as due to a complex of characters, such as the genetically determined internal (microscopic) structure of the cells and neurons and the organization to which these are subjected by experience, the abundance of the blood vessels, the character or their walls, and the efficiency of the drainage.[90]

The attack of the successors of Boas, who died in 1942, was concentric on the target. In 1940, Ruth Benedict strived to sharply separate race from racism, "poles apart": the first being a matter for scientific study, the second "an unproved assumption of the biological and perpetual superiority of one human group over another". In particular, organic differences in the brain could never by themselves answer the question of superiority and inferiority. People with brains of similar structure might use them very differently, due to the various experiences in which their minds have habitually functioned. People with organically inferior brains were born in all races, whereas no race as a whole could be condemned to feeble-mindedness. Confusing hereditary structure with learned functioning meant to committing a serious mistake. Less crude standards than the mere gross size of the brain had been introduced by

90 M. F. Ashley Montagu, *Man's Most Dangerous Myth: The Fallacy of Race* (New York: Columbia University Press, 1945²), pp. 198-199; G. Levin, "Racial and 'Inferiority' Characters in the Human Brain," *American Journal of Physical Anthropology*, 1937, 22:345-380. In 1949 the non-linear affair of the Unesco *Statements on Race* began, and Montagu was appointed – upon proposal by Claude Lévi-Strauss – *rapporteur* of the commission of experts charged with drafting a text.

FIGURE 4.15 Ad Reinhardt's drawing, in Ruth Benedict, Gene Weltfish, *The Races of Mankind* (New York 1943), p. 27.

psychological tests of racial superiority since the First World War, but their conception and administration were quite perplexing.[91]

In 1943 Ruth Benedict and Gene Weltfish – she too trained under Boas, and working at Columbia University – co-wrote a thirty-page popular booklet at the request of the United Service Organization (USO) to be handed out to the armed forces. Also with a series of simple and effective cartoon illustrations by the painter Ad Reinhardt, *The Races of Mankind* tried to counter, with the weapons of historical relativism, the racist views that had led to the then-ongoing war. Text and images unfold the tenets of Boasian culturalism: peoples of the earth are one family, a great inter-individual variability must be taken in account, education is all, history proves that progress in civilization is not the monopoly of one race or sub-race etc. A brief comment on the brain riddle was not spared:

91 Ruth Benedict, *Race: Science and Politics* (New York: Modern Age Books, 1940), pp. v-vi, 104-121.

> Because the brain is the thinking organ, some scientists have tried to find differences in the size and structure of the brain among different groups of people. In spite of these efforts, using the finest microscopes, the best scientists cannot tell from examining a brain to what group of people its owner belonged. The *average* size of the brain is different in different groups, but it has been proved over and over again that the size of the brain has nothing to do with intelligence. Some of the most brilliant men in the world have had very small brains. On the other hand, the world's largest brain belongs to an imbecile.[92]

As maintained by Benedict and Weltfish, the world was going through a "crucial moment of history", when all what science had learned in the past became needful, and supposedly condensed into the short space of the pamphlet. Significantly, members of Congress labeled it as "communistic", and its circulation in the Army was banned. After the war, *The Brotherhood of Man*, a ten-minute animation film, based on the booklet and directed by Robert Cannon, was shot in 1946 at the request of the United Auto Workers to help ease racial tensions. In the early 1950s, some of the people associated with the two products were picked on by the surveillance and repression of the House Committee on Un-American Activities, Joseph McCarthy's Senate Committee, and the US Senate Subcommittee on Internal Security.[93]

In a 1970 article, the South African paleoanthropologist Phillip V. Tobias still discussed the oft-repeated assertion that human races differed in quantity of brain substance and, especially, of gray matter in the cortex, with great disadvantage of the African lineage. Being an opponent of the apartheid regime, he scrutinized the evidence of that assertion, and ended up judging it inadequate, or, in some respects, totally lacking:

> We shall find that, even where differences in brain-size validly demonstrated among different populations, they would be unable to explain adequately variations in cerebral functions and achievement among living human beings. Furthermore, we shall find that such variations, if they do exist, are apparently not of much importance in modern man.[94]

---

92 Ruth Benedict, Gene Weltfish, *The Races of Mankind* (New York: The Public Affairs Committee, 1943), pp. 7-8.

93 Alice W. Campbell, "Influence and Controversy. *The Races of Mankind* and *The Brotherhood of Man*," *Social Welfare History Project*, 2018. <http://socialwelfare.library.vcu.edu/eras/wwii-1950s/influence-controversy-races-mankind-brotherhood-man/>.

94 Phillip V. Tobias, "Brain-size, Grey Matter and Race. Fact of Fiction?," *American Journal of Physical Anthropology*, 1970, 32:3-26, p. 3.

One of Tobias's targets was what the segregationist businessman Carleton Putnam had written in the early 1960s about the inherent anatomical causes of the evolutionary inferiority in Blacks, a tenet shared during the period of civil rights campaigns by other white "resistants", such as the racist anthropologist Donald W. Swan and the histologist and embryologist Wesley Critz George, an activist against racial integration, whom the Governor of Alabama John Patterson had commissioned with a report on *The Biology of the Race Problem*. Its Chapter IV took advantage of studies that in the twentieth century had insisted on the *Physical Basis for Intellectual and Behavior Differences*.[95]

For one thing, Tobias asked, "what facts do we have about the brain-size?", and specified how the cranial capacity did not represent a significant value, due to the amount of anatomical parts within the skull that were not properly pertaining to the hemispheres. Estimates of this proportion were highly variable (from 10 to 33.33%), and the ratio was not a constant figure within the adult lifetime of any individual, as the brain shrinks with age, illness, and other conditions. But also the measure of the real brain size would not be as easy to determine as it might sound, because of a large number of possible causes of variance: body weight and height, sex, age, nutritional state, environmental factors, source of the sample, occupational status, causes of death, lapse of time after death and treatment of the brain to be measured.

Considering all this, it was not at all puzzling that contradictory results connoted the vast scientific production amassed for over a century and a half. Tobias stressed that the order of variability stemming from only one or two of the possible aspects was greater than most of the supposed interracial differences, and that this circumstance would largely invalidate most of the comparisons made in the past. At the present state of knowledge, therefore, exploiting the allegedly smaller brain weight of Blacks as an explanation for their level of intelligence or kind of behavior was totally meaningless. Moreover, normal human beings existed with brain sizes three times that of others. Such a fantastic range remained an enigma:

95 See Carleton Putnam, *Race and Reason: A Yankee View* (Washington, DC: Public Affairs Press, 1961) and *Race and Reality: A Search for Solutions* (Washington, DC: Public Affairs Press, 1967); Donald W. Swan, "Juan Comas on «Scientific racism again?»: A scientific analysis," *Mankind Quarterly*, 1962, 2:231-245; Wesley C. George, *Race, Heredity, and Civilisation: Human Progress and the Race Problem* (London: Britons Publishing Society, 1961) and *The Biology of the Race Problem* (n. p., 1962), pp. 25-35. See George Lewis, "«Scientific Certainty»: Wesley Critz George, Racial Science and Organized White Resistance in North Caroline, 1954-1962," *Journal of American Studies*, 2004, *38*:227-247; John P. Jackson, Jr., *Science for Segregation: Race, Law, and the Case against Brown v. Board of Education* (New York-London: New York University Press, 2005).

> But within our species, we do not know what the microscopic or cellular basis is of varying brain-sizes. If we do not know that simple *physico-physical correlation*, how can we hope to make meaningful statements about the correlation between gross brain-size and cellular structure on one hand, and about psychical and behavioural attributes on the other? For the *physico-physical* correlation is basic to the *physico-psychical* association. We do not have the requisite information at either level. [...] We are back at square one. We must confess our ignorance of the functional meaning and value of different sized brains in modern human individuals.[96]

No wonder, after all, that in those turbulent years a scientific authority like Tobias still had to unmask old biases disguised as science.

96 Tobias, "Brain-Size, Grey Matter and Race" (cit. note 94), pp. 17-18.

Summary

1

Some time after Malpighi formulated his hypothesis (1665) of a *nigredo Aethiopum* set in the mucous layer that connects the tactile *papillae*, skin pigmentation began to stimulate scientific curiosity, to the point that in 1739 the Académie royale de sciences de Bordeaux offered a prize to address the old question of the color of Blacks. None of the competitors was judged worthy of the award, but the episode marked an ever-growing interest in differences – not only physical – between human groups. Soon speculation or even experiments were attempted about the influences of climate on the more superficial and visible body features – Buffon *docet* – eventually moving on to contemplate the more hidden and still little known organs of human anatomy, such as the central nervous system.

The corpses of some Blacks were examined shortly after the middle of the century, and in so doing the German anatomist Johann Friedrich Meckel announced his discovery that in their brains the color of the medullary substance was getting darker as the dissection proceeded downward. He also hazarded that the color of the subcuticular mucous membrane might derive, through the nerves, from the evaporation of the liquid that dyes the white matter. Since then, other anatomists confirmed or denied Meckel's thesis, which remained a reference point for decades. The real motive behind this kind of scientific curiosity was the vexed question of the relationship between brain and mind, tackled at the intersection with the other crucial issue – ever more pressing as European colonial expansion consolidated – of the varieties existing within the human species. It was that particular intersection that fostered the birth of craniology, which was bound to develop hugely in the following century.

A landmark is the dissection of Blacks performed in the early 1780s by the young anatomist Samuel Thomas Soemmerring, in the attempt to investigate their bodily *Verschiedenheit*. One of many things, he disproved Meckel's findings and observed a greater bulkiness in the nerves of the inferior cerebral surface, with the obvious implication of Blacks having a smaller brain. His conclusions were so ambiguous that they aroused concern in his friend Johann Friedrich Blumenbach – a collector and classifier of skulls – for the excessive distance between Africans and other human varieties that Soemmerring's work might suggest. It is no coincidence that the popular philosopher Christoph Meiners exploited it in his hammering campaign to fuel opposition between Whites and Blacks. Against him – and also demolishing Kant's abstract concept of race – Georg Forster claimed that he had acquired direct knowledge of distant peoples when he took part in Captain Cook's second voyage around the world (1772-1775).

In his 1787 Virginia report, Thomas Jefferson painted an unflattering portrait of American blacks, yet admitting that to demonstrate their physical and mental inferiority much study was still needed, with the aid of the anatomical knife, optical glasses, and analysis by fire or solvents. In a way, some sort of a research program, while others – like the English physician Charles White – envisaged a gradation from the brain of the European to the polypus, and proximity of the "Negro" type to the ape.

2

Although some signs had appeared until then, it was in the first half of the nineteenth century that the wide spreading of Gall's doctrine helped give a strong impulse not only to the efforts to unveil connections between brain and mind, but also to make the study of the organic causes of the so-called "national characters" increasingly crucial. A sort of Pandora's box opened up, even regardless of any affiliation with the phrenological faith. A new natural history of man, which then made its first tests, radically changed Buffon's approach, while the Napoleonic decree of 1802 – revoking the previous abolition of slavery – is symptomatic, not only for France, of a rapidly changing cultural landscape. It became possible to bluntly say that human races differ in most parts of the body and significantly in the governing organ constituted by the brain, as well as, correlatively, in their respective levels of development, intelligence, attitudes and behavior. Without hesitation, the observed differences were organized in such a way as to outline a hierarchical order. Virey's popular writings are good evidence of this trend, which took hold in the first decades of the century.

There were also sporadic pronouncements aimed at defending populations condemned as irreparably inferior – like the activity of the abbé Grégoire and the anatomical study by Friedrich Tiedemann – or more moderate and cautious views, but the mainstream favored anatomical research, systematically conducted to detect any revealing clue in the structure and weight of the brain. Given the scarcity of its dissectible specimens – especially from outside Europe – and the difficulty of handling them properly, it was necessary to resort to skulls, which could provide an indirect (albeit problematic) estimate of the brain weight. Since the early 1840s, the ratio of the width to the length of the head, known as cephalic index, had been measured on skulls for classification purposes, and gave also rise to guesswork and controversies on the correlated quantity of brain. Measuring was a supposedly essential practice for eliminating subjective biases and mistakes, and craniology set up an ideal field in which to take advantage of the benefits of calculation. Throughout the century, a frenzy for amassing skulls caught on, collections multiplied in Europe and in the United States and, piece by piece, got wider – suffice to think of Morton's initiative in Philadelphia.

The brain shape too was put under observation, since the mere volume (or weight) did not seem enough to effectively explain the functional qualities of the organ. The relationships between its various parts, and especially the degree of complication in the convolutions – long neglected by science – were brought into play. Anterior and superior lobes seemed to be in charge of intellectual faculties. Likewise, by virtue of what comparative anatomy had been finding, more tortuous cortical gyri and deeper sulci ended up being connected with a better performance of the central nervous system. As soon as a certain amount of data had been gathered and analyzed, and when both brain size and shape had been given a relevant role in the advancement and destiny of races, the first doubts and contradictions began to appear on the horizon.

3

The second half of the nineteenth century witnessed both the greatest efforts being made to racialize brains or skulls, and frequent displays of perplexity about the value of the results. For a while, Paris was one of the main operational centers, thanks to the Société d'Anthropologie founded in 1859 by Paul Broca, who was at the head of a school that tried to give an identity to a new discipline. The weight, volume, and shape of the brain attracted much research and discussions, especially in the period that saw the European public opinion take sides with opposing fronts of the American Civil War. In 1860, the anti-materialist Rudolph Wagner, heir of Blumenbach's chair in Göttingen, stressed the lack of constant relations between encephalic mass and intelligence, while soon after Broca and the anatomist Gratiolet disagreed about the meaning to be given to brain volume, in individuals and races.

A veritable international enterprise then involved a few generations of players. Since 1863, the Anthropological Society of London – which barred membership to women – had felt called to staunchly undermine the principle of unity of the human species and to denigrate those varieties that were reputedly inferior. Charles Darwin remained loyal to the preexisting Ethnological Society, which advocated monogenism. It should also be mentioned, though, that Darwinians differed little from their opponents in linking brain size and morphology with mental development. In this sense, a speech given by Thomas Huxley in 1865 – the Civil War having just ended – is exemplary, in affirming the insuperable gap between Blacks and their "bigger-brained and smaller-jawed rivals". However, right then, the Italian anatomist Calori dissected the brain of a Guinean man in Bologna and came to deny the attributions of diversity that had been increasingly emphasized until then. His was certainly a lone voice, furthermore referring to a single case, and yet he testified to the periodical emergence of a maverick position.

Likewise, from time to time, the issue of the female brain resurfaced, almost always associated to the belief that its smaller size – compared to the male one – implied physical and intellectual inferiority. The average capacity of a black male brain seemed slightly exceeding that of the European female, showing how race and gender had come to share a discursive space. Within this, nevertheless, it was often repeated that too little or inept work had been done in measuring brains, and that data were insufficient, thus producing unreliable outcomes. Not without reason, in 1878 an authority like Virchow made an appeal for a "comparative encephalology", which was considered an urgent desideratum for science. But, also that year, in his guide to the study of convolutions, Carlo Giacomini affirmed the scarce significance of sexual or racial differences on the cortical surface and advised not to swear by the superiority of the European cortex. Shortly beforehand a Société d'autopsie mutuelle had been founded in Paris, to facilitate the dissection of brains belonging to individuals who considered themselves excellent. The search for cerebral explanations of 'genius' lasted throughout the century and beyond. No matter how unsuccessful it might have been, international toiling with brains, skulls, and heads in relation to intelligence continued unabated, forming "a literature by itself", as the American neurologist Donaldson noted in 1895.

4

Head shape underwent far-reaching changes in European immigrants: this is what Boas claimed in 1911-12, after three years measuring thousands of them in New York. Apparently, the new environment could quickly alter their physical features, and it was not in the size of the brain that something interesting could be discovered, rather in its morphology. Be as it may, at the dawn of the twentieth century, studies on outstanding brains had regained strength, accomplihed – among others – by Gustaf Retzius in Sweden and by Edward Anthony Spitzka in the United States, both of them also interested in 'national' or racial brains.

In France, Vaschide and Pelletier tirelessly visited schools, looking for physical signs of intelligence in a large number of students, while Pearson undertook something on the same lines in London, although without brilliant results. More biometrical research was carried out at that time, while the idiosyncratic American context still bred the confidence that the key to the riddle laid in the brain. Since 1904, and for a quarter of a century, the anatomist Robert Bennett Bean thought he had found it, despite the criticisms received from many sides, for instance from some of the speakers at a meeting held in New York by the newly founded National Association for the Advancement of Colored People (1909).

For a couple of years, the young Sergio Sergi was given the opportunity of studying brains in von Waldeyer's Berlin anatomical institute, brains that came from the Herero ethnic group, recently exterminated by German troops. Area by area, he accurately described a long series of morphological peculiarities. Not surprisingly, Rudolf Martin still defined a clear hierarchy of cranial capacities in his *Lehrbuch der Anthropologie* (1914), with the Europeans at the top with the highest average of 1450 cm^3. However, Martin, inspired by Sergi, ruled out the existence of any disparity between higher and lower races in the weight of the frontal lobes. That was also the period in which histology might have provided new tools, and a particular viewpoint, except that most of the research still focused on macroscopic morphology, also by paying attention to populations that had been generally overlooked until then: Asiatic, American Indians, Australians. In 1931, Sergi complained that variations in the minute structure of the brain, race by race, still remained *terra incognita*, while shortly thereafter some microscopic research was performed by F. W. Vint on the brains of colonial subjects, in Kenya, which yielded knotty results.

To realize how much the topic was still a source of interest in the 1920s, just browse Fritz Kahn's popular and ultra-illustrated *Das Leben des Menschen*, where a table shows the progression of the facial angle from the pig, through the apes and four human races, to the European. But Kahn did not think the quality of mental faculties could possibly be inferred from the brain weight – on which he had written a chapter. According to him, for the sake of science, and in the wake of medieval methods, all the statistics hoarded in the past deserved to be burnt in a big fire. The deconstruction of the field was underway elsewhere too. Boasians like Melville J. Herskovits, Ashley Montagu, Ruth Benedict and Gene Weltfish began to dismantle long-held beliefs, piece by piece, in the interwar period and during the war. It was 1970 when the South African paleoanthropologist Phillip v. Tobias unmasked them again, as an activist for the eradication of apartheid, and just after the decade of the civil rights movement in the United States.

In a letter written to the Editor of the *Journal of the African Society* (1934), the biologist Lancelot Hogben had already expressed boredom for the "threadbare issue concerning the existence of racial differences with respect to the cubic capacity of the skull." He was quite right, given that for over a century the "issue" had been taking up an enormous amount of scientific energy, continually renewed in the face of disappointing results, incongruities, and admitted mistakes. But it is precisely this astonishing continuity that poses a historical problem: why did skulls and brains remain so stubbornly the object of anatomical and anthropological investigation during for such a long time? How could the related practices take shape – the hunt for ever-greater numbers of specimens, to be gauged in various ways, and the drive to fill countless tables with data or produce plates of visual representations? This book has tried to give a possible answer, which takes into account both the increasingly obsessive racial discourse and the contemporary metamorphosis of the brain into a hopefully revealing organ.

Bibliography

Ackermann, Jacob Fidelis, *Dissertatio inauguralis anatomica de discrimine sexuum praeter genitalia* (Moguntiae: Hered. Haeffner, 1788)

Ackermann, Jacob Fidelis, *Ueber die körperliche Verschiedenheit des Mannes vom Weibe außer den Geschlechtstheilen. Uebersezt nebst einer Vorrede und einigen Bemerkungen von Joseph Wenzel* (Mainz: in der Winkoppischen Buchhandlung, 1788)

Actes du deuxième Congrès international d'anthropologie criminelle. Biologie et sociologie (Paris, août 1889) (Lyon-Paris: Storck-Masson, 1890)

Åhrén, Eva, "Figuring it out: Visualizations in the work of Swedish anatomists Anders and Gustaf Retzius, 1829-1921," *Nuncius. Journal of the Material and Visual History of Science*, 2017, *32*:166-211

Albertan-Coppola, Sylviane, "Des récits des voyageurs à l'*Histoire générale des voyages*," *Dix-huitième siècle*, 2012/1, *n. 44*: 165-181

Amadei, Giuseppe, "Studj sulle variazioni del peso cerebrale. Il peso del cervello in rapporto alla statura," *Archivio italiano per le malattie nervose e più particolarmente per le alienazioni mentali*, 1881, *18*:199-206

Ammon, Otto, *Die natürliche Auslese beim Menschen. Auf Grund der Ergebnisse der anthropologischen Untersuchungen der Wehrpflichtigen in Baden und anderer Materialien* (Jena: Gustav Fischer, 1893)

Anderson, Stephanie, "«Three Living Australians» and the Société d'Anthropologie de Paris, 1885," in Brownwen Douglas and Chris Ballard (eds.), *Oceania and the Science of Race* (Canberra: ANU E Press, 2008), pp. 229-255

Anthony, Raoul, "Rapport du Prix Fauvelle," *Bulletins et Mémoires de la Société d'anthropologie de Paris*, VI série, 1911, *2*:391-393

Arago, Jacques, *Promenade autour du monde, pendant les années 1817, 1818, 1819 et 1820, sur les corvettes du Roi l'Uranie et la Physicienne, commandés par M. Freycinet* (Paris: Leblanc, 1822)

Ariëns Kappers, Cornelius U., "Cerebral Localization and the Significance of Sulci," in *17th International Congress of Medicine. Anatomy and Embriology* (London: Frowde, Hodder & Stoughton, 1913), pp. 273-392

Ariëns Kappers, Cornelius U., "Indices for the anthropology of the brain applied to Chinese, dolicho- and brachycephalic Dutch, foetuses and neonati," *Koninklijke Akademie van Wetenschappen te Amsterdam. Proceedings of the Section of Sciences*, 1927, *30*: 81-94

Ariëns Kappers, Cornelius U., *The Evolution of the Nervous System in Invertebrates, Vertebrates and Man* (Haarlem: De Erven F. Bohn, 1929)

Augstein, Hannah Franziska, *James Cowles Prichard's Anthropology: Remaking the Science of Man in Early Nineteenth Century Britain* (Amsterdam-Atlanta: Rodopi, 1999)

Bache, Meade R., "Reactions Time with Reference to Race," *The Psychological Review*, 1895, *2*:475-486

Bachman, John, *The Doctrine of the Unity of the Human Race Examined on the Principles of Science* (Charleston: C. Canning, 1850)

Baillarger, Jules, "De l'étendue de la surface du cerveau et de ses rapports avec le développement de l'intelligence," *Annales medico-psychologiques*, 1853, *5*:1-9

Baillarger, Jules, "Recherches sur la structure de la couche corticale des circonvolutions du cerveau," *Mémoires de l'Académie Royale de Médecine*, 1840, *8*: 149-183

Barber, John Warner, *A History of the Amistad Captives: Being a Circumstantial Account of the Capture of the Spanish Schooner Amistad, by the Africans on Board; their Voyage, and Capture near Long Island, New York; with Biographical Sketches of Each of the Surviving Africans. Also, an Account of the Trials had on their Case etc.* (New Haven, CT: E.L. & J.W. Barber, 1840)

Bean, Robert Bennett, "Some Racial Peculiarities of the Negro Brain," *The American Journal of Anatomy*, 1906, *5*:353-431

Bean, Robert Bennett, *The Racial Anatomy of the Philippine Islanders. Introducing New Methods of Anthropology and Showing their Application to the Filipinos with a Classification of Human Ears and a Scheme for the Heredity of Anatomical Characters in Man* (Philadelphia and London: J. B. Lippincott, 1910)

Bean, Robert Bennett, "A Racial Peculiarity in the Pole of the Temporal Lobe of the Negro Brain," *The Anatomical Record*, 1914, *8*:479-491

Bean, Robert Bennett, *The Races of Man. Differentiation and Dispersal of Man* (New York: The University Society, 1932)

Benedict, Ruth and Weltfish, Gene, *The Races of Mankind* (New York: The Public Affairs Committee, 1943)

Benedict, Ruth, *Race: Science and Politics* (New York: Modern Age Books, 1940)

Benedikt, "Sulla questione dei cervelli dei delinquenti. Lettera aperta al prof. Carlo Giacomini in Torino (con note del prof. Giacomini)," *L'Osservatore. Gazzetta delle cliniche di Torino*, 1883, *19*:141-143, 156-159, 174-176

Bergmann, Gottlob Heinrich, "Einige Bemerkungen über das Gewicht des Gehirns nebst drei Tabellen," *Allgemeine Zeitschrift für Psychiatrie und psychisch-gerichtliche Medicin*, 1852, *9*:361-376

Bericht über die Zusammenkunft einiger Anthropologen im September 1861 in Göttingen zum zwecke gemeinsamer besprechungen, erstattet von Karl Ernst von Baer und Rudolph Wagner (Leipzig: Leopold Voss, 1861)

Berner, Senta, *Die fünf Anatomen Meckel* (Freiburg im Brisgau: Universität, Medizinische Fakultät, Diss., 1963)

[Bernier, François], "Nouvelle division de la Terre par les différentes Especes ou Races qui l'habitent, envoyée par un fameux Voyageur à N. l'abbé de la ***** à peu près en ce termes," *Journal des sçavans*, 1684, pp. 133-140

Bernon, Thomas, "La science de races: la Société Ethnologique de Paris et le tournant colonial (1839-1848)," *La Révolution française. Cahiers de l'Institut d'histoire de la Révolution française*, 2018, *15*:1-56

Binet, Alfred et Vaschide, Nicolas, "Historique des recherches sur les rapports de l'intelligence avec la grandeur et la forme de la tête," *L'Année psychologique*, 1898, *5*:245-298

Bischoff, Theodor L. W. von, *Gedächtnissrede auf Friedrich Tiedemann. Vorgetragen in der öffentlichen Sitzung der k. Akademie der Wissenschaften am 28. November 1861, als am allerhöchsten Geburtstage Sr. Majestät des Königs Maximilian II. Von Bayern* (Münich: Verlag der k. Akademie, 1861)

Bischoff, Theodor L. W. von, "Die Grosshirnwindungen des Menschen mit Berücksichtigung ihrer Entwicklung bei dem Fötus und ihrer Anordnung bei den Affen," *Abhandlungen der mathematisch-physikalischen Classe der königlich bayerischen Akademie der Wissenschaften*, 1870, *10*:387-497

Bischoff, Theodor L. W. von, *Das Studium und die Ausübung der Medicin durch Frauen* (München: Literarisch-artistische Anstalt Th. Riedel, 1872)

Bischoff, Theodor L. W. von, *Das Hirngewicht des Menschen. Eine Studie* (Bonn: P. Neusser, 1880)

Blakeman John; Lee, Alice and Pearson, Karl, "A Study of the Biometric Constants of English Brain-Weights, and their Relationships to External Physical Measurements," *Biometrika*, 1905, *4*:124-160

Blanchard, Émile-Théophile, *Note sur la conformation particulière de la tête observée dans le Limousin* (Limoges: Chapoluaud Frères, 1859)

Blanckaert, Claude, "J.-J. Virey, Observateur de l'Homme," in Claude Benichou et Claude Blanckaert (sous la direction de), *Julien-Joseph Virey naturaliste et anthropologue* (Paris: Vrin, 1988)

Blanckaert, Claude, *De la race à l'évolution. Paul Broca et l'anthropologie française (1850-1900)* (Paris: L'Harmattan, 2009)

Blanckaert, Claude (coordonné par), *La Vénus hottentotte entre Barnum et Muséum* (Paris: Publications Scientifiques du Muséum d'Histoire naturelle, 2013)

Bloch, Adolphe, "L'intelligence est-elle en rapport avec le volume du cerveau?," *Revue d'anthropologie*, III série, 1885, *14*:577-619

Blumenbach, Johann Friedrich, *De generis humani varietate nativa. Illustris facultatis medicae consensus pro gradu doctoris medicinae disputavit d. XVI Sept. MDCCLXXV* (Gottingae: typis Frid. Andr. Rosenbuschii, 1775)

Blumenbach, Johann Friedrich, [on Soemmerring] in *Göttingische Anzeigen von Gehlerten. Der erste Band auf das Jahr 1785*, pp. 108-111

Blumenbach, Johann Friedrich, [on Soemmerring] in *Göttingische Anzeigen von Gehlerten. Der erste Band auf das Jahr 1786*, pp. 302-303

Blumenbach, Johann Friedrich, "Einige naturhistorische Bemerkungen bey Gelegenheit einer Schweizerreise," *Magazin für das Neueste aus der Physik und Naturgeschichte*, 1787, *4/3*: 1-12

Blumenbach, Johann Friedrich, "Vorrede," in *Sammlung seltener und merkwürdiger Reisegeschichten* (Memmingen: Seyler, 1789), vol. I

Blumenbach, Johann Friedrich, *De generis humani varietate nativa. Editio tertia: premissa est epistola ad virum perillustrem Iosephum Banks…*(Gottingae: Apud Vandenhoeck et Ruprecht, 1795)

Blumenbach, Johann Friedrich, *Handbuch der Naturgeschichte. Sechste Auflage* (Göttingen: Johann Christian Dieterich, 1799)

Boas, Franz, "The Limitations of the Comparative Method of Anthropology," *Science*, new series, 1896, *4*:901-908

Boas, Franz, "The Cephalic Index," *American Anthropologist*, 1899, *1/3*:448-461

Boas, Franz, *The Mind of Primitive Man. A Course f Lectures Delivered before the Lowell Institute, Boston, Mass., and the National University of Mexico, 1910-1911* (New York: The Macmillan Company, 1911)

Boas, Franz, "Change in the Bodily Form of Descendants of Immigrants," *American Anthropologist*, 1912, *14*: 530-562

Boas, Franz, *Change in Bodily Form of Descendants of Immigrants* (New York: Columbia University Press, 1912)

Bödeker, Hans Erich, "Aufklärerische ethnologische Praxis: Johann Reinhold Forster und Georg Forster," in Id., Peter Hanns Reill und Jürgen Schlumbohm (Hrsg.), *Wissenschaft als kulturelle Praxis, 1750-1900* (Göttingen: Vandenhoeck & Ruprecht, 1999), pp. 227-253

Borck, Cornelius, "Communicating the Modern Body: Fritz Kahn's Popular Images of Human Physiology as an Industrialized World," *Canadian Journal of Communication*, 2007, *32*: 495-520

Borsarelli, Antonio, "Della quantità di fosforo che si trova nella materia del cervello e del ventricolo dell'uomo e di alcuni altri animali, ed in differenti età," *Annali di chimica applicata alla medicina*, 1861, *32*: 354-357

Bory de Saint Vincent, Jean-Baptiste, *Essais sur les Isles Fortunées et l'antique Atlantide, ou Précis de l'Histoire générale de l'Archipel des Canaries* (Paris: Baudouin, 1803)

Bory de Saint Vincent, Jean-Baptiste, *Voyage dans les quatre principales îles des mers d'Afrique, fait par ordre du gouvernement, pendant les années neuf et dix de la République (1801-1802), avec l'Histoire de la Traversée du Capitaine Baudin jusqu'àu Port Louis de l'Île Maurice* (Paris: F. Buisson, 1804)

[Bory de Saint Vincent, Jean-Baptiste], "Homme. *Homo*," in *Dictionnaire classique d'histoire naturelle. Tome huitième. H-INV.* (Paris: Rey et Gravier, 1825), pp. 269-346

Bory de Saint Vincent, Jean-Baptiste, *L'Homme (Homo.) Essai zoologique sur le genre humain. 2e Édition enrichie d'une carte nouvelle pour l'intelligence de la distribution des espèces d'hommes à la surface du globe terrestre* (Paris, Rey et Gravier, 1827)

Boyd, Robert, "Tables of the Weights of the Human Body and Internal Organs in the Sane and Insane of both Sexes at various Ages, arranged from 2614 post-mortem examinations," *Philosophical Transactions of the Royal Society of London for the Year MDCCCLXI*, 1862, *151*:241-262

Bra, Charles Marie Joseph, *Étude sur le poids de l'encéphale dans les maladies mentales* (Paris: Adrien Delahaye et Émile Lecrosnier, 1882)

Brace, C. Loring, *"Race" is a Four-Letter Word. The Genesis of the Concept* (New York-Oxford: Oxford University Press, 2005)

Branson, Susan, "Phrenology and the Science of Race in Antebellum America," *Early American Studies: An Interdisciplinary Journal*, 2017, *15*:164-193

Broca, Paul, *Recherches sur l'hybridité animale en général et sur l'hybridité humaine en particulier, considérées dans leurs rapports avec la question de la pluralité des espèces humaines* (Paris: Imprimerie de J. Claye, 1860)

Broca, Paul, "Études sur le cerveau d'un nègre," *Bulletins de la Société d'Anthropologie de Paris*, 1860, *1*:53-5

Broca, Paul, *Sur le volume et la forme du cerveau suivant les individus et suivant les races* (Paris: Typographie Hennuyer, 1861)

Broca, Paul, "Revue critique. Topographie cranio-cérébrale. Position relative des diverses parties du crâne et des diverses parties du cerveau," *Revue d'Anthropologie*, 1876, *5*:278-279

Broca, Paul, "Sur la topographie cranio-cérébrale et sur les rapports anatomiques du crâne et du cerveau," *Revue d'Anthropologie*, 1876, *5*:193-248

Broca, Paul, "Sur la fausseté des résultats céphalométriques obtenus à l'aide du conformateur des chapeliers," *Bulletins et mémoires de la Société d'Anthropologie de Paris*, III série, 1879, *2*:101-106

Broca, Paul, [Discours d'ouverture], in *Congrès international des sciences anthropologiques tenu à Paris du 16 au 21 août 1878* (Paris: Imprimerie Nationale, 1880), pp. 17-23

Brodmann, Korbinian, *Vergleichende Lokalisationslehre der Grosshirnrinde in ihren Prinzipien dargestellt auf Grund des Zellenbaues* (Leipzig: J. A. Barth, 1909)

Büchner, Ludwig, *Kraft und Stoff. Empirisch-naturphilosophischen Studien. In allgemein-verständlicher Darstellung* (Frankfurt a. M.: Meidinger Sohn & Cie., 1855)

Buffon, Georges-Louis Leclerc de, *Histoire naturelle, générale et particulière, avec la description du Cabinet du roy* (Paris: De l'Imprimerie royale, 1749), vol. III

Buffon, Georges-Louis Leclerc de, *Histoire naturelle, générale et particulière, avec la description du Cabinet du roy* (Paris: De l'Imprimerie royale, 1766), vol. XIV

Burrell, Brian, "The Strange Fate of Whitman's Brain," *Walt Whitman Quarterly Review*, 2003, *20*:107-133

Burt, Cyril, "The Native Brain – Intelligence Tests," *Times*, 1933, December 15

Caldwell, Charles, *Elements of Phrenology* (Lexington, KY: Thomas T. Skillman, 1824)

Caldwell, Charles, *Thoughts on the Original Unity of the Human Race* (New York: E. Bliss, 1830)

Calori, Luigi, "Cervello di un negro della Guinea illustrato con otto tavole litografiche, Mem. letta nella sessione 9 Novembre 1865," *Memorie della Accademia delle Scienze dell'Istituto di Bologna*, 1865, *5*:177-212

Calori, Luigi, "Del tipo brachicefalo negli Italiani odierni. Memoria," *Memorie dell'Accademia delle Scienze dell'Istituto di Bologna*, serie II, 1868, *8*:205-234

Calori, Luigi, "Del cervello nei due tipi brachicefalico e dolicocefalico italiani. Memoria," *Memorie dell'Accademia delle Scienze dell'Istituto di Bologna*, serie II, 1870, *10*:35-152

Campbell, Alice W., "Influence and Controversy. *The Races of Mankind* and *The Brotherhood of Man*," *Social Welfare History Project*, 2018. <http://socialwelfare.library.vcu.edu/eras/wwii-1950s/influence-controversy-races-mankind-brotherhood-man/>

Campbell, Chloe, *Race and Empire: Eugenics in Colonial Kenya* (Manchester: Manchester University Press, 2007)

[Camper, Petrus], *Verhandeling van Petrus Camper, over het natuurlijk verschil der wezenstrekken in menschen van onderscheiden landaart en ouderdom; over het schoon in antyke beelden en gesneedene steenen. Gevolgd door een voorstel van eene nieuwe manier om hoofden van allerleye menschen met zekerheid te tekenen* (Utrecht: B. Wild en J. Altheer, 1791)

[Camper, Peter] *Peter Camper über den natürlichen Unterschied des Gesichtszüge in Menschen verschiedener Gegenden und verschiedenen Alters…* (Berlin: in der Vossischen Buchhandlung, 1792)

Camper, Stephen T. and Gavrus, Delia (eds.), *The History of the Brain and Mind Sciences* (Rochester, University of Rochester Press, 2017)

[Carlyle, Thomas], "Characteristics," in *The Edinburgh Review*, 1831, *54*:351-383

Carus, Carl Gustav, *System der Physiologie umfassend das Allgemeine der Physiologie, die physiologische Geschichte der Menschheit, die des Menschen und die der einzelnen organische Systeme im Menschen, für Naturforscher und Aerzte* (Dresden und Leipzig: Gerhard Fleischer, 1838)

Carus, Carl Gustav, *Denkschrift zum hundertjährigen Geburtsfeste Goethe's: Über ungleiche Befähigung der verschiedenen Menscheitsstämme für höhere geistige Entwicklung* (Leipzig: Brockhaus, 1849)

Carus, Carl Gustav, *Symbolik der menschlichen Gestalt. Ein Handbuch zur Menschenkenntnis* (Leipzig: Brockhaus, 1853)

[Chambers, Robert], *Vestiges of the Natural History of Creation* (London: John Churchill, 1844)

Chappey, Jean-Luc, *La Société des Observateurs de l'homme. Des anthropologues au temps de Bonaparte* (Paris: Société des études robespierristes, 2002)

Choudhury, Suparna and Slaby, Jan (eds.), *Critical Neuroscience: A Handbook of the Social and Cultural Contexts of Neuroscience* (Oxford: Wiley-Blackwell, 2011)

Chudzinski, Théophile, "Sur les trois encéphales des Esquimaux morts de la variole, du 13 au 16 janvier 1881, dans le service de M. Andrieux à l'hôpital Saint-Louis," *Bulletins de la Société d'Anthropologie de Paris*, 1881, III série, *4*:312-318

Chudzinski, Théophile et Duval Mathias, "Description morphologique du cerveau de Gambetta," *Bulletins et mémoires de la Société d'anthropologie de Paris*, III série, 1886, *9*: 129-152

Church, Henry Ward, "Corneille de Pauw, and the Controversy over his *Recherches philosophiques sur les Américains*," *PMLA*, 1936, *51*:178-206

Clapham, Crochley, "The Weight of the Brain in the Insane," *West Riding Lunatic Asylum Medical Reports*, 1873, *3*:285-298; 1876, *6*:11-26

Clapham, Crochley, "On the Brainweights of some Chinese and Pelew Islanders," *The Journal of the Anthropological Institute of Great Britain and Ireland*, 1878, *7*:89-94

Clapham, Crochley, "On the Weight of the Brain and its Component Parts in the Insane," *Brain. A Journal of Neurology*, 1879, *1*:504-518; 1880, *2*:42-67

Clendinning, John, "Report on the Experience of the St. Marylebone Infirmary, since 1827, with respect to Admissions, Duration of Treatment, Mortality, and other Statistical Results, according to Age and Sex," *Journal of the Statistical Society of London*, 1844, *7*: 292-310

Coburn, Kathleen, Harding, Anthony John, *The Notebooks of Samuel Taylor Coleridge. Volume 5 1827-1834. Notes* (Princeton: Princeton University Press, 2002)

Coleridge, Samuel Thomas, *Marginalia. Edited by George Whalley* (London-Princeton: Routledge and Princeton University Press, 1980)

Coleridge, Samuel Thomas, *Shorter Works and Fragments. Edited by H. J. Jackson and J. R. De J. Jackson* (London-Princeton: Routledge and Princeton University Press, 1995)

Combe, Andrew, "Remarks on the Fallacy of Professor Tiedemann's Comparison of the Negro Brain and Intellect with those of the European," *The Phrenological Journal and Magazine of Moral Science*, 1837-1838, *11*:13-22

Combe, George, "On the Coincidence between the Natural Talents and Dispositions of Nations and the Development of their Brains," *The Phrenological Journal and Miscellany*, 1824, *2/1*:1-19

Combe, George, *The Constitution of Man Considered in Relation to External Objects* (Edinburgh: John Anderson Jun., 1828)

Combe, George, "Phrenological Remarks on the Relation between the Natural Talents and Dispositions of Nations, and the Developments of their Brains," in Morton, *Crania Americana*, 1839, pp. 269-291

Combe, George, *Notes on the United States of North America during a Phrenological Visit in 1838-9-40* (Edinburgh: MacLachlan, Stewart, & Co., 1841)

Congrès international des sciences anthropologiques tenu à Paris du 16 au 21 août 1878 (Paris: Imprimerie Nationale, 1880)

Cox, Robert, "Remarks on the Skulls and Character of the Thugs," *The Phrenological Journal and Miscellany*, 1834, *8*:524-530

Curran, Andrew S., *The Anatomy of Blackness. Science & Slavery in an Age of Enlightenment* (Baltimore: The Johns Hopkins University Press, 2011)

Curran, Andrew S., "Buffon et l'histoire naturelle des Africains," *Dix-huitième siècle*, 2012/1, *n. 44*:183-199,

Cuvier, Georges, "Extrait d'observations faites sur le cadavre d'une femme connue à Paris et à Londres sous le nom de *Vénus hottentotte*," *Mémoires du Muséum d'Histoire naturelle*, 1817, *3*: 259-274

Cuvier, Georges, "Sur un nouveau rapprochement à établir entre les classes qui composent le Règne Animal," in *Annales du Muséum d'Histoire naturelle*, 1812, *19*:73-84

Cuvier, Georges, *Le Règne animal distribué d'après son organisation, pour servir de base à l'histoire naturelle des animaux et d'introduction à l'anatomie comparée* (Paris: Deterville, 1817)

Darwin, Charles, *The Descent of Man, and Selection in Relation to Sex* (London: John Murray, 1871)

Davis, Joseph Barnard and Thurnam, John, *Crania Britannica. Delineations and Descriptions of the Skulls of the Aboriginal and Early Inhabitants of the British Islands: with Notices of their Other Remains* (London: Printed for the Subscribers, 1865)

Davis, Joseph Barnard, *Thesaurum Craniorum. Catalogue of the Skulls of the Various Races of Man* (London: Printed for the Subscribers, 1867)

[Davis, Joseph Barnard] J.B.D., "The Brain of a Negro of Guinea," *The Anthropological Review*, 1868, *6*:279-285

Davis, Joseph Barnard, "Contributions towards Determining the Weight of the Brain in Different Races of Man," *Philosophical Transactions of the Royal Society of London*, 1868, *158*:505-527

Davis, Joseph Barnard, "On the Weight of the Brain of the Negro," *The Anthropological Review*, 1869, *7*:190-192

Davis, Joseph Barnard, *Supplement to Thesaurum Craniorum. Catalogue of the Skulls of the Various Races of Man* (London: Printed for the Subscribers, 1875)

[De Pauw, Cornelis] Mr. de P.***, *Recherches philosophiques sur les Américains, ou Mémoires intéressants pour servir à l'Histoire de l'Espèce humaine* (Berlin: George Jacques Decker, 1768-1769)

[De Quincey, Thomas], "Notes from the Pocket Book of a late Opium Eater. English Physiology," *The London Magazine*, June 1824

[De Quincey, Thomas], "Style No. IV," *Blackwoods Edinburgh Magazine*, February 1841

Debierre, Charles, "De l'influence du travail cérébral sur le volume et la forme du crâne," *Bulletin de la Société d'anthropologie de Lyon*, 1884, *3/2*:243-264

Della Rocca, Mattia and Pogliano, Claudio (eds.), "Different Histories from 20th Century Neuroscience," *Nuncius. Journal of the Material and Visual History of Science*, 2017, *32/2*:253-500

Deniker, Joseph, *The Races of Man. An Outline of Anthropology and Ethnography* (London: Walter Scott, 1900)

Desmoulins, Louis-Antoine, "De l'état du système nerveux sous ses rapports de volume et de masse dans le marasme non sénile, et de l'influence de cet état sur les fonctions nerveuses," *Journal de physique, de chimie, d'histoire naturelle et des arts*, 1820, *90*:442-452

Desmoulins, Louis-Antoine, *Histoire naturelle des races humaines du nord-est de l'Europe, de l'Asie boréale et orientale, et de l'Afrique australe, d'après des recherches spéciales d'antiquités, de physiologie, d'anatomie et de zoologie, appliquée à la recherche des origines des anciens peuples, à la science éthymologique, à la critique de l'histoire etc.* (Paris: Méquignon-Marvis, 1826)

Dewey, John, "Address,", in *Proceedings of the National Negro Conference 1909. New York, May 31 and June 1* (New York: The Conference, 1909), pp. 71-73

Dias, Nélia, "La Société d'autopsie mutuelle et le dévouement absolu aux progrès de l'anthropologie," *Gradhiva*, 1991, *n.10*: 26-36

Diderot, Denis, "Humaine espèce," in *Encyclopédie ou Dictionnaire raisonné des sciences, des arts et des métiers, par une société des gens de lettres* (Neufchastel: Samuel Faulche & Compagnie, 1765), 8, pp. 346-348

[Diderot, Denis], *Oeuvres inédites de Diderot. Le neveu de Rameau. Voyage de Hollande* (Paris: Brière, 1821)

Donaldson, Henry H., "Anatomical observations on the brain and several sense organs of the blind deaf mute, Laura Dewey Bridgman," *American Journal of Psychology*, 1891, *3-4*:293-342, 248-294

Donaldson, Henry H., *The Growth of the Brain: A Study of the Nervous System in Relation to Education* (London-New York: Walter Scott and Charles Scribner's Sons, 1895)

Dougherty, Frank Peter William, *Bibliographie der Werke und Schriften von Johann Friedrich Blumenbach nebst ihren Übersetzungen und Digitalisierungen. Bibliography of Johann Friedrich Blumenbach. Durchgesehen, ergänzt und mit Anmerkungen und einer Einleitung herausgegeben von Norbert Klatt. Mit einem Geleitwort von Prof. Dr. Dr. Heinz Schott* (Göttingen: Norbert Klatt, 2009)

Down, John Langdon, "Observations on an Ethnic Classifications of Idiots," *London Hospital Reports*, 1866, *3*:259-262

Dubois, Eugène, "On the Significance of the Large Cranial Capacity of Homo Neanderthalensis," *Koninklijke Akademie van Wetenschappen te Amsterdam. Proceedings of the Section of Sciences*, 1923, *23*:1271-1288

Dumont, Franz, "«Teutschlands Hippokrates». Der Anatom Samuel Thomas Soemmerring in seinen Mainzer Jahren (1784-1792/97)," in Id. (*et alii*), *Moguntia medica. Das medizinische Mainz. Vom Mittelater bis ins 20. Jahrhundert* (Wiesbaden: Wyličil, 2002), pp. 55-64

Durkheim, Émile, *De la division du travail social* (Paris: Félix Alcan, 1893)

Economo, Constantin von, *The Cytoarchitectonics of the Human Cerebral Cortex* (Oxford: Oxford University Press, 1929)

Eilers, Miriam, "Fritz Kahns *Das Leben des Menschen.* Zur Produktion und Transkription eines populären Werks," *NTM Zeitschrift für Geschichte der Wissenschaft, Technik und Medizin*, 2015, *23*: 1-31

Enke, Ulrike, "Soemmerings erste Professur am Collegium Carolinum zu Kassel", in Manfred Wenzel (Hrsg.), *Samuel Thomas Soemmerring in Kassel, 1779-1784: Beiträge zur Wissenschaftsgeschichte der Goethezeit* (Stuttgart-New York: Gustav Fischer, 1994), pp. 75-141

Evans, Gavin, *Black Brain White Brain. Is Intelligence Skin Deep?* (London: Thistle, 2014)

Eysenck, Hans, *The IQ Argument: Race, Intelligence, and Education* (New York: Library Press, 1971

Fabian, Ann, *The Skull Collectors. Race, Science, and America's Unburied Dead* (Chicago-London: The University of Chicago Press, 2010)

Farrand, Livingston, "Race Differentiation – Race Characteristics," in *Proceedings of the National Negro Conference 1909. New York, May 31 and June 1* (New York: The Conference, 1909), pp. 14-21

Fee, Elizabeth, "Nineteenth-Century Craniology: the Study of the Female Skull," *Bulletin of the History of Medicine*, 1979, *53*:415-433

Feijóo y Montenegro, Benito Jerónimo, *Teatro critico universal, ò Discursos varios en todo genero de materias, para desengaño de errores communes* (Madrid: Imprenta de Francisco del Hierro, 1736)

Ferrière, Hervé, *Bory de Saint-Vincent: l'évolution d'un voyageur naturaliste* (Paris: Syllepse, 2009)

Finger, Stanley, Eling, Paul, *Franz Joseph Gall. Naturalist of the Mind. Visionary of the Brain* (Oxford: Oxford University Press, 2019)

Fishberg, Maurice, "Physical Anthropology of the Jews. I. – The Cephalic Index," *American Anthropologist*, 1902, *4*:684-706

Fishberg, Maurice, *The Jews: A Study of Race and Environment* (London-New York: The Walter Scott Publishing Co.- Charles Scribner's Sons, 1911)

Flourens, Pierre, "Éloge historique de Frédéric Tiedemann, l'un des huit associés étrangers de l'Académie," *Gazette médicale de Paris*, 1861, *31*:815-827

de reconnoître plusieurs dispositions intellectuelles et morales de l'homme et des animaux, par la configuration de leurs tête (Paris: Schoell, 1810)

Gall, Franz Joseph et Spurzheim, Johann Gaspar, *Des dispositions innées de l'âme et de l'esprit, du materialisme, du fatalisme et de la liberté morale, avec une réflexion sur l'éducation et sur la legislation criminelle* (Paris: Schoell, 1811)

Gall, Franz Joseph, *Anatomie et physiologie du système nerveux en général et du cerveau en particulier, avec des observations sur la possibilité de reconnoître plusieurs dispositions intellectuelles et morales de l'homme et des animaux, par la configuration de leurs têtes* (Paris: N. Maze, 1819), vol. IV

Galton, Francis, "On Head Growth in Students at the University of Cambridge", *The Journal of the Anthropological Institute of Great Britain and Ireland*, 1889, *18*:155-156

Gascoigne, John, "Blumenbach, Banks, and the Beginnings of Anthropology in Germany," in Nicolaas Rupke (ed.), *Göttingen and the Development of the Natural Sciences* (Göttingen: Wallstein, 2002), pp. 86-98

George, Wesley C., *Race, Heredity, and Civilisation: Human Progress and the Race Problem* (London: Britons Publishing Society, 1961)

George, Wesley C., *The Biology of the Race Problem* (n. p., 1962)

Gerbi, Antonello, *The Dispute of the New World. The History of a Polemic, 1750-1900. Revised and enlarged edition translated by Jeremy Moyle* (Pittsburgh: University of Pittsburgh Press, 1973)

Giacomini, Carlo, "Nuovo processo per la conservazione del cervello," *L'Osservatore. Gazzetta delle cliniche di Torino*, 1878, *14*:421-425, 433-440

Giacomini, Carlo, "Variétés des circonvolutions cérébrales chez l'homme," *Archives Italiennes de Biologie*, 1882, *1*:1-64

Giacomini, Carlo, *Guida allo studio delle circonvoluzioni cerebrali dell'uomo* (Torino: Loescher, 1884), 2nd edition

Giacomini, Carlo, "Annotazioni sull'anatomia del negro," *Atti della R. Accademia delle scienze di Torino*, 1886-87, *22*:465-483

Giacomini, Carlo, "Annotazioni sull'anatomia del negro," *Giornale dell'Accademia di medicina di Torino*, serie III, 1878, *24*:454-470, 506-519; 1882, *30*:729-805; 1884, *32*:462-483; 1892, *40*:17-64; serie IV, 1897, *3*:649-672

Gladstone, Reginald J., "A Study of the Relations of the Brain to the Size of the Head," *Biometrika*, 1905, *4*:105-123

Gliddon, George G., "The Monogenists and the Polygenists; being an exposition of the doctrines of schools professing to sustain dogmatically the unity or the diversity of human races etc.," in Nott, Josiah C. and Gliddon, George G, *Indigenous Races of the Earth; or, Few Chapters of Ethnological Enquiry; including Monographs on Special Departments of Philology, Iconography, Cranioscopy, Palaentology, Pathology, Archaeology, Comparative Geography, and Natural History: contributed by Alfred*

Maury, Francis Pulszky and J. Aitken Meigs. Presenting fresh Investigations, documents, and Materials (Philadelphia: J. B. Lippincott & Co., 1857)

Gobineau, Arthur de, *The Moral and Intellectual Diversity of Races, with Particular Reference to their Respective Influence in the Civil and Political History of Mankind ...: with an Analytical Introduction and Copious Historical Notes by H. Hotz. To which is added an Appendix Containing a Summary of the Latest Scientific Facts upon the Question of Unity or Plurality of Species. By J. C. Nott* (Philadelphia: J. B. Lippincott & Co., 1856)

Goldsmith, Oliver, *A History of the Earth and Animated Nature in Four Volumes* (London: Henry Fisher, 1824)

Goldstein Sepinwall, Alyssa, *The Abbé Grégoire and the French Revolution. The Making of Modern Universalism* (Berkeley: University of California Press, 2005)

Goldstein, Jürgen, *Georg Forster. Zwischen Freiheit und Naturgewalt* (Berlin: Matthes & Seitz, 2015)

Gordon, Henry Laing, "The Mental Capacity of the African," *Journal of the Royal African Society*, 1934, *33*:226-242

Gordon, Henry Laing, "The Native Brain – Observations in Kenya – A Comparison with Europeans," *Times*, 1933, December 8

Gosse, Louis-André, *Essai sur les déformations artificielles du crâne* (Paris: J.-B. Baillière, 1855)

Gould, Stephen Jay, *Ontogeny and Phylogeny* (Cambridge, Mass.: The Belknap Press of Harvard University Press, 1977)

Gould, Stephen J., *The Panda's Thumb. More Reflections in Natural History* (New York-London: W. W. Morton & Company, 1980)

Gould, Stephen Jay, *The Mismeasure of Man* (New York-London: Norton & Company, 1996, revised and expanded edition)

Gould, Stephen J., " The Great Physiologist of Heidelberg," *Natural History*, 1999, *108/7-8*:26-29, 62-64, 68-70

Gratiolet, Pierre, *Mémoire sur les plis cérébraux de l'homme et des primatès* (sic) (Paris: Arthus Bertrand, 1854)

Gratiolet, Pierre, "Mémoire sur le développement de la forme du crâne de l'homme, et sur quelques variations qu'on observe dans la marche de l'ossification de ses sutures," *Comptes rendus hebdomadaires des séances de l'Académie des sciences*, 1856, *47*:428-431

Gratiolet, Pierre, "Sur la forme et la cavité crânienne d'un Totonaque, avec réflexions sur la signification du volume de l'encéphale," *Bulletins de la Société d'Anthropologie de Paris*, 1861, *2/1*:66-81

Gravlee, Clarence C.; Bernard, H. Russell, Leonard and William R., "Boas's *Changes in Bodily Form*: The Immigrant Study, Cranial Plasticity, and Boas's Physical Anthropology," *American Anthropologist*, 2003, *105*:326-332

Grégoire, Henri, *De la Littérature des Nègres, ou Recherches sur leurs facultés intellectuelles, leurs qualités morales et leur littérature; suivies de Notices sur les ouvrages des Nègres qui se sont distingués dans les Sciences, les Lettres et les Arts* (Paris: Maradan, 1808)

Grégoire, Henri, *De la noblesse de la peau, ou du préjudice des blancs contre la couleur des Africains et celle de leurs descendans noirs et sang-mêlés* (Paris: Baudouin Frères, 1826)

Gross, Charles G., "Hippocampus Minor and Man's Place in Nature: a Case Study in the Social Construction of Neuroanatomy," *Hippocampus*, 1993, *3*:403-416

Gubler, Adolphe, "Sur la coloration noirâtre des centres nerveux chez le individus de race blanche, remarquables par l'abondance du pigment extérieur," *Mémoires de la Société d'Anthropologie de Paris*, 1860-63, *1*:57-60

Guenebault, J. H., *Natural History of the Negro Race. Translated from the French* (Charleston, S.C.: D. J. Dowling, 1837)

Haeckel, Ernst, *Natürliche Schöpfungsgeschichte. Gemeinverständliche wissenschaftliche Vorträge über die Entwickelungslehre im Allgemeinen und dejenige von Darwin, Goethe und Lamarck im Besonderen...Zweite, verbesserte und vermehrte Auflage* (Berlin: Georg Reimer, 1870)

Hagner, Michael, *Homo cerebralis. Der Wandel vom Seelenorgan zum Gehirn* (Berlin: Berlin Verlag, 1997)

Hagner, Michael (Hrsg,), *Ecce Cortex: Beiträge zur Geschichte des modernen Gehirns* (Göttingen: Wallstein, 1999)

Hagner, Michael, *Geniale Gehirne. Zur Geschichte der Elitegehirnforschung* (Göttingen: Wallstein, 2004)

Haines, Duane, "Spitzka and Spitzka on the brains of the assassins of presidents," *Journal of the History of the Neurosciences*, 1995, *4*:236-266

Haldane, John B. S., "The Native Brain," *Times*, 1933, December 19

[Haller, Albrecht von], *Epistolarum ab eruditis viris ad Alb. Hallerum scriptarum Pars 1. Latinae. Vol. 11 – Epistolae 195 ad 404* (Bernae: Sumptibus Societatis Typographicae, 1773)

Haller, Alberts von, *Grundriss der Physiologie für Vorlesungen...durch Herrn Hofrath Soemmerring in Mainz* (Berlin: Hauder und Spener, 1788)

Hamilton, Keith and Salmon, Patrick (eds.), *Slavery, Diplomacy and Empire: Britain and the Suppression of the Slave Trade, 1807–1975* (Brighton and Portland: Academic Press, 2009)

Hamilton, William, *Sir William Hamilton and Phrenology. An exposition of Phrenology; shewing the complete inefficacy of the objections lately advanced in the Royal Society, and the real ground on which the system ought to be assailed* (Edinburgh-London: Hunter and Duncan, 1826)

Hamilton, William, *An Account of Experiments on the Weight and Relative Proportions of the Brain, Cerebellum, and Tuber Annulare in Man and Animals, under Various Circumstances of Age, Sex, Country, etc.* in Monro, Alexander *The Anatomy of the Brain: with some Observations of its Functions* (Edinburgh: J. Carfrae, 1831), pp. 4-8

Hamilton, William, "Remarks on Morton's Tables on the Size of the Brain," *The Edinburgh New Philosophical Journal*, 1849-1850, *48*:330-333

Hamilton, William, *Lectures on Metaphysics and Logic, edited by the Rev. Henry L. Mansel and John Veitch* (Boston: Gould and Lincoln, 1860)

Harris, George, "On the Distinctions, Mental and Moral, occasioned by the Differences of Sex," *Anthropological Review*, 1869, *7*:clxxxix-cxcv

Hauger, Otto, "Der Gehirnreichtum der Australier und anderer Hominiden, beurteilt nach ihrem Skelet," *Anatomische Hefte*, 1921, *59*:579-617

Hecht, Jennifer Michael, *Scientific Modernity, Atheism, and Anthropology in France* (New York: Columbia University Press, 2003)

Helo, Ari and Onuf, Peter S., "Jefferson, Morality, and the Problem of Slavery," *The William and Mary Quarterly*, 2003, *60*: 583-614

Herder, Johann Gottfried, *Ideen zur Philosophie der Geschichte der Menschheit* (Riga: Hartknoch, 1785), zweiter Theil

Hrdlička Aleš, *An Eskimo Brain* (New York: The Knickerbocker Press, 1901)

Herskovits, Melville J., *The American Negro. A Study in Racial Crossing* (New York: A. A. Knopf, 1928)

Herskovits, Melville J.; Cameron, Vivian K. and Smith, Harriet, "The Physical Form of the Mississippi Negro," *American Journal of Physical Anthropology*, 1931, *16*:193-201

Herskovits, Melville J., *The Myth of the Negro Past* (New York-London: Harper & Brothers, 1941)

Hervé, Georges, "Du poids de l'encéphale," *Revue d'Anthropologie*, II série, 1881, *10/4*: 681-698

Hildebrand, Reinhard, "Attic perfection in anatomy: Bernhard Siegfrid Albinus (21697-1770) and Samuel Thomas Soemmerring (1755-1830)," *Annals of Anatomy*, 2005, *187*: 555-573

Hildebrand, Reinhard, "Petrus Camper in his relationship to Samuel Thomas Soemmerring and other German scientists of the *Goethezeit*," in van Klaas Berkel und Bart Ramakers (eds.), *Petrus Camper in Context. Science, the Arts, and Society in the Eighteenth-Century Dutch Republic* (Hilversum: Verloren, 2015)

Hirmscher, Christoph, *Louis Agassiz. Creator of American Science* (Boston-New York: Houghton Mifflin Harcourt, 2013)

Hoffman, Michel A. and Johnson, John. I., "The C.U. Ariëns Kappers Brain Collection," *Annals of the New York Academy of Sciences*, 2011, *1225 S1*:E64-E84

Hogben, Lancelot, "To the Editor," *Journal of the Royal African Society*, 1934, *33*:432

[Home, Henry, Lord Kames], *Sketches of the History of Man. In two Volumes* (Edinburgh: W. Creech, London: W. Strahan and T. Cadell, 1774)

Hope, Thomas, *Essay on the Origin and Prospects of Man* (London: John Murray, 1831)

Horsman, Reginald, *Josiah Nott of Mobile: Southerner, Physician, and Racial Theorist* (Baton Rouge: Louisiana State University Press, 1987)

Hrdlička, Aleš, *Anthropological Investigations on One Thousand White and Colored Children of Both Sexes: the Inmates of the New York Juvenile Asylum, with Additional Notes on One Hundred Colored Children of the New York Colored Orphan Asylum* (New York-Albany: Wynkoop Hallenbeck Crawford Co., [1899])

Hunt, James, "Introductory Address on the Study of Anthropology, Delivered before the Anthropological Society of London, February 24th, 1863," *The Anthropological Review*, 1863, *1*:1-20

Hunt, James, *On the Negro's Place in Nature* (*Read before the Anthropological Society of London, November 17th, 1863*) (London: Trübner & Co., 1863)

Hunt, James, *The Negro's Place in Nature: A Paper Read before the London Anthropological Society* (New York: Van Evrie, Horton & Co., 1864)

Hunt, Sanford B., "The Negro as a Soldier," *The Anthropological Review*, 1869, *7*:40-54

Hurpin, Gérard, "Claude-Nicolas Le Cat ou de la notoriété médicale au XVIIIème siècle," *Histoire des sciences médicales*, 2001, *25*: 151-162

Huschke, Emil, *Schädel, Hirn und Seele des Menschen und der Thiere nach Alter, Geschlecht und Raçe. Nebst sechs Steintafeln mit photographischen Abbildungen* (Jena: Friedrich Mauke, 1854)

Huteau, Michel, "Un météore de la psychologie français: Nicolae Vaschide (1874-1907), *Bulletin de psychologie*, 2008/2, *n. 494*:173-199

Huxley, Julian, "The Native Brain: Size and Growth," *Times*, 1933, December 18

Huxley, Thomas H., "On the Zoological Relations of Man with the Lower Animals," *Natural History Review*, 1861: 67-84

Huxley, Thomas H., "Emancipation – Black and White," *Collected Essays. 3. Science & Education* (London: Macmillan and Co., 1893), pp. 66-75

Jackson, John P., Jr., *Science and Segregation: Race, Law, and the Case against Brown v. Board of Education* (New York-London: New York University Press, 2005)

Jacobs, Joseph, *The Jewish Question 1875-1884. Bibliographical Hand-List* (London: Trübner and Co., 1885)

Jacobs, Joseph, "On the Racial Characteristics of Modern Jews," *The Journal of the Anthropological Institute of Great Britain and Ireland*, 1886, *15*:23-62

Janjua, Rashid M. (et al.), "The Legacy of Johann Friedrich Meckel the Elder (1724-1774): A 4-Generation Dinasty of Anatomists," *Neurosurgery*, 2010, *66*: 758-771

Jefferson, Thomas, *On the State of Virginia. Illustrated with a Map, including the States of Virginia, Maryland, Delaware and Pennsylvania* (London: John Stockdale, 1787)

Jenkins, Bill, "Phrenology, Heredity and Progress in George Combe's *Constitution of Man*", *British Journal for the History of Science*, 2015, *48*: 455-473

Jensen, Arthur, "How Much Can We Boost IQ and Achievement?," *Harvard Educational Review*, 1969, *39/1*:1-123

Jones, Jeannette Eileen, "«On the Brain of the Negro». Race, Abolitionism and Friedrich Tiedemann's Scientific Discourse on the African Diaspora," in Mischa Honeck, Martin Klimke and Anne Kuhlmann (eds.), *Germany and the Black Diaspora. Points of Contact, 1250-1914*, (New York-Oxford: Berghahn, 2013)

Kahn, Fritz, *Das Leben des Menschen. Eine volksthümliche Anatomie, Biologie, Physiologie und Entwicklungsgeschichte des Menschen* (Stuttgart: Kosmos Franckh'sche Verlagshandlung, 1927), II

Kahn, Fritz, *Das Leben des Menschen. Eine volksthümliche Anatomie, Biologie, Physiologie und Entwicklungsgeschichte des Menschen* (Stuttgart: Kosmos Franckh'sche Verlagshandlung, 1929), IV

Kant, Immanuel, "Bestimmung des Begriffs einer Menschenrace," *Berlinische Monatsschrift*, 1785, *16*: 390-417

Kaplan, Jonathan M., "Race, IQ, and the Search for Statistical Signals Associated with So-Called 'x'- Factors: Environments, Racism, and the 'Hereditarian Hypothesis'," *Biology and Philosophy*, 2014, *30*:1-17

Karafyllis, Nicole C. and Ulshöfer, Gotlind (eds.), *Sexualized Brains. Scientific Modeling of Emotional Intelligence from a Cultural Perspective* (Cambridge, Mass.: The MIT Press, 2008)

Keegan, J. Jay, "A Study of a Plains Indian Brain," *The Journal of Comparative Neurology*, 1916, *26*:403-420

Keegan, J. Jay, "The Indian Brain," *American Journal of Physical Anthropology*, 1920, *3*:25-62

Keevak, Michael, *Becoming Yellow: A Short History of Racial Thinking* (Princeton: Princeton University Press, 2011)

Kelch, Wilhelm Gottlieb, *Ueber den Schädel Kants. Ein Beytrag zu Galls Hirn- und Schädellehre* (Königsberg: Friedrich Nicolovius, 1804)

Kitson, Peter J., *Romantic Literature, Race, and Colonial Encounter* (New York: Palgrave Macmillan, 2007)

Knox, "Inquiry into the Origin and Characteristic Differences of the Native Races inhabiting the Extra-Tropical Part of Southern Africa," *Memoirs of the Wernerian Natural History Society*, 1823-24, *5*:206-219

Knox, Robert, *The Races of Man. A Fragment* (Philadelphia: Lea & Blanchard, 1850)

Kohlbrugge, Jacob H. F., *Die Gehirnfurchen der Javanen. Eine vergleichend-anatomische Studie* (Amsterdam: Johannes Müller, 1906)

Kohlbrugge, Jacob H. F., "Untersuchungen über Großhirnfurchen der Menschenrassen," *Zeitschrift für Morphologie und Anthropologie*, 1908, *11*:596-609

Kohlbrugge, Jacob H. F., *Die Gehirnfurchen Malayischer Völker verglichen mit denen der Australier und Europäer. Ein Beitrag zur Evolutionslehre* (Amsterdam: Johannes Müller, 1909)

Kohlbrugge, Jacob H. F., "Gehirn und Kultur," *Biologisches Centralblatt*, 1911, *31*:248-256, 309-316

Kohlbrugge, Jacob H. F., "Le cerveau suivant les races," *Bulletins et Mémoires de la Société d'Anthropologie de Paris*, VII série, 1935, *6*:61-84

Kössler, Reinhart, *Namibia and Germany. Negotiating the Past* (Windhoeck: University of Namibia Press, 2015)

Kroke, Claudia, *Johann Friedrich Blumenbach. Bibliographie seiner Schriften* (Göttingen: Universitätsverlag, 2010).

Kurz, Eugen, "Das Chinesengehirn," *Zeitschrift für Anatomie und Entwicklungsgeschichte*, 1924, *72*:199-382

Kurz, Eugen, "Zwei Chinesengehirne. Ein Beitrag zur Rassenanatomie," *Zeitschrift für Morphologie und Anthropologie*, 1913, *16*:281-328

Lacassagne, Alexandre et Cliquet, "De l'influence du travail intellectual sur le volume et la forme de la tête," *Annales d'hygiène publique et de médecine légale*, II série, 1878, *1*:50-72

Lagier, Raphaël, "Un outsider de la fondation de l'anthropologie: Georg Forster," *Revue d'Histoire des Sciences humaines*, 2006, *14*: 137-152

Lawrence, William, *Lectures on Physiology, Zoology, and the Natural History of Man, delivered at the Royal College of Surgeons* (London: J. Smith, 1822)

Le Bon, Gustave, "Recherches anatomiques et mathématiques sur les lois des variations du volume du cerveau," *Revue d'Anthropologie*, II série, 1879, *2*:27-104

Le Bon, Gustave, "Psychologie. L'influence de la race dans l'histoire," *Revue scientifique* (*Revue rose*), III série, 1888, *15*:525-532

Le Bon, Gustave, *Les premières civilisations* (Paris: Flammarion, 1889)

Le Cat, Claude-Nicolas, *Traité des sens. Nouvelle edition* (Amsterdam: J. Wetstein, 1744)

Le Cat, Claude-Nicolas, *Traité de la couleur de la peau humaine en général, de celle des nègres en particulier, et de la métamorphose d'une de ces couleurs en l'autre, soit de naissance, soit accidentellement* (Amsterdam, 1765), pp. 51-62

Leakey, Louis, "The Native Brain: is Size a True Guide? Reasons for Stunted Growth," *Times*, 1933, December 13

Lee, Alice; Lewenz, Marie A. and Pearson, Karl, "On the Correlation of the Mental and Physical Characters in Man. Part II," *Proceedings of the Royal Society of London*, 1903, *71*:106-114

Lélut, Louis Francisque, *Du démon de Socrate, spécimen d'une application de la science psychologique à celle de l'histoire* (Paris: Trinquart, 1836)

Lélut, Louis Francisque, "Du poids du cerveau dans ses rapports avec le développement de l'intelligence," *Gazette médicale de Paris*, 1837, *5*:146-149

Lemercier, Jean-Pierre, "Claude-Nicolas Le Cat et l'Académie des Sciences, Belles-Lettres et Arts de Rouen," *Histoire des sciences médicales*, 2001, *25*:163-168

Lepenies, Wolf, *Autoren und Wissenschaftler im 18. Jahrhundert. Buffon, Linné, Winckelmann, Georg Forster, Erasmus Darwin* (München-Wien: Carl Hanser, 1988)

Letourneau, Charles, "Rapport sur le prix Godard, année 1879," *Bulletins de la Société d'Anthropologie de Paris*, III série, 1879, *2*:373-386

Leuret, François et Gratiolet, Pierre, *Anatomie comparée du système nerveux considéré dans ses rapports avec l'intelligence* (Paris: J.-B. Baillière, 1839-1857)

Leuret, François, *Anatomie comparée du système nerveux considéré dans ses rapports avec l'intelligence* (Paris: J.-B. Baillière, 1839)

Levin, G., "Racial and 'Inferiority' Characters in the Human Brain," *American Journal of Physical Anthropology*, 1937, *22*:345-380

Lewis, George, "«Scientific Certainty»: Wesley Critz George, Racial Science and Organized White Resistance in North Caroline, 1954-1962," *Journal of American Studies*, 2004, *38*:227-247

Lewis, Jason E. [et al.], "The Mismeasure of Science: Stephen Jay Gould versus Samuel George Morton on Skulls and Bias," *PLoS Biol*, 2011, *9*(*6*): e1001071. <https://doi.org/10.1371/journal.pbio.1001071>

Loeb, Jacques, *Comparative Physiology of the Brain and Comparative Psychology* (New York: G.P. Putnam's Sons, 1900)

Lombroso, Cesare, *L'uomo bianco e l'uomo di colore. Letture sull'origine e le varietà delle razze umane* (Padova: Tip.Editrice F. Sacchetto, 1871)

Lombroso, Cesare, *Nuovi studii sul genio. Volume II. Origine e natura dei genii* (Milano-Palermo-Napoli: Sandron, 1902)

[Long, Edward], *The History of Jamaica. Or, General Survey of the Antient and Modern State of that Island: with Reflections on its Situation, Settlements, Inhabitants, Climate, Products, Commerce, Laws, and Government* (London: T. Lowndes, 1774)

Loreti, Francesco, *Contributo alla storia dello "Studio" anatomico della Università di Torino. Carlo Giacomini (1840-1898)* (Torino: Accademia delle Scienze, 1963)

Lotter, Friedrich, "Christoph Meiners und die Lehre von der unterschiedlichen Wertigkeit der Menschenrassen," in Hartmut Boockmann und Hermann Wellenreuther (Hrsg.), *Geschichtswissenschaft in Göttingen. Eine Vorlesungsreihe* (Göttingen: Vandenhoeck & Ruprecht, 1987)

Lucae, Johann Christian Gustav, *Zur Morphologie der Rassen-Schädel. Einleitende Bemerkungen und Beiträge. Ein Sendschreiben an Carl Ernst v. Baer. Mit zwölf Tafeln* (Frankfurt a.M.: Heinrich Ludwig Brünner, 1861)

Luschan, Felix von, "Die anthropologische Stellung der Juden," *Correspondenz-Blatt der Deutschen Gesellschaft für Anthropologie, Ethnologie und Urgeschichte*, 1892, *3*: 94-102

Luyendijk-Elshout, Antoine M., "«Les beaux esprit se rencontrent». Petrus Camper und Samuel Soemmerring," in Gunter Mann und Franz Dumont (Hrsg.), *Samuel*

Thomas Soemmerring und die Gelehrten der Goethezeit (Stuttgart-New York: Gustav Fischer, 1985)

Lyell, Charles, *The Geological Evidences of the Antiquity of Man with Remarks on Theories of the Origin of Species by Variation* (London: John Murray, 1863)

Lyell, Katharine, *The Life, Letters and Journals of Sir Charles Lyell* (London: John Murray, 1881)

M.***[Pierre Barrère], *Dissertation sur la cause physique de la couleur des nègres, de la couleur de leurs cheveux, et de la degeneration de l'un et de l'autre* (Paris: Pierre-Guillaume Simon, 1741)

MacDonald, Carlos F., "The Trial, Execution, Autopsy and Mental Status of Leon F. Czolgosz, Alias Fred Nieman, the Assassin of President McKinley. With a Report of the Post-Mortem Examination by Edward Anthony Spitzka," *The Journal of Mental Pathology*, 1901-1902, *1*:179-209

Mall, Franklin P., "On Several Anatomical Characters of the Human Brain, said to vary According to Race and Sex, with Especial Reference to the Weight of Frontal Lobes," *The American Journal of Anatomy*, 1909, *9*:1-32

Malpighi, Marcello, *De externo tactus organo anatomica observatio* (Neapoli: apud Aegidium Longum, 1665), pp. 20-21

Mandressi, Rafael, *Félix Vicq d'Azyr: l'anatomie, l'État, la médecine*, in *Bibliothèque numérique Médic@*. 2006

Manouvrier, Léonce, "Recherches d'anatomie comparative et d'anatomie philosophique sur les caractères du crâne et du cerveau," *Bulletin de la Société Zoologique de France*, 1882, *7*:113-229

Manouvrier, Léonce, "Sur l'interpretation du poids de l'encéphale et ses applications," *Comptes rendus hebdomadaires des séances de l'Academie des Sciences*, 1882, *94*: 143-145

Manouvrier, Léonce, "Sur la valeur de la taille et du poids du corps comme terme de comparaison entre la masse de l'encéphale et la masse du corps," *Bulletin de la Société d'Anthropologie de Paris*, III série, 1882, *5*: 85-105

Manouvrier, Léonce, "Variétés: l'internat en mèdecine des femmes," *Revue scientifique*, 1884, *3*:592-597

Manouvrier, Léonce, "Recherches d'anatomie comparative et d'anatomie philosophique sur les caractères du crâne et du cerveau. Sur l'interprétation de la quantité dans l'encéphale et dans le cerveau en particulier," *Mémoires de la Société d'Anthropologie de Paris*, II série, 1888, *3*:137-326

Manouvrier, Léonce, "Indications anatomique et physiologiques relatives aux attributions naturelles de la femme," in *Congrès français et international du droits de femmes* (Paris: F. Dentu, 1889), pp. 41-51

Manouvrier, Léonce, "Le cerveau d'un Fuégien," *Bulletins de la Société d'Anthropologie de Paris*, 1894, IV série, *5*:595-614

Manouvrier, Léonce, "§ II – Morphologie générale – Anatomie comparée – Rapports du poids et de la forme du cerveau avec l'intelligence," in *Dictionnaire de physiologie par Charles Richet. Tome II B-C* (Paris: Félix Alcan, 1897), pp. 670-745

Manouvrier, Léonce, "Considérations sur l'hypermégalie cérébrale," *Revue de l'École d'anthropologie de Paris*, 1902, *12*:391-414

Manouvrier, Léonce, "Conclusions générales sur l'anthropologie des sexes et applications sociales," *Revue de l'École d'anthropologie de Paris*, 1903, *13*:405-423

Mantegazza, Paolo, "Dei caratteri gerarchici del cranio umano. Studi di critica craniologica," *Archivio per l'Antropologia e la Etnologia*, 1875, *5*:32-81

Mantegazza, Paolo, "La riforma craniologica," *Archivio per l'Antropologia e la Etnologia*, 1880, *10*:117-137

Mantegazza, Paolo, "Di alcune recenti proposte di riforma della craniologia," *Archivio per l'Antropologia e la Etnologia*, 1893, *23*:45-55

Marchand, Felix, *Ueber das Hirngewicht des Menschen* (Leipzig: Teubner, 1902)

Marino, Mario, "Noch etwas über die Menschenrassen. Eine Lektüre der Kant-Herder-Forster Kontroverse," in Simone De Angelis, Florian Gelzer, Lucas Marco Gisi (Hrsg.), *Natur, Naturrecht und Geschichte. Aspekte eines fundamentalen Begründungsdiskurses der Neuzeit (1600-1900)* (Heidelberg: Universitätsverlag Winter, 2010), pp. 393-413

Marshall, John, "On the Brain of a Bushwoman; and on the Brains of two Idiots of European Descent," *Philosophical Transactions of the Royal Society of London*, 1864, *154*: 501-558

Marshall, John, "On the Influence of Stature on the Weight of the Encephalon and its Parts in Man," *Proceedings of the Royal Society of London*, 1874, *23*:564-565

Martin, Rudolf, *Lehrbuch der Anthropologie in systematischer Darstellung mit besonderer Berücksichtigung der anthropologischen Methoden* (Jena: Gustav Fischer, 1914)

Martin, Rudolf, *Zur physischen Anthropologie der Feuerländer* (Braunschweig: Friedrich Vieweg und Sohn, 1893)

Matiegka, Heinrich, *Über das Hirngewicht, die Schädelkapazität und die Kopfform, sowie deren Beziehung zur psychischen Tätigkeit des Menschen. 1. Über das Hirngewicht des Menschen* (Prag: Verlag der Königl. Böhmischen Gesellschaft der Wissenschaften, 1902)

Maudsley, Henry, *The Physiology and Pathology of the Mind* (New York: Appleton and Company, 1867)

Mazliak, Paul, *Félix Vicq d'Azyr, créateur révolutionnaire de l'anatomie comparée* (Paris: Hermann, 2017)

Mazzolini, Renato G., "Skin Color and the Origin of Physical Anthropology," in Susanne Lettow (ed.), *Reproduction, Race, and Gender in Philosophy and Early Life Sciences* (Albany, NY: State University of New York Press, 2014)

McCulloch, Jock, *Colonial Psychiatry and the African Mind* (Cambridge: Cambridge University Press, 1995)

McGrigor Allan, James, "On the Real Differences in the Minds of Men and Women," *Anthropological Review*, 1869, *7*:cxcv-ccxix

Meckel, Johann Friedrich, *Tractatus anatomico-physiologicus de quinto pare nervorum cerebri, duabus figurarum tabulis illustratus* (Gottingae: Apud Abram Vandenhoeck, 1748)

Meckel, Johann Friedrich, "Recherches anatomiques. I. Sur la nature de l'epiderme, et du réseau, qu'on appelle malpighien; II. Sur la diversité de couleur dans la substance médullaire du cerveau des nègres; III. Description d'une maladie particulière du peritoine," *Histoire de l'Académie Royale des Sciences et Belles Lettres. Année MDCCLIII* (Berlin: chez Haud et Spener, 1755), pp. 79-113

Meckel, Johan Friedrich, "Nouvelles observations sur l'epiderme et le cerveau des Nègres," *Mémoires de l'Académie Royale des Sciences et Belles-Lettres. Classe de philosophie expérimentale*, 1757, *13*: 61-71

Meckel, Johann Friedrich, "Observations anatomiques sur la glande pinéale, sur la cloison transparente, et sur l'origin du nerf de la septième paire", *Histoire de l'Académie Royale des Sciences et Belles-Lettres. Année MDCCLXV* (Berlin: chez Haude et Spener, 1767), pp. 91-101

Meijer, Miriam Claude, *Race and Aesthetics in the Anthropology of Petrus Camper (1722-1749)* (Amsterdam-Atlanta: Rodopi, 1999)

Meijer, Miriam Claude, "Cranial Varieties in the Human and Orangutan Species", in Nicolas Bancel, Thomas David and Dominic Thomas (eds.), *The Invention of Race. Scientific and Popular Representations* (New York: Routledge, 2014)

Meiners, Christoph, *Grundriß der Geschichte der Menschheit* (Lemgo: im Verlage der Meyerschen Buchhandlung, 1785)

Meiners, Christoph, "Ueber die grosse Verschiedenheit der Biegsamkeit und Unbiegsamkeit, der Härte und Weichheit der verschiedenen Stämme und Racen der Menschen", *Göttingisches Historisches Magazin*, 1787, *1*: 210-246

Meiners, Christoph, "Ueber die sympathetische Reizbarkeit, und einige daraus zu erklärende Erscheinungen in den schwächern Völkern," *Göttingisches Historisches Magazin*, 1788, *2*:40-56

Meiners, Christoph, "Ueber die Natur des Afrikanischen Neger, und die davon abhangende Befreyung, oder Einschränkung der Schwarzen," *Göttingisches Historisches Magazin*, 1790, *6*: 385-456

Meiners, Christoph, "Ueber die Natur der Germanischen und übrigen Celtischen Völker", *Göttingisches Historisches Magazin*, 1791, *8*: 1-48, 67-124

Michael John S., "A New Look at Morton's Craniological Research," *Current Anthropology*, 1988, *29*: 349-354

Michael, John S. and Mitchell, Paul Wolff, "Bias, Brains, and Skulls. Tracing the Legacy of Scientific Racism in the Nineteenth-Century Works of Samuel George Morton and Friedrich Tiedemann," in Jamie A. Thomas and Christina Jackson (eds.), *Embodied Difference. Divergent Bodies in Public Discourse* (Lanham MD: Lexington Books, 2019)

Miclouho-Maclay, Nikolai von, (and Virchow, Rudolf), "Anthropologische Notizen, gesammelt auf einer Reise in West-Mikronesien und Nord-Melanesien im Jahre 1876," *Zeitschrift für Ethnologie*, 1878, *10*:101-119

Mill, John Stuart, *The Subjection of Women* (London: Longmans, Green, Reader, and Dyer, 1869

Mingazzini, Giovanni, *Il cervello in relazione con i fenomeni psichici* (*Studio sulla morfologia degli emisferi cerebrali dell'uomo*) (Torino: Bocca, 1895)

Mingazzini, Giovanni, *Il cervello e i fenomeni mentali. Discorso inaugurale* (Roma: Pallotta, 1912)

Mitchell, Paul Wolff, "The fault in his seeds: Lost notes to the case of bias in Samuel George Morton's cranial race science," *PLoS Biol*, 2018, *16*(*10*): e2007008. <https://doi.org/10.1371/journal.pbio.2007008>

Mochi, Aldobrandino, "Sui rapporti tra lo sviluppo intellettuale e la morfologia craniense. Tesi per la libera docenza in Antropologia," *Archivio per l'Antropologia e la Etnologia*, 1904, *34*:83-142

Moebius, Paul Julius, *Ueber die Anlage zur Mathematik* (Leipzig: Barth, 1900)

Moebius, Paul Julius, *Ueber den physiologischen Schwachsinn des Weibes* (Halle: Marhold, 1900)

Moebius, Paul Julius, *Geschlecht und Kopfgrösse* (Halle: Marhold, 1903)

Moebius, Paul Julius, "Ueber den Schädel eines Mathematikern," in *Ausgewählte Werke* (Leipzig: Barth, 1905), Bd. 7, pp. 211-221

Mogilner, Marina, *Homo Imperii. A History of Physical Anthropology in Russia* (Lincoln, University of Nebraska Press, 2013)

Moleschott, Jacob, *Georg Forster, der Naturforscher des Volks* (Frankfurt am Main: Eidinger Sohn & Cie., 1857)

Montagu, Ashley M. F., *Man's Most Dangerous Myth: The Fallacy of Race* (New York: Columbia University Press, 1945[2])

Montaldo, Silvano, "The Relics of Two 19th-Century Scientists," in Marco Beretta, Maria Conforti and Paolo Mazzarello (eds.), *Savant Relics. Brains and Remains of Scientists*, (Sagamore Beach: Science History Publications, 2016)

Montesquieu, Charles Louis de Secondat, baron de, *Mélanges inédits de Montesquieu publiés par le Baron de Montesquieu* (Bordeaux-Paris: G. Gounouilhou-J. Rouam, 1892), pp. 109-148

Montessori, Maria, "Sui caratteri antropometrici in relazione alle gerarchie intellettuali dei fanciulli nelle scuole. Ricerche di antropologia pedagogica," *Archivio per l'Antropologia e la Etnologia*, 1904, *34*:243-297

Morris-Reich, Amos, "Photography in Economies of Demonstration: The Idea of the Jews as a Mixed-Race People," *Jewish Social Studies: History, Culture, Society*, 2013, *20*:150-183

Morselli, Enrico, "Il peso specifico dell'encefalo negli alienati. Studio critico e sperimentale," *Rivista sperimentale di Freniatria e di Medicina legale*, 1882, *8*: 58-82, 206-247

Morselli, Enrico, "Studi di Antropologia patologica sulla pazzia – I. Sul peso comparativo dei due emisferi cerebrali negli alienati," *La Psichiatria, la neuropatologia e le scienze affini*, 1887, *4*:279-301

Morselli, Enrico, "Studi di Antropologia patologica sulla pazzia – I. Sul peso dell'encefalo in rapporto ai caratteri craniometrici negli alienati," *Rivista sperimentale di Freniatria e di Medicina legale*, 1887, *13*: 365-388

Morton, Samuel George, *Crania Americana, or a Comparative View of the Skulls of Various Aboriginal Nations of North and South America: to which is prefixed an Essay on the Varieties of the Human Species Illustrated by Seventy-eight Plates and a Colored Map* (Philadelphia: J. Dobson; London: Simpkin, Marshall & Co., 1839)

Morton, Samuel George, "Observations on the Size of the Brain in various Races and Families of Man," *The Edinburgh New Philosophical Journal*, 1849-1850, *48*:262-265

Nicobar Islands," *Proceedings of the American Philosophical Society*, 1908, *47*: 51-58

Nicolucci, Giustiniano, "La stirpe ligure in Italia ne' tempi antichi, e ne' moderni. Memoria letta nella tornata del 6 ottobre 1863," *Società Reale di Napoli. Rendiconto dell'Accademia delle scienze fisiche e matematiche*, 1863, *2*:225-234

Nicolucci, Giustiniano, "Sul peso del cervello dell'uomo. Studii," *Società Reale di Napoli. Rendiconto dell'Accademia delle scienze fisiche e matematiche*, 1881, *20*:22-30, 40-59

Nott, Josiah C. and Gliddon, George G., *Types of Mankind: or, Ethnological Researches, based upon the Ancient Monuments, Paintings, Sculptures, and Crania of Races, and upon their Natural, Geographical, Philological, and Biblical History: Illustrated by selections from the Inedited Papers of Samuel George Morton, M.D., and by Additional Contributions from Prof. L. Agassiz, LL.D.; W. Usher, M.D.; and Prof. H. S. Patterson, M.D.* (Philadelphia: Lippincott, Grambo & Co., 1854)

Oehler-Klein, Sigrid, "Anatomie und Kunstgeschichte. Soemmerrings *Rede über die Schönheit der antiken Kinderköpfe* vor der Société des Antiquités in Kassel (1779)," in Manfred Wenzel (Hrsg.), *Samuel Thomas Soemmerring in Kassel, 1779-1784: Beiträge zur Wissenschaftsgeschichte der Goethezeit* (Stuttgart-New York: Gustav Fischer, 1994), pp. 189-239

Oehler-Klein, Sigrid, "Der «Mohr» auf der «niedrigeren Staffel am Throne der Menschheit»? Georg Forsters Rezeption der Anthropologie Soemmerrings," in Horst Dippel und Helmut Scheuer (Hrsg.), *Georg-Forster-Studien III* (Kassel: Kassel University Press, 1999), pp. 119-166

Oehler-Klein, Sigrid, "Mordsinn und Diebsinn. Franz Joseph Galls Erklärung des Verbrechens und die Anfänge der Neuropsychologie zu Beginn des 19. Jahrhunderts," in Roger Fayet (Hrsg.), *Die Anatomie des Bösen: Ein Schnitt durch Körper, Moral und Geschichte* (Baden: Hier & Jetzt, 2008)

Onuf, Peter S., *The Mind of Thomas Jefferson* (Charlottesville: University of Virginia Press, 2007)

Oppenheimer, Jane M., *New Aspects of John and William Hunter. I. Everard Home and the Destruction of John Hunter Manuscripts. II. William Hunter and his* Contemporaries (New York: Henry Schuman, 1946)

Owen, Richard, "On the Characters, Principles of Division, and Primary Groups of the Class Mammalia," *Journal of the Proceedings of the Linnean Society*, 1858, *2*:1-37

Owen, Richard, *On the Classification and Geographical Distribution of the Mammalia, being the Lecture on Sir Robert Reade's Foundation, Delivered before the University of Cambridge, in the Senate-House, May 10, 1859* (London: John Parker, 1859)

Panese, Francesco, "The creation of the 'Negro' at the Turn of the Nineteenth Century. Petrus Camper, Johann Friedrich Blumenbach, and Julien-Joseph Virey," in Nicolas Bancel, Thomas David and Dominic Thomas (eds.), *The Invention of Race* (New York: Routledge, 2014)

Parchappe, Maximien, *Recherches sur l'encéphale, sa structure, ses fonctions et ses maladies. Premier mémoire. Du volume de la Tête et de l'Encéphale chez l'Homme* (Paris: Librairie des sciences médicales de Just Rouvier et E. Le Bouvier, 1836)

Parker, A. J., "Cerebral Convolutions of the Negro Brain," *Proceedings of the Academy of Natural Sciences of Philadelphia*, 1878, *30*:11-15

Parker, A. J., "Simian Characters in Negro Brains," *Proceedings of the Academy of Natural Sciences of Philadelphia*, 1878, *30*:339-340

Peacock, Thomas B., "Tables of the Weights of the Brain and of some of the Organs of the Human Body," *The Monthly Journal of Medical Science*, 1846-47, *7*: 101-110, 166-178

Peacock, Thomas B., "On the Weight of the Brain in the Negro," *Memoirs read before the Anthropological Society of London*, 1963-64, *1*: 65-71

Pearl, Raymond, "Biometrical Studies on Man. I. Variation and Correlation in Brain-Weight," *Biometrika*, 1905, *4*:13-104

Pearson, Karl, "On the Correlation of Intellectual Ability with the Size and Shape of the Head (Preliminary Notice)," *Proceedings of the Royal Society of London*, 1902, *69*: 333-342

Pearson, Karl, "Biometrical Studies on Man. I. Variation and Correlation in Brain-Weight," *Biometrika*, 1905, *4*:13-104

Pearson, Karl, "On the Relationship of Intelligence to Size and Shape of the Head, and to Other Physical and Mental Characters," *Biometrika*, 1906, *5*:105-146

Pelletier, Madeleine, "Recherches sur les indices pondéraux du crâne et des principaux os longs d'une série de squelettes japonais," *Bulletins de la Sociétè d'Anthropologie de* Paris, V série, 1900, *1*:514-529

Pierre Flourens, "Éloge historique de Jean-Frédéric Blumenbach", in *Mémoires de l'Académie royale des sciences de l'Institut de France*, 1847:*21*, pp. i-xxi

Pigozzi, Marinella e Ruggeri, Alessandro, (a cura di), *Cesare Bettini 1814-1885. Disegnatore e modellatore anatomico, pittore e litografo Bolognese* (Bologna: Edizioni TIP.LE.CO., 2010)

Platner, Ernst, *Anthropologie für Aerzte und Weltweise* (Leipzig: Dyck, 1772)

Plischke, Hans, *Johann Friedrich Blumenbachs Einfluss aud die Entdeckungsreisenden seiner Zeit* (Göttingen: Vandenhoeck & Ruprecht, 1937)

Pogliano, Claudio, "Between Form and Function: A New Science of Man," in Pietro Corsi (ed.), *The Enchanted Loom. Chapters in the History of Neuroscience* (New York-Oxford, Oxford University Press, 1991), pp. 144-203

Pogliano, Claudio, *L'ossessione della razza. Antropologia e genetica nel XX secolo* (Pisa: Edizioni della Normale, 2005)

Pogliano, Claudio, *Storie di cervelli, Dall'antichità al Novecento* (Milano: Editrice Bibliografica, 2017)

Poignant, Roslyn, *Professional Savages: Captive Lives and Western Spectacle* (New Haven: Yale University Press, 2004)

Pont, Jean-Claude, et al. (édité par), *Carl Vogt. Science, philosophie et politique (1817-1895). Actes du colloque de mai 1995* (Genève: Georg, 1998)

Popkin, Jeremy D. and Popkin, Richard H. (eds.), *The Abbé Grégoire and his World* (Dordrecht: Kluwer, 2000)

Poskett, James, "National Types: the Transatlantic Publication and Reception of *Crania Americana*," *History of Science*, 2015, *53*:264-295

Poskett, James, *Materials of the Mind. Phrenology, Race, and the Global History of Science, 1815-1920* (Chicago: University of Chicago Press, 2019)

Pouliquen, Yves, *Félix Vicq d'Azyr, les Lumières et la Révolution* (Paris: Odile Jacob, 2009)

Poullain de La Barre, François, *De l'egalité de deux sexes, discours physique et moral, où l'on voit l'importance de se défaire de préjugez* (Paris: Jean Du Puis, 1673)

Poynter, Charles W. M. and Keegan, J. Jay, "A Study of the American Negro Brain," *The Journal of Comparative Neurology*, 1915, *25*:183-212

Poynter, Charles W. M., "Some Conclusions Based on Studies in Cerebral Anthropology," *American Anthropologist*, 1917, *19*:495-502

Pozzi, Samuel, "Du poids du cerveau suivant les races et suivant les individus," *Revue d'anthropologie*, 1878, II série, *1*:277-285

Forster, Georg, *A Voyage round the World in His Britannic Majesty's Sloop, Resolution, commanded by Capt. James Cook, during the Years 1772, 3, 4, and 5* (London: B. White, J. Robson, P. Elmsly, and G. Robinson, 1777)

Forster, Georg, "Noch etwas über die Menschenraßen," *Der Teutsche Merkur*, 1786, *4*: 57-86, 150-166

Forster, Georg, *James Cook, der Entdecker* (1787), in Id., *Kleine Schriften. Ein Beytrag zur Völker- und Länderkunde, Naturgeschichte und Philosophie des Lebens* (Leipzig: Paul Gotthelf Kummer, 1789)

Forster, Georg, *Werke in vier Bänden* (Leipzig: Insel Verlag, 1968)

Forster, Georg, *Werke. Briefe 1784-Juni 1787* (Berlin: De Gruyter, 1978), Band 14

Forster, Georg, *Werke. Briefe 1790-1791* (Berlin: De Gruyter, 1980), Band 16

[Forster, Georg], "Leitfaden zu einer künftigen Geschichte der Menschheit," *Neues Deutsches Museum*, 1789, *1/3*: 269-283

Forster, Thomas, "Observations on a new system of phrenology, or the anatomy and the physiology of the brain, of Drs. Gall and Spurzheim," *Philosophical Magazine & Journal*, 1815: *45*:44-50

Foville, Achille, *Déformation du crâne résultant de la méthode la plus générale de couvrir la tête des enfans: Influence des vêtemens sur nos organes* (Paris: Madame Prevost-Crucius, 1834

Fowler, Lorenzo Niles, "Phrenological Development of Joseph Cinquez, alias Ginqua," *The American Phrenological Journal and* Miscellany, 1840, *2*:136-138

Fowler, Orson Squire, "Hereditary Descent: its Laws and Facts, Illustrated and Applied to the Improvement of Mankind; with Hints to Woman etc.," *The American Phrenological Journal and Miscellany*, 1843, *5*:385-650

[Freier, Gustav], *Leben und Thaten des Freiherrn Quinctius Heymeran von Flaming* (Berlin: In der Vossischen Buchhandlung, 1795-1796)

Fulford, Tim, Lee Debbie, and Kitson, Peter J., *Literature, Science and Exploration in the Romantic Era. Bodies of Knowledge* (Cambridge: Cambridge University Press, 2004)

Gainot, Bernard, "L'abbé Grégoire et la place des Noirs dans l'histoire universelle," *Gradhiva. Revue d'anthropologie et d'histoire des arts*, 2009, *10*: 22-39

[Gall, Franz Joseph], "Des Herrn Dr. F. J. Gall Schreiben über seinen bereits geendigten Prodromus über die Verrichtungen des Gehirns der Menschen und der Thiere an Herrn Jos. Fr. von Retzer," *Der Neue Teutscher Merkur*, 1798, *3*:310-335

Gall, Franz Joseph et Spurzheim, Johann Gaspar, *Recherches sur le système nerveux en général et sur celui du cerveau en particulier. Mémoire présenté à l'Institut de France, le 14 mars 1808; suivi d'Observations sur le rapport qui en a été fait à cette compagnie par ses commissaires* (Paris: Schoell-Nicole, 1809)

Gall, Franz Joseph et Spurzheim, Johann Gaspar, *Anatomie et physiologie du système nerveux en général et du cerveau en particulier, avec des observations sur la possibilité*

Prichard, James Cowles, *Disputatio inauguralis de generis humani varietate* (Edinburgi: Abernethy & Walker), 1808

Prichard, James Cowles, *Researches into the Physical History of Man* (London: John and Arthur Arch, 1813)

Prichard, James Cowles, *Researches into the Physical History of Mankind. Second edition in two volumes* (London: John and Arthur Arch, 1826)

Prichard, James Cowles, *Researches into the Physical History of Mankind. Third Edition* (London: Sherwood, Gilbert, and Piper, 1836)

Pruner-Bey, Franz Ignaz, "Mémoire sur les Négres," *Mémoires de la Société d'Anthropologie de Paris*, 1860-63, *1*: 293-336

Putnam, Carleton, *Race and Reason: A Yankee View* (Washington, DC: Public Affairs Press, 1961)

Putnam, Carleton, *Race and Reality: A Search for Solutions* (Washington, DC: Public Affairs Press, 1967)

Quatrefages, Armand de, *Rapport sur les progrès de l'anthropologie* (Paris: A l'Imprimerie imperial, 1867)

Querner, Hans, "Samuel Thomas Soemmerring und Johann Georg Forster – eine Freundschaft," in Gunter Mann, und Franz Dumont (Hrsg.), *Samuel Thomas Soemmerring und die Gelehrten der Goethezeit* (Stuttgart-New York: Gustav Fischer, 1985), pp. 229-245

Quetelet, Adolphe, *Sur l'homme et le développement de ses facultés, ou Essai de physique sociale* (Paris: Bachelier, 1835)

Quoy, Jean René Constant et Gaimard Joseph Paul, "Observations sur la Constitution physique des Papous qui habitant les îles Rawak et Vaigiou," *Annales des sciences naturelles*, 1826, *7*:27-38

Quoy, Jean René Constant et Gaimard Joseph Paul, *Voyage autour du monde, entrepris par ordre du Roi* [...] *exécuté sur les corvettes l'Uranie et la Physicienne pendant les années 1817, 1818, 1819 et 1820* [...] *Zoologie* (Paris: Pillet aîné, 1824)

Rediker, Marcus, *The Amistad Rebellion. An Atlantic Odissey of Slavery and Freedom* (New York: Viking, 2012)

Renneville, Marc, *Le langage des crânes. Une histoire de la phrenologie* (Paris: Sanofi-Synthélabo, 2000)

Retzer, Joseph Friedrich von, "Antwort an Hernn Doktor Gall", *Der Neue Teutscher Merkur*, 1798:*3*, pp. 332-335

Retzius, Anders, "Mémoire sur les formes du crâne des habitants du Nord," *Annales des sciences naturelles*, 1846, *6*:133-172

Retzius, Anders, "Ueber die Form des Knochengerüstes des Kopfes bei den verschiedenen Völkern. Vorgetragenen in der vierten Versammlung der scandiv. Naturforscher," *Archiv für Anatomie, Physiologie und wissenschaftliche Medicin*, 1848, 263-284

Retzius, Gustaf, "Das Gehirn des Astronomen Hugo Gyldéns," *Biologische Untersuchungen*, 1898, *8*:1-22

Retzius, Gustaf, "Das Gehirn des Mathematikers Sonja Kovalevski," *Biologische Untersuchungen*, 1900, *9*:1-16

Retzius, Gustaf, "Ueber das Hirngewicht der Schweden," *Biologische Untersuchungen*, 1900, *9*:51-68

Retzius, Gustaf, *The So-Called North European Race of Mankind. A Review of and Views on the Development of Some Anthropological Questions* (*The Huxley Lecture for 1909*) (London: Royal Anthropological Institute, 1909)

Richards, Evelleen, "Huxley and woman's place in science: The 'woman question' and the control of Victorian anthropology," in James R. Moore (ed.), *History, Humanity and Evolution. Essays for John C. Greene* (Cambridge: Cambridge University Press, 1989), pp. 253-284

Richards, Graham, *'Race', Racism and Psychology* (London-New York: Routledge, 1997)

Riese, Walther, "The 150th Anniversary of S. T. Soemmerring's *Organ of the Soul*," *Bulletin of the History of Medicine*, 1946, *20*: 310-321

Ripley, William Z., *The Races of Europe. A Sociological Study* (*Lowell Institute Lectures*) (New York: D. Appleton & Co., 1899)

Rippon, Gina, *The Gendered Brain. The New Neuroscience that Shatters the Myth of the Female Brain* (London: The Bodely Head, 2019)

Robert le jeune [Robert, Louis-Joseph-Marie], *Mégalanthropogénésie, ou l'Art de faire des Enfans d'esprit, qui deviennent de grands-hommes; Suivie des traits physiognomoniques propres à les faire reconnaître, décrits par Lavater, et du meilleur mode de génération* (Paris: Debray-Bailleul, 1801),

Roque, Ricardo, "Race and the Mobility of Human as Things," *Science, Technology, & Human Values*, 2014, *39*:607-617

Roque, Ricardo, "Introduction. Anthropologie et materialité de la race," in *Revue d'histoire des sciences humaines*, 2015, *n. 27*:7-27

Rose, Nikolas and Abi-Rached, Joelle M., *Neuro: The New Brain Sciences and the Management of the Mind* (Princeton: Princeton University Press, 2013)

Rosner, Lisa, *The Anatomy Murders Being the True and Spectacular History of Edinburgh's Notorious Burke and Hare and of the Man of Science Who Abetted Them in the Commission of Their Most Heinous Crimes* (Philadelphia: University of Pennsylvania Press, 2010)

Rousseau, Jean-Jacques, *Émile, ou de l'éducation* (La Haye: Jean Néaulme, 1762)

Roussel, Pierre, *Système physique et moral de la femme, ou Tableaux philosophique de la Constitution, de l'État organique, du Tempérament, des Moeurs, & des Fonctions propres au Sexe* (Paris: Vincent, 1775)

[Roussel, Pierre], "Bibliothèque universelle des dames. Physique de l'homme. Tome premier," (Paris: Rue et Hötel Serpente, 1787)

Roussel Pierre, *Système physique et moral de la femme, suivi du Système physique et moral de l'homme, et d'un fragment sur la sensibilité* (Paris: Crapart, Caille et Ravier, 1805)

Rüdinger, Nikolaus, "Beitrag zur Anatomie der Affenspalte und der Interparietalfurche beim Menschen nach Race, Geschlecht und Individualität," in *Beiträge zur Anatomie und Embryologie. Als Festgabe Jacob Henle zum 4. April 1882 dargebracht von seinen Schülern* (Bonn: Cohen & Sohn, 1882), pp. 186-198

Rüdinger, Nikolaus, "Über die Hirne von Zwillingen," *Verhandlungen der Anatomischen Gesellschaft auf der 8. Versammlung in Strassburg*, 1894, *8*:177

Ruggendorfer, Peter, Szemethy, Hubert D. (Hg.), *Felix von Luschan. Leben und Wirken eines Universalgelehrten* (Wien-Köln-Weimar: Böhlau Verlag, 2009)

Rupke, Nicolaas A., *Richard Owen. Biology without Darwin, a Revised Edition* (Chicago-London, The University of Chicago Press, 2009)

Rupke, Nicolaas and Lauer, Gerhard (eds.), *Johann Friedrich Blumenbach. Race and Natural History, 1750-1850* (New York: Routledge, 2019)

Rupp-Eisenreich, Britta, "Christoph Meiners' «New Science»", in Nicolas Bancel, Thomas David and Dominic Thomas (eds.), *The Invention of Race. Scientific and Popular Representations* (New York: Routledge, 2014), pp. 68-83

Sahmland, Irmtraut, "Soemmerring als Freimaurer und Rosenkreuzer in Kassel", in Manfred Wenzel (Hrsg.), *Samuel Thomas Soemmerring in Kassel, 1779-1784: Beiträge zur Wissenschaftsgeschichte der Goethezeit* (Stuttgart-New York: Gustav Fischer, 1994), pp. 353-422

Santorini, Giandomenico, *Observationes anatomicae* (Venetiis: Apud Jo. Baptistam Recurti, 1724)

Sappey, Constance, "Recherches sur le volume et la capacité du crâne, sur le volume et le poids de l'encéphale, comparés chez l'homme et chez la femme, lues à la Societé de Biologie, dans sa séance du 18 mai 1861," *Comptes rendus des séances et Mémoires de la Société de Biologie*, 1862, *3*: 109-120

Sappol, Michael, *Body Modern. Fritz Kahn, Scientific Illustration, and the Homuncular Subject* (Minneapolis-London: University of Minneapolis Press, 2017)

Saumarez, Richard, *A New System of Physiology, Comprehending the Laws by which Animated Beings in General, and the Human Species in Particular, are Governed, in their Several States of Health and Disease* (London: T. Cox, 1798)

Schelling, Friedrich Wilhelm Joseph, "Einiges über die Schädellehre," *Morgenblatt für gebildete Stände*, 1807, *1/74*:293-294

Schiebinger, Londa, *The Mind Has No Sex? Women in the Origins of Modern Science* (Cambridge, Mass.: Harvard University Press, 1989)

Schiller, Johann Christoph Friedrich, *Versuch über den Zusammenhang der thierischen Natur des Menschen mit seiner geistigen. Eine Abhandlung welche in höchster Gegenwart Sr. Herzoglichen Durchlaucht, während den öffentlichen akademischen Prü-*

fungen vertheidigen wird Johann Christoph Friedrich Schiller Kandidat der Medizin in der Herzoglichen Militair-Akademie (Stuttgart: Christoph Friedrich Cotta, 1780)

Schiller's Werke. Nach den vorzüglichsten Quellen revidirte Ausgabe. Vierzehnter Theil. Kleinere prosaische Schriften (Berlin: Gustav Hempel, Berlin, 1869/70)

Schultze, Leonhard, *Zoologische und anthropologische Ergebnisse einer Forschungsreise im westlichen und zentralen Südafrika ausgeführt in den Jahren 1903-1905. Erster Band: Systematik und Tiergeographie* (Jena: Gustav Fischer, 1908)

Schwarz, Sabine, *Die anatomische Privatsammlung der Anatomenfamilie Meckel unter besonderer Berücksichtigung ihres präparationstechnischen Profils. Dissertation zur Erlangung des akademischen Grades Dr.med. vorgelegt dem Rat der Medizinischen Fakultät der Martin-Luther-Universität Halle-Wittenberg*, pp. 3-23. URL: <http://webdoc.sub.gwdg.de/ebook/m/2002/pub/medizin/00H110/prom.pdf>

Secord, James A., "Behind the Veil: Robert Chambers and *Vestiges*" in James R. Moore (ed.), *History, Humanity and Evolution. Essays for John C. Greene* (Cambridge: Cambridge University Press, 1989)

Secord, James A., *Visions of Science: Books and Readers at the Dawn of the Victorian Age* (Oxford: Oxford University Press, 2014)

Seligman, Edwin R. A., "Address," in *Proceedings of the National Negro Conference 1909. New York, May 31 and June 1* (New York: The Conference, 1909), pp. 67-70

Sera-Shriar, Efram, "Observing Human Difference. James Hunt, Thomas Huxley and Competing Disciplinary Strategies in the 1860's," *Annals of Science*, 2013, *70*:461-491

Sera-Shriar, Efram, *The Making of British Anthropology, 1813-1871* (London: Pickering & Chatto, 2013)

Sergi, Sergio, "Sulla craniologia degli Herero," *Bollettino della R. Accademia medica di Roma*, 1908, *34*:100-116

Sergi, Sergio,"Sulla morfologia del cervello degli Herero," *Atti della Società Romana di Antropologia*, 14, 1908, *14*: 71-74

Sergi, Sergio, *Cerebra Hererica* (Jena: Fischer, 1909)

Sergi, Sergio, *Note morfologiche sulla superficie metopica del lobo prefrontale in cervelli di Indiani e di Giapponesi* (Roma: Tip. Pallotta, 1912)

Sergi, Sergio, "Capacità del cranio," in *Enciclopedia Italiana* (Roma, Istituto della Enciclopedia Italiana, 1931), 11, pp. 790-791

Sergi, Sergio, "Il cervello nelle razze umane e negli antropomorfi," in *Enciclopedia Italiana* (Roma, Istituto della Enciclopedia Italiana, 1931), 9, pp. 845-858

Serres, Étienne, *Anatomie comparée du cerveau, dans les quatre classes des animaux vertébrés, appliquée à la physiologie et à la pathologie du système nerveux* (Paris: Gabon et Compagnie, 1824-1826)

Serres, Étienne, "Recherches d'anatomie transcendante, sur les Lois de l'Organogénie appliquées à l'anatomie pathologique," *Annales des sciences naturelles*, 1827, *11*:47-70

Serres, Étienne, *Recherches d'anatomie transcendante et pathologique. Traité des formations et déformations organique appliquées à l'anatomie de Ritta-Christina, et de la duplicité monstrueuse* (Paris: Baillière, 1832)

Shellshear, Joseph L., "The Occipital Lobe in the Brain of the Chinese with Special Reference to the Sulcus Lunatus," *Journal of Anatomy*, 1926, *61*:1-13

Shellshear, Joseph L., "The Primitive Features of the Cerebrum, with Special Reference to the Brain of the Bushwoman described by Marshall," *Philosophical Transactions of the Royal Society of London*, Series B, 1934, *223*:1-26

Shellshear, Joseph L., "The Brain of the Aboriginal Australian. A Study in Cerebral Morphology," *Philosophical Transactions of the Royal Society of London,* Series B, 1937, *227*: 293-409

Sims, John, "On Hypertrophy and Atrophy of the Brain," *Medico-Chirurgical Transactions*, 1835, *19*: 315-380

Skuncke, Marie-Christine, "Suède, Europe, Japon: le botaniste Carl Peter Thunberg sur le marché international", *La Révolution française*, 2018:*13*, URL : <http://journals.openedition.org/lrf/1928>; DOI : 10.4000/lrf.1928

Smail, Daniel Lord, *On Deep History and the Brain* (Berkeley: University of California Press, 2008)

Smith, C.U.M, "Coleridge's «Theory of Life»," *Journal of the History of Biology*, 1999, *32*: 31-50

Smith, Grafton Elliot, "The Morphology of the Retrocalcarine Region of the Cortex Cerebri," *Proceedings of the Royal Society of London*, 1904, *73*:59-65

Smith, Grafton Elliot, "The so-called «Affenspalte» in the Human (Egyptian) Brain," *Anatomischer Anzeiger*, 1904, *24*:74-83

Smith, Grafton Elliot, "Studies in the Morphology of the Human Brain, with Special Reference to that of the Egyptians. No. 1: The Occipital Region," *Records of the Egyptian Government School of Medicine*, 1904, *2*:123-173

Smith, Grafton Elliot, "A New Topographical Survey of the Human Cerebral Cortex, being an Account of the Distribution of the Anatomically Distinct Cortical Areas and their Relationship to the Cerebral Sulci", *Journal of Anatomy and Physiology*, 1907, *41*:237-254

Smith, Grafton Elliot, *The Evolution of Man: Essays* (London: Oxford University Press, 1924)

Smith, John David, "W.E.B. Du Bois, Felix von Luschan, and Racial Reform at the Fin de Siècle," *Amerikastudien/American Studies*, 2002, *47*: 23-38

Smith, Samuel Stanhope, *An Essay on the Causes of the Variety of Complexion and Figure in the Human Species. To which are Added Strictures on Lord Kaims's* Discourse on the Original Diversity of Mankind (Philadelphia: Robert Aitken, 1787)

Smyth, Thomas, *The Unity of the Human Races, proved to be the Doctrine of Scripture, Reason, and Science. With a Review of the Present Position and Theory of Professor Agassiz* (New York: George P. Putnam, 1850)

Soemmerring, Samuel Thomas, *De basi encephali et originibus nervorum cranio egredientium* (Goettingae: Apud Abr. Vandenhoeck Viduam, 1778)

Soemmerring, Samuel Thomas, *Ueber die körperliche Verschiedenheit des Negers vom Europäer* (Mainz: n. p., 1784)

Soemmerring, Samuel Thomas, *Ueber die körperliche Verschiedenheit des Negers vom Europäer* (Frankfurt-Mainz: Barrentrapp Sohn und Wenner, 1785)

Soemmerring, Samuel Thomas, *Über das Organ der Seele* (Königsberg: Nicolovius, 1796)

Soemmerring, Samuel Thomas, *Briefwechsel 1784-1792. Teil II Januar 1787 – Oktober 1792*, hrsg. von Franz Dumont (Stuttgart: Gustav Fischer, 1998)

Soemmerring, Samuel Thomas, *Ueber die körperliche Verschiedenheit des Negers vom Europäer bearbeitet und herausgegeben von Sigrid Oehler-Klein* (Stuttgart-New York: Gustav Fischer, 1998)

Sowervine, Charles, "Women's Brain, Man's Brain: Feminism and Anthropology in Late Nineteenth-Century France," *Women's History Review*, 2003, *12*:289-308

Sparks, Corey S. and. Jantz, Richard L.,"Changing Times, Changing Faces: Franz Boas's Immigrant Study in Modern Perspective," *American Anthropologist*, 2003, *105*: 333-337

Spencer, Herbert, *The Principles of Psychology* (London: Longman, Brown, Green, and Longmans, 1855)

Spencer, Herbert, "Progress: Its Law and Causes", *The Westminster Review*, 1857, *67*: 445-485

Sperino, Giuseppe, *L'encefalo dell'anatomico Carlo Giacomini* (Torino: Unione Tipografico-Editrice, 1900)

Spitzka, Edward Charles, "The Czolgosz Case," *The Philadelphia Medical Journal*, 1901, 8:693-695

Spitzka, Edward Anthony, "Contributions to the Encephalic Anatomy of the Races. First Paper: – Three Eskimo Brains, from Smith's Sound," *American Journal of Anatomy*, 1902-1903, *2*: 25-71

Spitzka, Edward Anthony, "Preliminary Note on the Brains of Natives of the Andaman and

Spitzka, Edward Anthony, "The Brain-Weight of the Japanese," *Science*, 1903, *18/455*: 371-373

Spitzka, Edward Anthony, "A Study of the Brains of Six Eminent Scientists and Scholars belonging to the American Anthropometric Society together with a Description of the Skull of Professor E. D. Cope," *Transactions of the American Philosophical Society*, 1908, *21*:175-308

Spry, Henry H., "Some Account of the Gang-Murderers of Central India, Commonly Called Thugs; Accompanying the Seven Skulls of Them," *The Phrenological Journal and Miscellany*, 1834, *8*:511-524

Spurzheim, Johann Gaspar, *Observations sur la phrénologie, ou la connaissance de l'homme moral et intellectuel, fondée sur les fonctions du système nerveux* (Paris: Treuttel & Würtz, 1818)

Spurzheim, Johann Gaspar, *Outlines of Phrenology, Being also a Manual of Reference for the Marked Bust* (Boston: Marsh, Capen and Lyon, 1832)

Spurzheim, Johann Gaspar, *Phrenology, in Connexion with the Study of Physiognomy. Illustration of Characters. First American Edition. To which is Prefixed a Biography of the Author, by Nahum Capen* (Boston: Marsh, Capen and Lyon, 1833)

Spurzheim, Johann Gaspar, *Phrenology, or the Doctrine of Mental Phenomena. Vol. I Physiological Part. Second American Edition, greatly Improved by the Author, from the Third London Edition* (Boston: Marsh, Capen and Lyon, 1833)

Spurzheim, Johann Gaspar, *A View of the Elementary Principles of Education, Founded on the Study of the Nature of Man* (Boston: Marsh, Capen and Lyon, 1835)

Stahnisch, Franz, "Über die Natur des weiblichen Gehirns. Geschlechterpolitik im Werk der Mainzer Anatomen Jacob Fidelis Ackermann (1765-1815), in Rüdiger Schultka und Josef N. Neumann, (Hrsg.), *Anatomie und anatomische Sammlungen in 18. Jahrhundert. Anlässlich der 250. Wiederkehr des Geburtstages von Philipp Friedrich Theodor Meckel (1755-1803)* (Berlin: LIT, 2007)

Statuts de la Société d'autopsie mutuelle," *Revue scientifique de la France et de l'étranger*, 1876-1877, *11*:527-528

Stephens, Lester D., *Science, Race, and Religion in the American South: John Bachman and the Charleston Circle of Naturalists, 1815-1895* (Chapel Hill: University of North Carolina Press, 2000)

Stetson, George R., "Some Memory Tests of Whites and Blacks," *The Psychological Review*, 1897, *4*:285-289

Stieda, Ludwig, "Über die Bedeutung der Hirnwindungen," *Correspondenz-Blatt der Deutschen Gesellschaft für Anthropologie, Ethnologie und Urgeschichte*, 1906, *37*:137-138

Stieda, Ludwig, "Das Gehirn eines Sprachkundigen," *Zeitschrift für Morphologie und Anthropologie*, 1907, *11*:83-138

Stocking, Jr., George W., *Victorian Anthropology* (London: The Free Press, 1987)

Swan, Donald W., "Juan Comas on «Scientific racism again?»: A scientific analysis," *Mankind Quarterly*, 1962, *2*:231-245

The Slave Trade Debate. Contemporary Writings For and Against. Introduction by John Pinfold (Oxford: Bodleian Libray, 2007)

Thulié, Henri, "Sur l'autopsie de Louis Asseline, membre de la Société d'Anthropologie de Paris," *Bulletin de la Société d'Anthropologie de Paris*, 1878, *1*:161-167

Thunberg, Carl Peter, *Resa uti Europa, Africa, Asia, Förrättad Åren 1770 – 1779* (Upsala: Joh. Edman, 1788)

Thurber, James, *Lantern and Lances* (London: Hamish Hamilton, 1961)

Thurnam, John, *The Weight of the Brain, and on the Circumstances Affecting it* (London: J. E. Adlard, 1866)

Tiedemann, Dietrich, "Beobachtungen über die Entwicklung der Seelenfähigkeiten bei Kindern," *Hessische Beiträge zur Gehlersamkeit und Kunst*, 1787, *2*:313-333, 486-502

Tiedemann, Friedrich, *Anatomie und Bildungsgeschichte des Gehirns im Foetus des Menschen nebst einer vergleichenden Darstellung des Hirnbaues in den Thieren* (Nürnberg: in der Steinischen Buchhandlung, 1816)

Tiedemann, Friedrich, "Anruf an die Humanität der Höheren Behörden der Gerechtigkeits-Pflege in Deutschland, veranlasst durch eine am 22. October 1827 in Heidelberg vollzogene Enthauptung," *Zeitschrift für Physiologie*, 1829, *3*: 283-288

Tiedemann, Friedrich, "On the Brain of the Negro, compared with that of the European and the Orang-Outang," *Philosophical Transactions of the Royal Society of London*, 1836, *126*: 497-527

Tiedemann, Friedrich, *Das Hirn des Negers mit dem des Europäers und Orang-Outangs verglichen* (Heldelberg: Karl Winter, 1837). Mit einer Einführung von Hans-Konrad Schmutz (Marburg: Basilisken-Presse, 1984)

Tilley, Helen, *Africa as a Living Laboratory: Empire, Development, and the Problem of Scientific Knowledge, 1870-1950* (Chicago and London: University of Chicago Press, 2011)

Tobias, Phillip V., "Brain-size, Grey Matter and Race. Fact of Fiction?," *American Journal of Physical Anthropology*, 1970, *32*:3-26

Tobias, Phillip V., *The brain in hominid evolution* (New York-London: Columbia University Press, 1971)

Topinard, Paul, "Sur les races indigènes de l'Australie," *Bulletins de la Société d'Anthropologie de Paris*, 1872, II série, *7*:211-327

Topinard, Paul, "Le poids du cerveau d'après les registres de Paul Broca," *Revue d'Anthropologie*, II série, 1882, *5*:1-30

Topinard, Paul, "Présentation de trois Australiens vivants," *Bulletins et mémoires de la Société d'Anthropologie de Paris*, III série, 1885, *8*:683-698

Topinard, Paul, *Éléments d'anthropologie générale* (Paris: Delahaye et Lecrosnier, 1885)

Topinard, Paul, "Le poids de l'encéphale d'après les registres de Paul Broca," *Mémoires de la Société d'Anthropologie de Paris*, II série, 1888, *3*:1-41

Tucker, William H., *The Funding of Scientific Racism. Wickliffe Draper and the Pioneer Fund* (Urbana, Ill.: University of Illinois Press, 2002)

Tucker, William H., *The Science and Politics of Racial Research* (Urbana, Ill.: University of Illinois Press, 1995)

Turgot, Robert-Jacques, "Plan du second discours sur l'histoire universelle, dont l'objet sera les progrès de l'esprit humain," in *Oeuvres de Turgot. Nouvelle édition classée par ordre de matières, avec les notes de Dupont de Nemours* (Paris: Guillaumin, 1844)

Turnbull, Paul, "«Rare Work amongst the Professors»: the Capture of Indigenous Skulls within Phrenological Knowledge in early colonial Australia," in Barbara Creed and Jeanette Horn (eds.), *Body Trade: Captivity, Cannibalism and Colonialism in the Pacific* (New York: Routledge, 2001), pp. 3-23

Turnbull, Paul, *Science, Museums and Collecting the Indigenous Dead in Colonial Australia* (London: Palgrave Macmillan, 2012)

u.s.f. [Soemmerring Samuel Thomas], review of Kelch's *Ueber den Schädel Kants, Göttingische gelehrte Anzeigen*, 1804, *3*:1457-1460

Van Berkel, Klaas and Ramakers, Bart (eds.), *Petrus Camper in Context. Science, the Arts, and Society in the Eighteenth-century Dutch Republic* (Hilversum: Verloren, 2015)

Van Evrie, John H., *Negroes and Negro Slavery. The First, an Inferior Race – The Latter, its Normal Condition. Introductory Number: Caused of Popular Delusion on the Subject* (Baltimore: John D. Toy, 1853)

Van Evrie, John H., *Negroes and Negro Slavery. The First, an Inferior Race – The Latter, its Normal Condition* (New York: Van Evrie, Horton & Co., 1861)

Van Hoorn, Tanja, *Dem Leibe abgelesen. Georg Forster im Kontext der physischen Anthropologie des 18. Jahrhunderts* (Tübingen: Niemeyer, 2004)

Van Wyhe, John, *Phrenology and the Origins of Victorian Scientific Naturalism* (Aldershot-Burlington: Ashgate, 2004)

Van Wyhe, John, "The Diffusion of Phrenology through Public Lecturing", in Aileen Fyfe, and Bernard Lightman (eds.), *Science in the Marketplace. Nineteenth-Century Sites and Experiences* (Chicago: University of Chicago Press, 2007),

Vaschide, Nicolas et Pelletier, Madeleine, "Recherches expérimentales sur les signes physiques de l'intelligence," *Revue de Philosophie*, 1902-1903, *3*:796-831; 1903-1904, *4*:168-195

Verga, Andrea, "Del peso e del volume dell'encefalo," in Id., *Studi anatomici sul cranio e sull'encefalo, psicologici e freniatrici* (Milano: Manini-Wiget, 1896), I, pp. 266-277

Verity, Robert, *Changes Produced in the Nervous System by Civilization, Considered according to the Evidence of Physiology and the Philosophy of History* (London: S. Highley, 1837)

Vicq d'Azyr, Félix, *Suite des recherches sur la structure du cerveau. Quatrième Mémoire sur la structure du cerveau des Animaux comparé avec celui de l'Homme*, in *Histoire de l'Académie Royale des Sciences. Année MDCCXLXXXIII. Avec les Mémoires de Mathématiques et de Physique* (Paris: De l'Imprimerie Royale, 1786)

Vicq d'Azyr, Félix, *Traité d'anatomie et de physiologie avec des planches coloriées. Représentant au naturel les divers organes de l'homme et des animaux* (Paris: François Didot l'Aîné, 1786)

Vicq d'Azyr, Félix, *Éloge de Pierre Camper*, in *Oeuvres de Pierre Camper, qui ont pour objet l'histoire naturelle, la physiologie et l'anatomie comparée* (Paris: H. J. Jansen, 1803), 1, pp. lxi-xci

Vidal, Fernando and Ortega, Francisco, *Being Brains. Making the Cerebral Subject* (New York: Fordham University Press, 2017)

Vint, F. W., "A Preliminary Note on the Cell Content of the Prefrontal Cortex of the East African Native," *East African Medical Journal*, 1932-33, *9*: 30-55

Vint, F. W., "The Brain of the Kenya Native," *Journal of Anatomy*, 1934, *68*/2:216-223

Virchow, Rudolf, "Création de laboratoires d'anthropologie dans les colonies," in *Congrès international des sciences anthropologiques tenu à Paris du 16 au 21 août 1878* (Paris: Imprimerie Nationale, 1880), pp. 222-224

Virchow, Rudolf, "Rassenbildung und Erblichkeit," in *Adolf Bastian als Festgruss zu seinem 70. Geburtstage 26. Juni 1896 gewidmet von seinen Freunden und Verehrern* (Berlin: Dietrich Reimer, 1896)

Virey, Julien-Joseph, *Histoire naturelle du genre humain, ou Recherches sur ses principaux Fondemens physiques et moraux; precedées d'un Discours sur la nature des êtres organiques, et sur l'ensemble de leurs physiologie. On y a joint une dissertation sur le sauvage de l'Aveyron* (Paris: F. Dufart, an IX [1800-1801])

Virey, Julien-Joseph], "Animal," in *Nouveau Dictionnaire d'Histoire naturelle appliquée aux Arts, principalement à l'Agriculture et à l'Economie rurale et domestique par une Société des Naturalistes et d'Agriculteurs* (Paris: Crapelet-Deterville, 1803), tome 1, pp. 419-466

Virey, Julien-Joseph, *L'Art de perfectionner l'homme, ou de la médecine spirituelle et morale* (Paris: Deterville, 1808)

Virey, Julien-Joseph, "Homme," in *Dictionnaire des sciences médicales, par une société de médecins et de chirurgiens* (Paris: Panckoucke, 1817), tome 21, pp. 191-344

Virey, Julien-Joseph, "Homme," in *Nouveau Dictionnaire d'Histoire naturelle, appliquée aux Arts, à l'Agriculture, à l'Économie rurale et domestique, à la Médecine, etc. par une Société de Naturalistes et d'Agriculteurs* (Paris: Deterville, 1817), tome 15, pp. 1-268

Virey, Julien-Joseph, "Nègre," in *Dictionnaire des sciences médicales, par une société de médecins et de chirurgiens* (Paris: Panckoucke, 1819), tome 35, pp. 378-432

Virey, Julien-Joseph, *De la femme sous ses rapports physiologique, moral et littéraire* (Paris: Crochard, 1823)

Virey, Julien-Joseph, *Histoire naturelle du genre humain, nouvelle édition augmentée et entièrement refondue. Tome premier* (Paris: Crochard, 1824); *Tome second et troisième* (Bruxelles: Aug. Wahlen, 1826)

Virey, Julien-Joseph, *De la physiologie dans ses rapports avec la philosophie* (Paris: Baillière, 1844)

Visser, Robert, "Die Rezeption der Anthropologie Petrus Campers (1770-1850)," in Gunter Mann und Franz Dumont (Hrsg.), *Die Natur des Menschen. Probleme der*

Physischen Anthropologie und Rassenkunde (1750-1850) (Stuttgart-New York: Gustav Fischer, 1990)

Vogt, Carl, *Vorlesungen über den Menschen, seine Stellung in der Schöpfung und in der Geschichte der Erde* (Giessen: J. Ricker'sche Buchhandlung, 1863)

Vogt, Carl, *Lectures on Man: his Place in Creation, and in the History of the Earth, edited by James Hunt* (London: Longman, Green, Longman, and Roberts, 1864)

Vogt, Carl, *Leçons sur l'homme, sa place dans la creation et dans l'histoire de la Terre* (Paris: C. Reinwald, 1865)

Voyage de La Pérouse autour du monde pendant les années 1785, 1786, 1787 et 1787, publié conformément au décret du 22 avril 1791, et rédigé par M. L. A. Milet-Mureau (Paris: de l'Imprimerie de la République, 1797)

Vv. Aa., "Themenschwerpunkt: Ernst Platner (1744-1818). Konstellationen der Aufklärung zwischen Philosophie, Medizin und Anthropologie," *Aufklärung*, 2007, *19*: 7-378

Vv. Aa."Focus: Neurohistory and the History of Science," *Isis*, 2014, *105*:100-154

Vv. Aa., *The Health and Physique of the Negro American. A Social Study made under the direction of Atlanta University by the Eleventh Atlanta Conference* (Atlanta, Georgia: The Atlanta University Press, 1906)

Wagner, Kim A., "Confessions of a Skull: Phrenology and Colonial Knowledge in Early Nineteenth-Century India," *History Workshop Journal*, 2010, *69*:27-51

Wagner, Rudolph, *Samuel Thomas Soemmerrings Leben und Verkehr mit seinen Zeitgenossen. Erste Abtheilung und zweite Abtheilung* (Leipzig: Voss, 1844)

Wagner, Rudolph, "Kritische und experimentelle Untersuchungen über die Functionen des Gehirns. Siebente Reihe. Über die angeblichen Verhältnisse des Gewichts und des Windungsreichthums des menschlichen Gehirns zur Intelligenz," *Nachrichten von der G. A. Universität und der Königl. Gesellschaft der Wissenschaften zu Göttingen*, 1860, *n.7*:65-80

Wagner, Rudolph, "Notiz über das Hirngewicht von Lord Byron mit einigen Berichtigungen und Bemerkungen über das Gehirn Cromwell's, Cuvier's und Dupuytrens, als Nachtrag zur siebenten Reihe der kritischen und experimentellen Untersuchungen über die Funktionen des Gehirns," *Nachrichten von der G. A. Universität und der Königl. Gesellschaft der Wissenschaften zu Göttingen*, 1860, *n.12*:125-130

Wagner, Rudolph, "Über das relative Hirngewicht der Hemisphären des grossen Gehirn's zum kleinen Gehirn und Hirnstamm mit besonderer Rücksicht auf geistige Begabung und Arbeit," *Nachrichten von der G. A. Universität und der Königl. Gesellschaft der Wissenschaften zu Göttingen*, 1860, *n.16*:176-180

Wagner, Rudolph, *Vorstudien zu einer wissenschaftlichen Morphologie und Physiologie des menschlichen Gehirns als Seelenorgan. Erste Abhandlung. Über die typischen Verschiedenheiten der Windungen der Hemisphären und über die Lehre vom Hirnge-*

wicht mit besondrer Rücksicht auf die Hirnbildung intelligenter Männer (Göttingen: Verlag der Dieterischen Buchhandlung, 1860)

Wagner, Rudolph, *Vorstudien zu einer wissenschaftlichen Morphologie und Physiologie des menschlichen Gehirns als Seelenorgan. Zweite Abhandlung. Über den Hirnbau der Mikrocephalen mit vergleichender Rücksicht auf den Bau des Gehirns der normalen Menschen und der Quadrumanen* (Göttingen: Verlag der Dieterischen Buchhandlung, 1862)

Waldeyer, Wilhelm von, "Über einige anthropologisch bemerkenswerthe Befunde an Negergehirnen," *Sitzungsberichte der Königlich Preussischen Akademie der Wissenschaften zu Berlin*, 1894, pp. 1213-1221

Waldeyer, Wilhelm von, "Gehirne südwestafricanischer Völker," *Sitzungsberichte der königlich Preussischen Akademie der Wissenschaften*, Erster Halbband, Januar 1906, pp. 3-8

Waldeyer, Wilhelm von, "Über Gehirne menschlicher Zwillings- und Dillingsfrüchte verschiedenen Geschlechtes," *Sitzungsberichte der Königlich Preussischen Akademie der Wissenschaften zu Berlin*, 1907, pp. 114-126

Wallace, Alfred R., "The Origin of Human Races and the Antiquity of Man deduced from the Theory of «Natural Selection»," *Journal of the Anthropological Society of London*, 1864, *2*:clviii-clxxxvii

[Wallace, Alfred R.], "Sir Charles Lyell on Geological Climates and the Origin of Species," *Quarterly Review*, 1869, *126*:359-394

Wallace, Alfred R., *Contributions to the Theory of Natural Selection. A Series of Essays* (London: Macmullan and Co., 1870)

[Walling, William English], "Preface," in *Proceedings of the National Negro Conference 1909. New York, May 31 and June 1* (New York: The Conference, 1909), pp. 9-13

Wartelle, Jean-Claude, "La Société d'Anthropologie de Paris de 1859 à 1920," *Revue d'histoire des sciences humaines*, 2004, *1/10*: 125-171

Weisbach, Augustin, "Die Gewichtsverhältnisse der Gehirne österreichischer Völker mit Rücksicht auf Körpergrösse, Alter, Geschlecht und Krankheiten," *Archiv für Anthropologie*, 1866, *1*:191-218, 285-319

Weisbach, Augustin, *Beiträge zur Kenntniss der Schädelformen österreichischer Völker* (Wien: Selbstverlag der k. k. Gesellschaft der Aerzte, 1867)

Welcker, Hermann, *Untersuchungen über Wachsthum und Bau des menschlichen Schädels. Erster Theil. Allgemeine Verhältnisse des Schädelwachsthums und Schädelbaues, normaler Schädel deutschen Stammes* (Leipzig: Wilhelm Engelmann, 1862)

Wenzel, Josephus et Carolus, *De penitiori structura cerebri hominis et brutorum* (Tubingae: Apud Cottam, 1812)

Whitaker, Harry and Jarema, Gonia, "The Split between Gall and Spurzheim," *Journal of the History of the Neurosciences*, 2017, *26/2*:216-223

White, Charles, *An Account of the Regular Gradation in Man, and in Different Animals and Vegetables; and from the Former to the Latter. Illustrated with Engravings Adapted to the Subject* (London: C. Dilly, 1799)

Wickens, Andrew P., *A History of the Brain. From Stone Age Surgery to Modern Neuroscience* (London-New York: Psychology Press, 2015)

Wilder, Burt G., "The Brain of the American Negro," in *Proceedings of the National Negro Conference 1909. New York, May 31 and June 1* (New York: The Conference, 1909), pp. 22-50

Wilder, Burt G., "American Negroes in the Civil War," in *Proceedings of the National Negro Conference 1909. New York, May 31 and June 1* (New York: The Conference, 1909), pp. 50-54

Wilson, Leonard G., "The Gorilla and the Question of Human Origins: The Brain Controversy," *The Journal of the History of Medicine and Allied Sciences*, 1996, *51*:184-207

Wolfgang Böker, "Zur Geschichte der Schädelsammlung Johann Friedrich Blumenbach," *Annals of the History and Philosophy of Biology* (2018), *23*:3-29

Wyman, Jeffries, [no title], *Proceedings of the Boston Society of Natural History*, 1862-1863, *9*:352-357

Wyman, Jeffries, "Observations on the Skeleton of a Hottentot," *The Anthropological Review*, 1865, *3*:330-335

Zammito, John H., *Kant, Herder, and the Birth of Anthropology* (Chicago and London: The University of Chicago Press, 2002)

Zimmerman, Andrew, *Anthropology and Antihumanism in Imperial Germany* (Chicago and London: The University of Chicago Press, 2001)

Index of Names

www.ingramcontent.com/pod-product-compliance
Lightning Source LLC
LaVergne TN
LVHW010639300926
849161LV00005B/993
* 9 7 8 9 0 0 4 4 2 9 3 3 8 *